ASTROPHYSICS OF PLANET FORMATION

Concise and self-contained, this textbook gives a graduate-level introduction to the physical processes that shape planetary systems, covering all stages of planet formation. Writing for readers with undergraduate backgrounds in physics, astronomy, and planetary science, Armitage begins with a description of the structure and evolution of protoplanetary disks, moves on to the formation of planetesimals, rocky, and giant planets, and concludes by describing the gravitational and gas dynamical evolution of planetary systems. He provides a self-contained account of the modern theory of planet formation and, for more advanced readers, carefully selected references to the research literature, noting areas where research is ongoing.

The second edition has been thoroughly revised to include observational results from NASA's *Kepler* mission, *ALMA* observations, and the *JUNO* mission to Jupiter, new theoretical ideas including pebble accretion, and an up-to-date understanding in areas such as disk evolution and planet migration.

PHILIP J. ARMITAGE is a professor in the Department of Physics and Astronomy at Stony Brook University and he leads the planet formation group at New York's Center for Computational Astrophysics. He teaches classes on planet formation to advanced undergraduate and graduate students, and has lectured on the topic at summer schools worldwide.

ASTROPHYSICS OF PLANET FORMATION

SECOND EDITION

PHILIP J. ARMITAGE

Stony Brook University and
Center for Computational Astrophysics

CAMBRIDGE
UNIVERSITY PRESS

University Printing House, Cambridge CB2 8BS, United Kingdom

One Liberty Plaza, 20th Floor, New York, NY 10006, USA

477 Williamstown Road, Port Melbourne, VIC 3207, Australia

314–321, 3rd Floor, Plot 3, Splendor Forum, Jasola District Centre, New Delhi – 110025, India

79 Anson Road, #06–04/06, Singapore 079906

Cambridge University Press is part of the University of Cambridge.

It furthers the University's mission by disseminating knowledge in the pursuit of education, learning, and research at the highest international levels of excellence.

www.cambridge.org
Information on this title: www.cambridge.org/9781108420501
DOI: 10.1017/9781108344227

© Philip J. Armitage 2009, 2020

First published 2009
Second edition 2020

A catalogue record for this publication is available from the British Library.

Library of Congress Cataloging-in-Publication Data
Names: Armitage, Philip J., 1971– author.
Title: Astrophysics of planet formation / Philip J. Armitage (Stony Brook University and Center for Computational Astrophysics).
Description: Second edition. | Cambridge : Cambridge University Press, 2020. | Includes bibliographical references and index.
Identifiers: LCCN 2019038227 (print) | LCCN 2019038228 (ebook) | ISBN 9781108420501 (hardback) | ISBN 9781108344227 (epub)
Subjects: LCSH: Planets–Origin. | Astrophysics.
Classification: LCC QB603.O74 A76 2020 (print) | LCC QB603.O74 (ebook) | DDC 5234–dc23
LC record available at https://lccn.loc.gov/2019038227
LC ebook record available at https://lccn.loc.gov/2019038228

ISBN 978-1-108-42050-1 Hardback

Contents

Preface   *page* xi

1   Observations of Planetary Systems   1
    1.1   Solar System Planets   2
    1.2   The Minimum Mass Solar Nebula   4
    1.3   Minor Bodies in the Solar System   6
    1.4   Radioactive Dating of the Solar System   9
          1.4.1   Lead–Lead Dating   12
          1.4.2   Dating with Short-Lived Radionuclides   13
    1.5   Ice Lines   14
    1.6   Meteoritic and Solar System Samples   16
    1.7   Exoplanet Detection Methods   18
          1.7.1   Direct Imaging   19
          1.7.2   Radial Velocity Searches   21
          1.7.3   Astrometry   26
          1.7.4   Transits   27
          1.7.5   Gravitational Microlensing   32
    1.8   Properties of Extrasolar Planets   34
          1.8.1   Parameter Space of Detections   35
          1.8.2   Orbital Properties   37
          1.8.3   Mass–Radius Relation   40
          1.8.4   Host Properties   42
    1.9   Habitability   44
    1.10 Further Reading   48

2   Protoplanetary Disk Structure   49
    2.1   Disks in the Context of Star Formation   49
    2.2   Observations of Protostellar Disks   51
          2.2.1   Accretion Rates and Lifetimes   52
          2.2.2   Inferences from the Dust Continuum   53
          2.2.3   Molecular Line Observations   55

	2.2.4	Transition Disks	56
	2.2.5	Disk Large-Scale Structure	56
2.3		Vertical Structure	57
2.4		Radial Force Balance	59
2.5		Radial Temperature Profile of Passive Disks	60
	2.5.1	Razor-Thin Disks	61
	2.5.2	Flared Disks	63
	2.5.3	Radiative Equilibrium Disks	65
	2.5.4	The Chiang–Goldreich Model	68
	2.5.5	Spectral Energy Distributions	69
2.6		Opacity	71
2.7		The Condensation Sequence	73
2.8		Ionization State of Protoplanetary Disks	75
	2.8.1	Thermal Ionization	75
	2.8.2	Nonthermal Ionization	76
2.9		Disk Large-Scale Structure	80
	2.9.1	Zonal Flows	80
	2.9.2	Vortices	81
	2.9.3	Ice Lines	83
2.10		Further Reading	85
3		Protoplanetary Disk Evolution	86
3.1		Observations of Disk Evolution	86
3.2		Surface Density Evolution of a Thin Disk	88
	3.2.1	The Viscous Time Scale	90
	3.2.2	Solutions to the Disk Evolution Equation	91
	3.2.3	Temperature Profile of Accreting Disks	95
3.3		Vertical Structure of Protoplanetary Disks	96
	3.3.1	The Central Temperature of Accreting Disks	97
	3.3.2	Shakura–Sunyaev α Prescription	98
	3.3.3	Vertically Averaged Solutions	100
3.4		Hydrodynamic Angular Momentum Transport	102
	3.4.1	The Rayleigh Criterion	102
	3.4.2	Self-Gravity	103
	3.4.3	Vertical Shear Instability	106
	3.4.4	Vortices	107
3.5		Magnetohydrodynamic Angular Momentum Transport	109
	3.5.1	Magnetorotational Instability	109
3.6		Effects of Partial Ionization and Dead Zones	114
	3.6.1	Ohmic Dead Zones	114
	3.6.2	Non-ideal MHD Terms	117
	3.6.3	Non-ideal Induction Equation	118

		3.6.4	Density and Temperature Dependence of Non-ideal Terms	123
		3.6.5	Application to Protoplanetary Disks	124
	3.7	Disk Winds		127
		3.7.1	Condition for Magnetic Wind Launching	128
		3.7.2	Net Flux Evolution	132
	3.8	Disk Dispersal		133
		3.8.1	Photoevaporation	134
		3.8.2	Viscous Evolution with Photoevaporation	135
	3.9	Magnetospheric Accretion		137
	3.10	Further Reading		140
4	Planetesimal Formation			141
	4.1	Aerodynamic Drag on Solid Particles		142
		4.1.1	Epstein Drag	142
		4.1.2	Stokes Drag	143
	4.2	Dust Settling		144
		4.2.1	Single Particle Settling with Coagulation	145
		4.2.2	Settling in the Presence of Turbulence	147
	4.3	Radial Drift of Solid Particles		149
		4.3.1	Radial Drift with Coagulation	153
		4.3.2	Particle Concentration at Pressure Maxima	154
		4.3.3	Particle Pile-up	155
		4.3.4	Turbulent Radial Diffusion	156
	4.4	Diffusion of Large Particles		158
	4.5	Particle Growth via Coagulation		160
		4.5.1	Collision Rates and Velocities	160
		4.5.2	Collision Outcomes	162
		4.5.3	Coagulation Equation	163
		4.5.4	Fragmentation-Limited Growth	165
	4.6	Gravitational Collapse of Planetesimals		166
		4.6.1	Gravitational Stability of a Particle Layer	167
		4.6.2	Application to Planetesimal Formation	172
		4.6.3	Self-Excited Turbulence	173
	4.7	Streaming Instability		176
		4.7.1	Linear Streaming Instability	176
		4.7.2	Streaming Model for Gravitational Collapse	177
	4.8	Pathways to Planetesimal Formation		178
	4.9	Further Reading		180
5	Terrestrial Planet Formation			181
	5.1	Physics of Collisions		181
		5.1.1	Gravitational Focusing	182

Contents

	5.1.2	Shear versus Dispersion Dominated Encounters	183
	5.1.3	Accretion versus Disruption	186
5.2		Statistical Models of Planetary Growth	190
	5.2.1	Approximate Treatment	191
	5.2.2	Shear and Dispersion Dominated Limits	193
	5.2.3	Isolation Mass	197
5.3		Velocity Dispersion	198
	5.3.1	Viscous Stirring	199
	5.3.2	Dynamical Friction	202
	5.3.3	Gas Drag	203
	5.3.4	Inelastic Collisions	204
5.4		Regimes of Planetesimal-Driven Growth	205
5.5		Coagulation Equation	207
5.6		Pebble Accretion	210
	5.6.1	Encounter Regimes	212
	5.6.2	Pebble Accretion Conditions	213
	5.6.3	Pebble Accretion Rates	216
	5.6.4	Relative Importance of Pebble Accretion	216
5.7		Final Assembly	217
5.8		Further Reading	219
6		Giant Planet Formation	220
6.1		Core Accretion	221
	6.1.1	Core/Envelope Structure	226
	6.1.2	Critical Core Mass	230
	6.1.3	Growth of Giant Planets	233
6.2		Constraints on the Interior Structure of Giant Planets	237
	6.2.1	Interior Structure from Gravity Field Measurements	238
	6.2.2	Internal Structure of Jupiter	239
6.3		Disk Instability	240
	6.3.1	Outcome of Gravitational Instability	241
	6.3.2	Cooling-Driven Fragmentation	242
	6.3.3	Disk Cooling Time Scale	243
	6.3.4	Infall-Driven Fragmentation	245
	6.3.5	Outcome of Disk Fragmentation	246
6.4		Further Reading	246
7		Early Evolution of Planetary Systems	247
7.1		Migration in Gaseous Disks	248
	7.1.1	Planet–Disk Torque in the Impulse Approximation	248
	7.1.2	Physics of Gas Disk Torques	251
	7.1.3	Torque Formulae	255
	7.1.4	Gas Disk Migration Regimes	257

	7.1.5	Gap Opening and Gap Depth	258
	7.1.6	Coupled Planet–Disk Evolution	261
	7.1.7	Eccentricity Evolution	264
7.2		Secular and Resonant Evolution	265
	7.2.1	Physics of an Eccentric Mean-Motion Resonance	266
	7.2.2	Example Definition of a Resonance	267
	7.2.3	Resonant Capture	270
	7.2.4	Kozai–Lidov Dynamics	272
	7.2.5	Secular Dynamics	275
7.3		Migration in Planetesimal Disks	276
	7.3.1	Application to Extrasolar Planetary Systems	280
7.4		Planetary System Stability	281
	7.4.1	Hill Stability	282
	7.4.2	Planet–Planet Scattering	287
	7.4.3	The Titius–Bode Law	289
7.5		Solar System Migration Models	289
	7.5.1	Early Theoretical Developments	290
	7.5.2	The Nice Model	291
	7.5.3	The Grand Tack Model	292
7.6		Debris Disks	293
	7.6.1	Collisional Cascades	293
	7.6.2	Debris Disk Evolution	298
	7.6.3	White Dwarf Debris Disks	299
7.7		Further Reading	300

Appendix A	Physical and Astronomical Constants	301
Appendix B	The Two-Body Problem	302
Appendix C	N-Body Methods	311
	References	319
	Index	329

Preface

The study of planet formation has a long history. The idea that the Solar System formed from a rotating disk of gas and dust – the *Nebula Hypothesis* – dates back to the writings of Kant, Laplace, and others in the eighteenth century. A quantitative description of terrestrial planet formation was already in place by the late 1960s, when Viktor Safronov published his now classic monograph *Evolution of the Protoplanetary Cloud and Formation of the Earth and the Planets*, while the main elements of the core accretion theory for gas giant planet formation were developed in the early 1980s. More recently, new observations have led to renewed interest in the problem. The most dramatic development has been the identification of extrasolar planets, first around a pulsar and subsequently in large numbers around main-sequence stars. These detections have allowed us to start to assess the Solar System's place amid an extraordinary diversity of extrasolar planetary systems. The advent of high resolution imaging of protoplanetary disks and the discovery of the Solar System's Kuiper Belt have been almost as influential, the former by providing direct information about the initial conditions for planet formation, the latter by highlighting the role of dynamics in the early evolution of planetary systems.

My goals in writing this text are to provide a concise introduction to the classical theory of planet formation and to more recent developments spurred by new observations. Inevitably, the range of topics covered is far from comprehensive. The emphasis is firmly on the *astrophysical* aspects of planet formation, including the physics of the protoplanetary disk, the agglomeration of dust into planetesimals and planets, and the dynamical interactions between those bodies and the disk and among themselves. The information that can be deduced from study of the chemical and geological makeup of Solar System bodies is discussed in places where that information is particularly pertinent, but this book is intended to complement rather than to replace textbooks on planetary science and cosmochemistry.

This book began as a graduate course that I taught at the University of Colorado in Boulder, for which the prerequisites were undergraduate classical physics and mathematical methods. The primary readership is beginning graduate students, but most of the text ought to be accessible to undergraduates who have had some exposure to Newtonian mechanics and fluid dynamics. Although the mathematical demands are relatively elementary, the text does not shy away from covering modern theoretical developments, including those where research is very much still ongoing. Especially in these areas I provide extensive references to the technical literature to enable interested readers to explore further.

The decade since the first edition was published has seen further dramatic advances. NASA's *Kepler* mission has revolutionized our understanding of the population of relatively small extrasolar planets, while high resolution images of protoplanetary disks with *ALMA* have identified a wealth of largely unexpected structure. The chapter on observations has required major revision. On the theory side there has been intense interest in several processes that were either unknown or under-appreciated (at least by me) ten years ago, including pebble accretion, disk winds, the streaming instability, and vortices. Those omissions have been remedied. I have also added reference material on dynamics, and thoroughly revised the existing text to reflect both new thinking in areas such as planetary migration and my own teaching preferences.

My understanding of planet formation has been shaped by the many collaborators that I have had the privilege to work with. I am indebted to them, to the students in Boulder and at various summer schools who have informed my thinking about how best to teach the subject, and to the colleagues who have provided feedback and encouragement. Lastly, my thanks to Dada, whose unwavering support brought this new edition to fruition.

1

Observations of Planetary Systems

Planets can be defined informally as large bodies, in orbit around a star, that are not massive enough to have ever derived a substantial fraction of their luminosity from nuclear fusion. This definition fixes the maximum mass of a planet to be at the deuterium burning threshold, which is approximately 13 Jupiter masses for solar composition objects (1 M_J = 1.899 × 10^{30} g). More massive objects are called brown dwarfs. The lower mass cut-off for what we call a planet is not as easily defined. For a predominantly icy body self-gravity overwhelms material strength when the diameter exceeds a few hundred km, leading to a hydrostatic shape that is near spherical in the absence of rapid rotation (the critical diameter is larger for rocky bodies). Planets (including dwarf planets) are defined as exceeding this threshold size. As planets get larger they typically become more interesting as individual objects; larger bodies retain more internal heat to power geological processes and can hold on to more significant atmospheres. As members of a planetary system the dynamical influence of massive bodies also acts to destabilize and clear out most neighboring orbits. These physical and dynamical characteristics can be used to sub-divide the class of planets, but we will not have cause to make such distinctions in this book. It is likely that some objects of planetary mass exist that are *not* bound to a central star, having formed either in isolation or following ejection from a planetary system. Such objects are normally called "planetary-mass objects" or "free-floating planets."

Complementary constraints on theories of planet formation come from observations of the Solar System and of extrasolar planetary systems. Space missions have yielded exquisitely detailed information on the surfaces (and in some cases interior structures) of the Solar System's planets and satellites, and an increasing number of its minor bodies. Some of the most fundamental facts about the Solar System are reviewed in this chapter, while other relevant observations are discussed subsequently in connection with related theoretical topics. By comparison with the Solar System our knowledge of individual extrasolar planetary systems is meager – in many cases it can be reduced to a handful of imperfectly known numbers characterizing the orbital properties of the planets – but this is compensated in part by the large and rapidly growing number of known systems. It is only by studying extrasolar planetary systems that we can make statistical studies of the range of outcomes of the planet formation process, and avoid bias introduced by the fact that the Solar System must necessarily be one of the subset of planetary systems that admit the existence of a habitable world.

1.1 Solar System Planets

The Solar System has eight planets. Jupiter and Saturn are gas giants composed primarily of hydrogen and helium, although their composition is substantially enhanced in heavier elements when compared to that of the Sun. Uranus and Neptune are ice giants, composed of water, ammonia, methane, silicates, and metals, atop which sit relatively low mass hydrogen and helium atmospheres. There are also four terrestrial planets, two of which (Earth and Venus) have quite similar masses. Mars is almost an order of magnitude less massive and Mercury is smaller still, though its density is anomalously high and similar to that of the Earth. There is more than an order of magnitude gap between the masses of the most massive terrestrial planets and the ice giants, and these two classes of planets have entirely distinct radii and structures. In addition there are a number of dwarf planets, including the trans-Neptunian objects Pluto, Eris, Haumea, and Makemake, and the asteroid Ceres. Many more dwarf planets of comparable size, and possibly even larger objects, remain to be discovered in the outer Solar System.

The orbital elements, masses and equatorial radii of the Solar System's planets are summarized in Table 1.1. With the exception of Mercury, the planets have almost circular, almost coplanar orbits. There is a small but significant misalignment of about $7°$ between the mean orbital plane of the planets and the solar equator. Architecturally, the most intriguing feature of the Solar System is that the giant and

Table 1.1 *The orbital elements (semi-major axis a, eccentricity e and inclination i), masses and equatorial radii of Solar System planets. The orbital elements are quoted for the J2000 epoch and are with respect to the mean ecliptic. Data from JPL.*

	a / AU	e	i / deg	M_p/g	R_p/cm
Mercury	0.3871	0.2056	7.00	3.302×10^{26}	2.440×10^8
Venus	0.7233	0.0068	3.39	4.869×10^{27}	6.052×10^8
Earth	1.000	0.0167	0.00	5.974×10^{27}	6.378×10^8
Mars	1.524	0.0934	1.85	6.419×10^{26}	3.396×10^8
Jupiter	5.203	0.0484	1.30	1.899×10^{30}	7.149×10^9
Saturn	9.537	0.0539	2.49	5.685×10^{29}	6.027×10^9
Uranus	19.19	0.0473	0.77	8.681×10^{28}	2.556×10^9
Neptune	30.07	0.0086	1.77	1.024×10^{29}	2.476×10^9

terrestrial planets are clearly segregated in orbital radius, with the giants only being found at large radii where the Solar Nebula (the disk of gas and dust from which the planets formed) would have been cool and icy.

The planets make a negligible contribution ($\simeq 0.13\%$) to the mass of the Solar System, which overwhelmingly resides in the Sun. The mass of the Sun, $M_\odot = 1.989 \times 10^{33}$ g, is made up of hydrogen (fraction by mass in the envelope $X = 0.73$), helium ($Y = 0.25$), and heavier elements (described in astronomical parlance as "metals," with $Z = 0.02$). One notes that even most of the condensible elements in the Solar System are in the Sun. This means that if a significant fraction of the current mass of the Sun passed through a disk during the formation epoch the process of planet formation need not be 100% efficient in converting solid material in the disk into planets. In contrast to the mass, most of the angular momentum of the Solar System is locked up in the orbital angular momentum of the planets. Assuming rigid rotation at angular velocity Ω, the solar angular momentum can be written as

$$J_\odot = k^2 M_\odot R_\odot^2 \Omega, \tag{1.1}$$

where $R_\odot = 6.96 \times 10^{10}$ cm is the solar radius. Taking $\Omega = 2.9 \times 10^{-6}$ s^{-1} (the solar rotation period is 25 dy), and adopting $k^2 \approx 0.1$ (roughly appropriate for a star with a radiative core), we obtain as an estimate for the solar angular momentum $J_\odot \sim 3 \times 10^{48}$ g cm^2 s^{-1}. For comparison, the orbital angular momentum associated with Jupiter's orbit at semi-major axis a is

$$J_J = M_J \sqrt{GM_\odot a} \simeq 2 \times 10^{50} \text{ g cm}^2 \text{ s}^{-1}. \tag{1.2}$$

Even this value is small compared to the typical angular momentum contained in molecular cloud cores that collapse to form low mass stars. We infer that substantial segregation of angular momentum and mass must have occurred during the star formation process.

The orbital radii of the planets do not exhibit any relationships that yield immediate clues as to their formation or early evolution. (We briefly mention the Titius–Bode law in Section 7.4.3, but this empirical relation is not thought to have any fundamental basis.) From a dynamical standpoint the most relevant fact is that although the planets orbit close enough to perturb each other's orbits, the perturbations are all nonresonant. Resonances occur when characteristic frequencies of two or more bodies display a near-exact commensurability. They adopt disproportionate importance in planetary dynamics because, in systems where the planets do not make close encounters, gravitational forces between the planets are generally much smaller (typically by a factor of 10^3 or more) than the dominant force from the star. These small perturbations are largely negligible unless special circumstances (i.e. a resonance) cause them to add up coherently over time. The simplest type of resonance, known as a *mean-motion resonance* (MMR), occurs when the periods P_1 and P_2 of two planets satisfy

$$\frac{P_1}{P_2} \simeq \frac{i}{j}, \tag{1.3}$$

where i and j are integers and use of the approximate equality sign denotes the fact that such resonances have a finite width. One can, of course, always find a pair of integers such that this equation is satisfied for arbitrary P_1 and P_2, so a more precise statement is that there are no dynamically important resonances among the major planets.[1] Nearest to resonance in the Solar System are Jupiter and Saturn, whose motion is affected by their proximity to a 5:2 mean-motion resonance known as the "great inequality" (the existence of this near resonance, though not its dynamical significance, was known even to Kepler). Among lower mass objects Pluto is one of a large class of Kuiper Belt Objects (KBOs) in 3:2 resonance with Neptune, and there are many examples of important resonances among satellites and in the asteroid belt.

1.2 The Minimum Mass Solar Nebula

The mass of the disk of gas and dust that formed the Solar System is unknown. However, it is possible to use the observed masses, orbital radii and compositions of the planets to derive a *lower limit* for the amount of material that must have been present, together with a crude idea as to how that material was distributed with distance from the Sun. This is called the "minimum mass Solar Nebula" (Weidenschilling,1977a). The procedure is simple:

(1) Starting from the observed (or inferred) masses of heavy elements such as iron in the planets, augment the mass of each planet with enough hydrogen and helium to bring the augmented mixture to solar composition.
(2) Divide the Solar System up into annuli, such that each annulus is centered on the current semi-major axis of a planet and extends halfway to the orbit of the neighboring planets.
(3) Imagine spreading the augmented mass for each planet across the area of its annulus. This yields a characteristic gas surface density Σ (units g cm^{-2}) at the location of each planet.

Following this scheme, out to the orbital radius of Neptune the derived surface density scales roughly as $\Sigma(r) \propto r^{-3/2}$. Since the procedure for constructing the

---

[1] Roughly speaking, a resonance is typically dynamically important if the integers i and j (or their difference) are small. Care is needed, however, since although the 121:118 mean-motion resonance between Saturn's moons Prometheus and Pandora formally satisfies this condition (since the *difference* is small) one would not immediately suspect that such an obscure commensurability would be significant.

distribution is somewhat arbitrary it is possible to obtain a number of different normalizations, but the most common value used is that quoted by Hayashi (1981):

$$\Sigma(r) = 1.7 \times 10^3 \left(\frac{r}{1 \text{ AU}}\right)^{-3/2} \text{ g cm}^{-2}. \tag{1.4}$$

Integrating this expression out to 30 AU the enclosed mass works out to be 0.01 $M_\odot$, which is comparable to the estimated masses of protoplanetary disks around other stars (though these have a wide spread). Hayashi (1981) also provided an estimate for the surface density of solid material as a function of radius in the disk:

$$\Sigma_s(\text{rock}) = 7.1 \left(\frac{r}{1 \text{ AU}}\right)^{-3/2} \text{ g cm}^{-2} \text{ for } r < 2.7 \text{ AU}, \tag{1.5}$$

$$\Sigma_s(\text{rock/ice}) = 30 \left(\frac{r}{1 \text{ AU}}\right)^{-3/2} \text{ g cm}^{-2} \text{ for } r > 2.7 \text{ AU}. \tag{1.6}$$

These distributions are shown in Fig. 1.1. The discontinuity in the solid surface density at 2.7 AU is due to the presence of icy material in the outer disk that would be destroyed in the hotter inner regions.

Although useful as an order of magnitude guide, the minimum mass Solar Nebula (as its name suggests) provides only an approximate lower limit to the amount of mass that must have been present in the Solar Nebula. As we will discuss later, it is very likely that both the gas and solid disks evolved substantially over time. There is no reason to believe that the minimum mass Solar Nebula reflects either the initial inventory of mass in the Solar Nebula, or the steady-state profile of the protoplanetary disk around the young Sun.

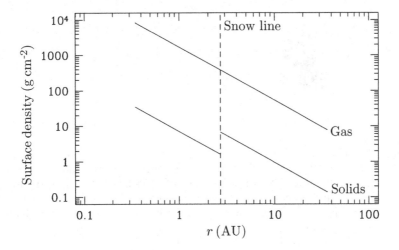

Figure 1.1 The surface density in gas (upper line) and solids (lower broken line) as a function of radius in Hayashi's minimum mass Solar Nebula. The dashed vertical line denotes the location of the snow line.

1.3 Minor Bodies in the Solar System

In addition to the planets, the Solar System contains a wealth of minor bodies: asteroids, Trans-Neptunian Objects (TNOs, including those in the Kuiper Belt), comets, and planetary satellites. The total mass in these reservoirs is now small.[2] The main asteroid belt has a mass of about $5 \times 10^{-4}\ M_\oplus$ (Petit *et al.*, 2001), while the more uncertain estimates for the Kuiper Belt are of the order of 0.1 $M_\oplus$ (Chiang *et al.*, 2007). Although dynamically unimportant, the distribution of minor bodies is extremely important for the clues it provides to the early history of the Solar System. As a very rough generalization the Solar System is dynamically full, in the sense that most locations where small bodies could stably orbit for billions of years are, in fact, populated. In the inner Solar System, the main reservoir is the main asteroid belt between Mars and Jupiter, while in the outer Solar System the Kuiper Belt is found beyond the orbit of Neptune.

Figure 1.2 shows the distribution of a sample of numbered asteroids in the inner Solar System, taken from the *Jet Propulsion Laboratory*'s small-body database. Most of the bodies in the main asteroid belt have semi-major axes a in the range between 2.1 and 3.3 AU. The distribution of a is by no means smooth, reflecting the crucial role of resonant dynamics in shaping the asteroid belt. The prominent regions, known as the Kirkwood (1867) gaps, where relatively few asteroids are

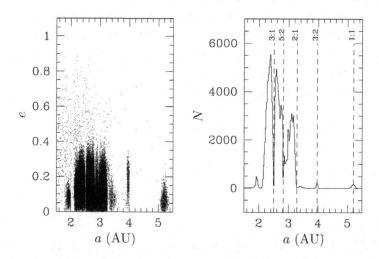

Figure 1.2 The orbital elements of a sample of numbered asteroids in the inner Solar System. The left-hand panel shows the semi-major axes a and eccentricity e of asteroids in the region between the orbits of Mars and Jupiter. The right-hand panel shows a histogram of the distribution of asteroids in semi-major axis. The locations of a handful of mean-motion resonances with Jupiter are marked by the dashed vertical lines.

---

[2] Indirect evidence suggests that the primordial asteroid and Kuiper belts were much more massive. A combination of dynamical ejection, and/or collisional grinding of bodies to dust that is then rapidly lost as a result of radiation pressure forces is likely to be responsible for their depletion.

found coincide with the locations of mean-motion resonances with Jupiter, most notably the 3:1 and 5:2 resonances. In addition to these locations – at which resonances with Jupiter are evidently depleting the population of minor bodies – there are *concentrations* of asteroids at both the co-orbital 1:1 resonance (the Trojan asteroids), and at the interior 3:2 resonance (the Hilda asteroids). This is a graphic demonstration of the fact that different resonances can either destabilize or protect asteroid orbits (for a thorough analysis of the dynamics involved the reader should consult Murray & Dermott, 1999). Also notable is that the asteroids, unlike the major planets, have a distribution of eccentricity e that extends to moderately large values. Between 2.1 and 3.3 AU the mean eccentricity of the numbered asteroids is $\langle e \rangle \simeq 0.14$. As a result, collisions in the asteroid belt today typically involve relative velocities that are large enough to be disruptive. Indeed, a number of asteroid families (Hirayama, 1918) are known, whose members share similar orbital elements (a, e, i). These asteroids are interpreted as debris from disruptive collisions taking place within the asteroid belt, in some cases relatively recently (within the last few Myr, e.g. Nesvorný *et al.*, 2002).

Figure 1.3 shows the distribution of a sample of outer Solar System bodies, maintained by the IAU's *Minor Planet Center*. Among the known planets outer Solar System bodies interact most strongly with Neptune, and to leading order they are classified based upon the nature of that interaction.

- *Resonant Kuiper Belt Objects (KBOs)* currently occupy mean-motion resonances with Neptune. The most common resonance is the 3:2 that is occupied by Pluto, and such objects are also called Plutinos. The eccentricity of some Plutinos – including Pluto itself – is large enough that their perihelion lies within the orbit of Neptune, and these objects depend upon their resonant configuration to avoid close encounters. The existence of this large population of moderately eccentric resonant bodies provided the original evidence for models in which Neptune migrated outward early in Solar System history.

- *Classical KBOs* orbit in a relatively narrow belt between Neptune's 3:2 and 2:1 MMRs (39.5AU $< a <$ 47.8 AU), and their number drops sharply toward the upper end of this range of semi-major axes (Trujillo & Brown, 2001). These objects are nonresonant and they have low enough eccentricity to avoid scattering encounters with Neptune. They can be divided into two sub-populations. The *cold* classical belt objects have lower inclinations $i < 2°$ (and generally also lower eccentricities) than the *hot* objects, which have $i > 6°$ (Dawson & Murray-Clay, 2012). (Objects with intermediate inclinations cannot be classified reliably using only orbital information.) The dynamical classification matches up to apparent physical differences that are inferred from measurements of the color and size distribution, which suggests that the cold and hot populations derive from distinct source populations.

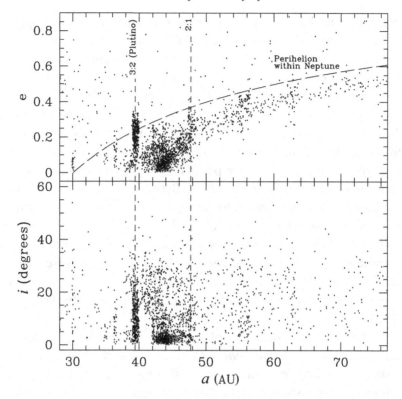

Figure 1.3 The distribution of eccentricity and inclination for a sample of minor bodies in the outer Solar System beyond the orbit of Neptune. The dashed vertical lines indicate the locations of mean-motion resonances with Neptune. Objects with eccentricity above the long-dashed line have perihelia that lie within the orbit of Neptune.

- The *scattering* population have perihelia $a(1 - e) \approx a_{Nep}$, where a_{Nep} is the semi-major axis of Neptune. These objects are in close dynamical contact with Neptune, and their orbits evolve as a result of the planet's perturbations.
- The *detached* population makes up the rest – objects on typically quite eccentric orbits that are not currently in dynamical contact with Neptune. Some of these objects are so detached that their orbits must have been established under different dynamical conditions earlier in Solar System history. A notable example is the large object Sedna, whose perihelion distance of 76 AU lies way beyond the orbit of Neptune.

Comets that approach the Sun are more easily accessible messengers from the outer Solar System. The Jupiter Family comets have low inclinations and are thought to originate from within the TNO reservoirs discussed above. Other comets, however, have a clearly distinct origin. In particular, among comets that are identified for the first time there is a population that has a broad inclination distribution and semi-major axes that cluster at $a \sim 2 \times 10^4$ AU, far beyond the Kuiper Belt. It was this evidence that led Jan Oort to postulate that the Sun is surrounded by a quasi-spherical reservoir of comets, now called the Oort cloud (Oort, 1950). The Oort cloud was established at an early epoch and delivers comets toward the inner Solar System over time as a consequence of Galactic tidal forces and perturbations from passing stars.

Planetary satellites in the Solar System also fall into several classes. The regular satellites of Jupiter, Saturn, Uranus, and Neptune have relatively tight prograde orbits that lie close to the equatorial plane of their respective planets. This suggests that these satellites formed from disks, analogous to the Solar Nebula itself, that surrounded the planets shortly after their formation. The total masses of the regular satellite systems are a relatively constant fraction (about 10^{-4}) of the mass of the host planet, with the largest satellite, Jupiter's moon Ganymede, having a mass of 0.025 $M_{\oplus}$. The presence of resonances between different satellite orbits – most notably the *Laplace resonance* that involves Io, Europa, and Ganymede (Io lies in 2:1 resonance with Europa, which in turn is in 2:1 resonance with Ganymede) – is striking. As in the case of Pluto's resonance with Neptune, the existence of these nontrivial configurations among the satellites provides evidence for past orbital evolution that was followed by resonant capture. Orbital migration within a primordial disk, or tidal interaction with the planet, are candidates for explaining these resonances.

The giant planets also possess extensive systems of irregular satellites, which are typically more distant and which do not share the common disk plane of the regular satellites. These satellites were probably captured by the giant planets from heliocentric orbits.

The sole example of a natural satellite of a terrestrial planet – the Earth's Moon – is distinctly different from any giant planet satellite. Relative to its planet it is much more massive (the Moon is more than 1% of the mass of the Earth), and its orbital angular momentum makes up most of the angular momentum of the Earth–Moon system. The Moon's composition is not the same as that of the Earth; there is less iron (resulting in a lower density than the uncompressed density of the Earth) and evidence for depletion of some volatile elements. Some aspects of the composition, in particular the ratios of stable isotopes of oxygen, are however essentially indistinguishable from those measured from terrestrial mantle samples. Qualitatively these properties are interpreted within models in which the Moon formed from the cooling of a heavy-element rich disk generated following a giant impact early in the Earth's history (Hartmann & Davis, 1975; Cameron & Ward, 1976), though some of the quantitative constraints remain challenging to reproduce. Pluto's large moon Charon may have formed in the aftermath of a similar impact.

1.4 Radioactive Dating of the Solar System

Determining the ages of individual stars from astronomical observations is a difficult exercise, and good constraints are normally only possible if the frequencies of stellar oscillations can be identified via photometric or spectroscopic data.

Much more accurate age determinations are possible for the Solar System, via radioactive dating of apparently pristine samples from meteorites.

It is worth clarifying at the outset how radioactive dating works, because it is not as simple as one might initially think. Consider a notional radioactive decay A $\rightarrow$ B that occurs with mean lifetime τ. After time t the abundance n_A of "A" is reduced from its initial value n_{A0} according to

$$n_A = n_{A0} e^{-t/\tau}, \tag{1.7}$$

while that of "B" increases,

$$n_B = n_{B0} + n_{A0}(1 - e^{-t/\tau}). \tag{1.8}$$

We can assume that τ is known precisely from laboratory measurements. However, it is clear that we cannot in general determine the age because we have three unknowns (t and the initial abundances of the two species) but only two observables (the current abundances of each species). Getting around this roadblock requires considering more complex decays and imposing assumptions about how the samples under consideration formed in the first place.

For a simple example that works we can look at a rock containing radioactive potassium (^{40}K) that solidifies from the vapor or liquid phases during the epoch of planet formation. One of the decay channels of ^{40}K is

$$^{40}\text{K} \rightarrow {}^{40}\text{Ar.} \tag{1.9}$$

This decay has a half-life of 1.25 Gyr and a branching ratio $\xi \approx 0.1$. (The branching ratio describes the probability that the radioactive isotope decays via a specific channel. In this case ξ is small because ^{40}K decays more often into ^{40}Ca.) If we assume that the rock, once it has solidified, traps the argon and that *there was no argon in the rock to start with*, then we have eliminated one of the generally unknown quantities and measuring the relative abundance of ^{40}Ar and ^{40}K suffices to determine the age. Quantitatively, if the parent isotope ^{40}K has an initial abundance $n_p(0)$ when the rock solidifies at time $t = 0$, then at later times the abundances of the parent isotope n_p and daughter isotope n_d are given by the usual exponential formulae that characterize radioactive decay:

$$n_p = n_p(0)e^{-t/\tau},$$
$$n_d = \xi n_p(0)\left[1 - e^{-t/\tau}\right], \tag{1.10}$$

where τ, the mean lifetime, is related to the half-life via $\tau = t_{1/2}/\ln 2$. The ratio of the daughter to parent abundance is

$$\frac{n_d}{n_p} = \xi\left(e^{t/\tau} - 1\right). \tag{1.11}$$

A laboratory measurement of the left-hand-side then fixes the age provided that the nuclear physics of the decay (the mean lifetime and the branching ratio) is accurately known. Notice that this method works to date the age of the rock (rather than the epoch when the radioactive potassium was formed) because minerals have distinct chemical compositions that differ – often dramatically so – from the average composition of the protoplanetary disk. In the example above, it is reasonable to assume that any ^{40}Ar atoms formed prior to the rock solidifying will not be incorporated into the rock, first because the argon will be diluted throughout the disk and second because it is an unreactive element that will not be part of the same minerals as potassium.

Radioactive dating is also possible in systems where we cannot safely assume that the initial abundance of the daughter isotope is negligible. The decay of rubidium 87 into strontium 87,

$$^{87}\text{Rb} \rightarrow {}^{87}\text{Sr}, \tag{1.12}$$

occurs with a half-life of 48.8 Gyr. Unlike argon, strontium is not a noble gas, and we cannot assume that the rock is initially devoid of strontium. If we denote the initial abundance of the daughter isotope as $n_d(0)$, then measurement of the ratio (n_d/n_p) yields a single constraint on two unknowns (the initial daughter abundance and the age) and dating appears impossible. Again, the varied chemical properties of rocks allow progress. Suppose we measure samples from two different minerals within the same rock, and compare the abundances of ^{87}Rb and ^{87}Sr not to each other, but to the abundance of a separate stable isotope of strontium ^{86}Sr. Since ^{86}Sr is chemically identical to the daughter isotope ^{87}Sr that we are interested in, it is reasonable to assume that the ratio ^{87}Sr/^{86}Sr was initially constant across samples. The ratio ^{87}Rb/^{86}Sr, on the other hand, can differ between samples. As the rock ages, the abundance of the parent isotope drops and that of the daughter increases. Quantitatively,

$$n_p = n_p(0)e^{-t/\tau},$$
$$n_d = n_d(0) + \xi n_p(0)\left[1 - e^{-t/\tau}\right]. \tag{1.13}$$

Eliminating $n_p(0)$ between these equations and dividing by the abundance n_{ds} of the second stable isotope of the daughter species (^{86}Sr in our example) we obtain

$$\left(\frac{n_d}{n_{ds}}\right) = \left(\frac{n_d(0)}{n_{ds}}\right) + \xi\left(\frac{n_p}{n_{ds}}\right)\left[e^{t/\tau} - 1\right]. \tag{1.14}$$

The first term on the right-hand-side is a constant. We can then plot the relative abundances of the parent isotope (n_p/n_{ds}) and the daughter isotope (n_d/n_{ds}) from different samples on a ratio–ratio plot called an isochron diagram, such as the one shown schematically in Fig. 1.4. Inspection of Eq. (1.14) shows that we should expect the points from different samples to lie on a straight line whose slope (together with independent knowledge of the mean lifetime) fixes the age. Two samples are in principle sufficient to yield an age determination, but additional data

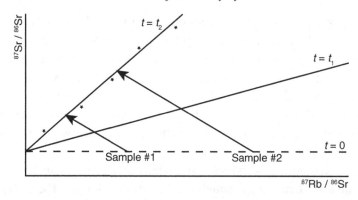

Figure 1.4 Ratio–ratio plot for dating rocks using the radioactive decay $^{87}Rb \rightarrow ^{87}Sr$. The abundance of these isotopes is plotted relative to the abundance of a separate stable isotope of strontium ^{86}Sr. When the rock solidifies, different samples contain identical ratios of $^{87}Sr/^{86}Sr$, but different ratios of $^{87}Rb/^{86}Sr$. The ratios of different samples track a steepening straight line as the rock ages.

provide a check against possible systematic errors. If the points fail to lie on a straight line something is wrong.

1.4.1 Lead–Lead Dating

The most important system for absolute determination of the age of Solar System samples, lead–lead dating, is based upon the measurement of lead and uranium isotopes. Lead has four stable and naturally occurring isotopes with mass numbers 204, 206, 207, and 208. Of these, ^{204}Pb preserves its primordial abundance, while the abundance of the heavier isotopes increases over time because these are daughter isotopes from the decay of uranium and thorium. The parent/daughter relationships are

$$^{238}U \rightarrow ^{206}Pb \quad (t_{1/2} = 4.47 \times 10^9 \text{ yr}),$$
$$^{235}U \rightarrow ^{207}Pb \quad (t_{1/2} = 7.04 \times 10^8 \text{ yr}), \qquad (1.15)$$
$$^{232}Th \rightarrow ^{208}Pb \quad (t_{1/2} = 1.40 \times 10^{10} \text{ yr}).$$

The relatively high abundance of the parent isotopes, coupled with the favorable half-lives, mean that this system is well suited to deliver ages of high accuracy and precision.

There are several ways to derive dates using the above system. Limiting attention to the two uranium decays gives a pair of equations analogous to Eq. (1.14):

$$\left(\frac{^{206}Pb}{^{204}Pb}\right) = \left(\frac{^{206}Pb}{^{204}Pb}\right)_0 + \left(\frac{^{238}U}{^{204}Pb}\right)\left[e^{t/\tau_1} - 1\right], \qquad (1.16)$$

$$\left(\frac{^{207}Pb}{^{204}Pb}\right) = \left(\frac{^{207}Pb}{^{204}Pb}\right)_0 + \left(\frac{^{235}U}{^{204}Pb}\right)\left[e^{t/\tau_2} - 1\right]. \qquad (1.17)$$

Here the ratios presented without subscripts are the present-day (measurable) values, while those with the subscript "0" refer to the initial values at the time when the system became closed. The mean lifetimes τ_1 and τ_2 refer to the decay of ^{238}U and ^{235}U. At this point we could, following the logic above, make two ratio–ratio plots (e.g. from the first equation ^{206}Pb/^{204}Pb against ^{238}U/^{204}Pb) and derive two independent ages from a single sample of primitive material. It turns out, however, that we can do a better job at reducing practical sources of error (associated, for example, with terrestrial contamination) by considering the two uranium/lead systems jointly. By taking the ratio of Eq. (1.17) to Eq. (1.16) we obtain a form in which the left-hand-side depends only on initial or current lead isotope ratios, while the right-hand-side depends only on the current uranium isotope ratio:

$$\frac{(^{207}\text{Pb}/^{204}\text{Pb}) - (^{207}\text{Pb}/^{204}\text{Pb})_0}{(^{206}\text{Pb}/^{204}\text{Pb}) - (^{206}\text{Pb}/^{204}\text{Pb})_0} = \left(\frac{^{235}\text{U}}{^{238}\text{U}}\right)\left(\frac{e^{t/\tau_1} - 1}{e^{t/\tau_2} - 1}\right). \tag{1.18}$$

Multiplying first by the denominator and then by $(^{204}\text{Pb}/^{206}\text{Pb})$ shows that a ratio–ratio plot of $(^{204}\text{Pb}/^{206}\text{Pb})$ versus $(^{207}\text{Pb}/^{206}\text{Pb})$ ought to give a straight line:

$$\left(\frac{^{207}\text{Pb}}{^{206}\text{Pb}}\right) = \underbrace{\left(\frac{^{235}\text{U}}{^{238}\text{U}}\right)\left(\frac{e^{t/\tau_1} - 1}{e^{t/\tau_2} - 1}\right)}_{\text{intercept}} + b(t)\left(\frac{^{204}\text{Pb}}{^{206}\text{Pb}}\right). \tag{1.19}$$

(Here $b(t)$ is an age-dependent constant whose exact form is not important for our purposes.) Measuring the intercept from a ratio–ratio plot involving the three lead isotopes, along with a current determination of the uranium isotope ratio, yields a measure of the sample age.

Connelly *et al.* (2017) review both the theory and the practical difficulties encountered in lead–lead dating of Solar System samples. Astonishingly, age determinations using this method are able to resolve events at the dawn of the Solar System, 4.57 Gyr ago, with a precision that is of the order of 0.2 Myr.

1.4.2 Dating with Short-Lived Radionuclides

Absolute chronometry provides an estimated age for the earliest Solar System solids of 4.5673 ± 0.0002 Gyr (Connelly *et al.*, 2017). This is the measurable quantity that, by convention, is described as the "age of the Solar System." Knowing the age of the Sun is useful for calibrating solar evolution models, but the exact absolute age is otherwise superfluous for studies of planet formation. More valuable are constraints on the time scale of critical phases of the planet formation process. For example, it is of interest to know whether the formation of the km-sized bodies called planetesimals was sudden or spread out over many Myr. Questions of this kind, which involve only *relative* ages, can of course be addressed given sufficiently accurate absolute chronometry. A complementary approach determines only relative ages from the decay products of short-lived isotopes. It is well established that the Solar Nebula initially contained radioactive species with very short half-lives, including ^{26}Al which has a half-life of only 0.72 Myr. The likely origin of these isotopes is nucleosynthesis within the cores of massive stars followed by ejection into the surrounding medium via either a supernova explosion or Wolf–Rayet stellar winds. In either case the implication is that the Sun formed in proximity to a rich star forming region (or within a cluster, now dissolved) that also contained massive stars.

Heating due to the radioactive decay of ^{26}Al is important for the thermal evolution of small bodies forming within protoplanetary disks, and the ionization produced from these decays may also be physically significant. For dating, the key point is that we can learn something about the relative ages of samples by measuring the abundance of daughter isotopes even when all of the parent species has long since decayed. For example, we could imagine measuring the abundance of ^{129}Xe, formed via the decay,

$$^{129}\text{I} \rightarrow \ ^{129}\text{Xe}, \tag{1.20}$$

and comparing it to the abundance of a stable isotope of iodine ^{127}I. The ^{129}I decay has a a half-life of only 17 Myr. Accordingly, we would expect that samples that form early would incorporate a higher $(^{129}\text{I}/^{127}\text{I})$ ratio, and this would lead to a higher $(^{129}\text{Xe}/^{127}\text{I})$ ratio that is preserved at late times after all of the radioactive iodine is gone.

The main caveat to be borne in mind with radioactive dating (both absolute and relative) is that it depends upon there being no other processes besides radioactive decay that alter the abundance of either the parent or daughter isotopes. The radioactive age of a rock measures the moment it solidified only if there has been no diffusion, or other alteration of the rock, during the intervening period. Even if the rock itself is pristine, high energy particles (cosmic rays) can induce nuclear reactions that may change the abundances of some critical isotopes. Relative chronometry additionally relies on the assumed spatial homogeneity of the parent species within the disk. Because the half-lives of the species used for relative dating are short, there may not have been enough time for the short-lived parents to have been mixed throughout the gas that forms the disk, and the assumption of homogeneity can reasonably be questioned.

1.5 Ice Lines

Liquid water is not stable under the low pressure conditions of the gaseous protoplanetary disk. Water ice, which is abundant at low temperature, sublimates directly to vapor in regions of the disk where the temperature exceeds about 150–170 K. The radial location in the disk beyond which the temperature falls below this value is called the snow line. Analogous "ice lines" exist for other common chemical species. The carbon monoxide (CO) ice line corresponds to a temperature of 23–28 K, while the carbon dioxide (CO_2) ice line is predicted to be at 60–72 K.

Theoretically, the location of the snow line is expected to change over time, due to variations in the stellar luminosity and in the rate at which gas is being accreted through the disk (as we will discuss in Chapter 3, accretion can be an important source of heat). For a solar mass, solar luminosity star, accreting at an observationally typical rate, the snow line is predicted to lie between 1 and 2 AU. Reasonable variations of these parameters lead to snow lines between about 0.5 and 5 AU (Garaud & Lin, 2007). Observationally, the location of the snow line is difficult to pin down from astronomical data on protoplanetary disks, and hence the main constraints come from laboratory measurement of meteoritic samples. Meteorites of the class known as carbonaceous chondrites, which are water-rich, have properties (such as their reflectance spectra) that match those of asteroids found only in the outer asteroid belt beyond about 2.5 AU. Conversely, the ordinary and enstatite chondrites, which contain negligible amounts of water, appear to originate from the inner asteroid belt at a distance from the Sun of around 2 AU. Consistent with this, the dwarf planet Ceres in the outer asteroid belt is known to have both surface and sub-surface water, and may contain a substantial interior water reservoir. The empirical conclusion is that compositional variations in the asteroid belt preserve evidence for a snow line in the early Solar Nebula at about $a \simeq 2.7$ AU.

The location of the snow line has important consequences for the composition and habitability of terrestrial planets. The standard definition of the "habitable zone" for terrestrial planets is based on requiring that liquid water is stable on the surface. Under Earth's atmospheric pressure, that translates into a surface temperature between 273 K $< T <$ 373 K. As noted above, however, the lower pressure in the disk gas means that prior to planet formation water would have existed only in the vapor phase at temperatures above $T \simeq 150$–170 K. This leads to a somewhat surprising prediction; planets that formed in situ within the habitable zone are typically assembled from planetesimals that formed interior to the snow line in a region of the disk that would have been too hot for water-rich minerals to condense. Solar System evidence for a snow line at 2.7 AU supports this scenario for the Earth.

The idea that habitable planets form inside the snow line is consistent not just with Solar System meteoritic evidence, but with the first-order composition of the Earth. Surface appearances to the contrary, the Earth is not a water world. The mass of water in the ocean, atmosphere, and crust of the Earth is just $2.8 \times 10^{-4} M_\oplus$ (where the Earth mass, $M_\oplus = 5.974 \times 10^{27}$ g). The amount of extra water locked in the mantle is uncertain (especially early in the Earth's history) and could be substantial, but it is clear that the total water fraction is orders of magnitude below that found in ice-rich bodies beyond the snow line.

Several models have been proposed to explain where the Earth's water came from. The leading hypothesis is that water was delivered to the Earth from a reservoir of small bodies (asteroids, comets, or a now vanished class of objects) beyond the snow line. Alternatively, Solar System material at 1 AU may have contained a small admixture of water (in the form of water molecules that can bind to some grain surfaces at relatively high temperature), and not have been as dry as is assumed in the standard model. The isotopic makeup of the Earth's water provides a constraint. The reference value for the deuterium to hydrogen (D/H) ratio of water in the Earth's ocean is 156 parts per million (ppm), in broad agreement with the mean 159 ± 10 ppm measured in carbonaceous chondrites (Morbidelli *et al.*, 2000). Data on comets are more limited, but values measured for Oort cloud comets are substantially higher (around 300 ppm). Hartogh *et al.* (2011) reported an Earth-like value of ≈ 160 ppm for the Jupiter family comet 103P/Hartley 2, but a much higher ratio of ≈ 530 pm was measured for 67P/Churyumov–Gerasimenko (the comet, also from the Jupiter family, which was the target of the *Rosetta* mission; Altwegg *et al.*, 2015).

Given the admittedly inconclusive evidence, the most likely source for the bulk of the Earth's water is asteroids from the outer part of the main belt. The mass of asteroids required is substantial. If our water arrived via asteroids with compositions similar to the carbonaceous chondrites, which have mass fractions of water of $\approx 10\%$, the total mass required would have been a few $10^{-3} M_\oplus$. Although this mass is negligible on the scale of the impacts that assembled the Earth and formed the Moon, it still exceeds the total mass in the present-day asteroid belt by an order of magnitude.

1.6 Meteoritic and Solar System Samples

Unique but sometimes hard to interpret information about conditions in the Solar Nebula comes from the study of Solar System samples, recovered from meteorites and from a handful of sample return missions. Meteorites originate from the asteroid belt, the Moon, and Mars, and have a composition that reflects the structure of their parent bodies. Undifferentiated bodies, which are generally smaller, never became hot enough to form a distinct core/mantle/crust structure. They give rise to chondritic meteorites, which once the most volatile elements are excluded have a bulk chemical composition that is similar to that of the Sun. Differentiated bodies, by contrast, are the progenitors of iron meteorites (made up of iron–nickel alloys) and achondrites, stony meteorites that are mostly made up of igneous rocks.

The chondritic meteorites are divided into numerous classes and are not, for the most part, entirely pristine. Their minerals often show evidence for changes that occurred on the parent body due to the action of heat and/or water. The carbonaceous chondrites are among the least altered (or most "primitive"), and together with samples returned directly from comets and asteroids represent the closest approximation to early Solar System material that can be studied in the laboratory.

The rock of chondritic meteorites is a mixture of three quite distinct components: refractory inclusions, chondrules, and matrix. The refractory inclusions – calcium, aluminium-rich inclusions (CAIs), and amoeboid olivine aggregates – have a chemical composition and structure that suggest that they formed at temperatures in excess of 1350 K. Both absolute and relative dating support the idea that CAIs are the oldest known Solar System materials, with an age estimated at 4.5673 ± 0.0002 Gyr (Connelly *et al.*, 2017). The dispersion in the inferred ages of CAIs is small, suggesting that they may have formed under high temperature disk conditions, presumably close to the Sun, during the earliest stage of disk evolution.

Chondrules are a second high temperature component of chondrites. Chondrules are typically 0.1–1 mm spheres of igneous rock that make up a variable but often large fraction of chondritic meteorites (as much as 80% of the volume of ordinary and enstatite chondrites, and 0 to 70% of carbonaceous chondrites; Nittler & Ciesla, 2016). The most basic fact about chondrules, indicated by their spherical shape and mineral texture, is that they formed due to cooling of initially molten precursors. A great deal is known experimentally about what happens as igneous rocks cool and crystallize, and as a consequence there are surprisingly informative constraints on the conditions under which chondrules formed (Desch *et al.*, 2012; Nittler & Ciesla, 2016). It is generally accepted that chondrule precursors – the solid particles or material that were melted to become chondrules – formed under relatively low temperature conditions, because chondrules contain FeS, which would not be present at $T \gtrsim 650$ K. The precursor material was then heated rapidly, on a time scale of at most minutes, to a temperature of about 2000 K. Volatile elements, such as Na and K, may have been lost while the chondrule was molten, but did not experience the mass-dependent isotope fractionation that would be expected if they were surrounded by a low–pressure gas. This implies formation in a relatively large high density environment where evaporation and fractionation are suppressed by the development of a volatile-rich vapor. After melting, the various mineral textures observed in chondrules suggest cooling at rates between about 10 K per hour and 10^4 K per hour. Even the top end of this range is orders of magnitude below the cooling rate expected for a single small sphere radiating freely to space, implying again that the formation regions were large enough so that radiative transfer effects slowed the cooling rate. Both absolute and relative chronometry show that chondrules did not form before CAIs. Some chondrules may have formed at roughly the same time as refractory inclusions, but others formed several Myr later.

The third component of chondritic meteorites is matrix, small micrometer-sized particles of minerals such as crystalline magnesian olivine, pyroxene, and amorphous ferromagnesian silicates. Matrix materials contain a larger fraction of volatile elements, consistent with formation under lower temperature conditions than those needed for either CAI or chondrule formation. The average composition of chondrules and matrix is, in a colloquial sense, "complementary," in that the former is depleted and the latter enriched in volatiles. The hypothesis of chemical *complementarity* holds that this relationship also holds for individual chondrites.

Complementarity in this stricter sense would provide evidence that chondrules and matrix formed from a common reservoir of disk gas.

Both CAIs and chondrules are found within meteorites whose parent bodies orbited within the asteroid belt at orbital radii not far from the inferred location of the snow line. At this radii there is almost an order of magnitude gulf between the expected disk temperatures and those implicated in the formation of these high temperature materials. For chondrules two main classes of models have been proposed to explain how the required high temperature can be realized at 3 AU from the Sun. One holds that chondrules formed as pre-existing solid particles passed through shock waves in the gas disk. The shock waves could have been either large-scale spiral shocks in the disk (although it is hard to see how these would form), or more localized bow shocks around orbiting planetary embryos. Alternatively, chondrules may have formed out of the cooling debris of planetesimal impacts. For CAIs the standard scenario is that these materials formed within the disk, probably closer to the Sun, and were transported radially before being incorporated into larger solid bodies.

The mystery of how CAIs and (especially) chondrules formed is one of the most important open problems in meteoritics and cosmochemistry. The wider relevance for planet formation cannot be determined without first knowing the answer. All that we know for sure is that the small mass currently in undifferentiated asteroids is made up of a high fraction by mass of chondrules. If the formation mechanism for chondrules involved long-lived disk processes then it is reasonable to assume that a large fraction of the solid material that forms terrestrial planets was once processed through a chondrule-like phase. Chondrules would then be generic precursors of planets. If, on the other hand, chondrules form from a process such as late planetesimal collisions, they are bystanders of only peripheral interest for understanding the main phases of planet formation.

Results from the *Stardust* sample return mission from the Jupiter family comet Wild 2 raise district but thematically related questions. Analysis of 1–10 μm particles collected from the comet showed that many of them are made up of olivine or pyroxene. These are crystalline silicates that can be produced from amorphous precursors via annealing at temperatures of 800 K and above (Brownlee *et al.*, 2006). This result implies that even the very cold regions of the disk where comets formed were somehow polluted with a fraction of material processed through high temperatures.

1.7 Exoplanet Detection Methods

The first extrasolar planetary system to be discovered was identified by Alex Wolszczan and Dale Frail via precision timing of pulses from the millisecond pulsar PSR1257+12 (Wolszczan & Frail, 1992). The system contains at least three planets on nearly circular orbits within 0.5 AU of the neutron star, with the outer two planets having masses close to 4 $M_\oplus$ and the inner planet having a mass 0.02 $M_\oplus$ which is comparable to the mass of the Moon. Millisecond pulsars are a type of rapidly rotating neutron star that form in supernova explosions, and the mass loss during the explosion would unbind any pre-existing planets. The planets in the PSR1257+12 system must therefore have originated within a disk formed subsequent to the explosion. The observation that pulsar planets are not common (Kerr *et al.*, 2015) implies that the conditions leading to their formation could involve moderately rare events.

The precision required to find Jupiter mass planets in relatively short-period orbits via monitoring of the radial velocity of their host stars first became available in the 1980s (Campbell *et al.*, 1988). The first accepted detection, of 51 Peg b, a Jupiter mass body orbiting a solar-type star, was made using this technique by Michel Mayor and Didier Queloz in 1995 (Mayor & Queloz, 1995). The first two decades of exoplanet discovery were dominated by detections made via the radial velocity and transit methods, with smaller numbers of planets being found using microlensing and direct imaging techniques. An additional technique, astrometry, is expected to play a growing role in future planet discovery. Based on existing data it is known that planets have a broad range of masses, from sub-Earth mass up to the deuterium burning limit, and orbit all the way from 0.01 AU to 100 AU from their host star. No single planet discovery method is sensitive across this entire parameter space, large fractions of which remain unexplored by any method. A synthesis of multiple surveys, each with its own selection function in planetary mass, radius, and orbital period, is needed to build up an understanding of the population of extrasolar planetary systems.

1.7.1 Direct Imaging

The most straightforward way to detect extrasolar planets is to image the planet as a source of light that is spatially separated from the stellar emission. The main difficulty is the extreme contrast between a planet and its host star. A planet of radius R_p, orbital radius a, and albedo A intercepts and reflects a fraction f of the incident starlight given by

$$f = \left(\frac{\pi R_p^2}{4\pi a^2}\right) A = 1.4 \times 10^{-10} \left(\frac{A}{0.3}\right) \left(\frac{R_p}{R_\oplus}\right)^2 \left(\frac{a}{1\,\text{AU}}\right)^{-2}. \qquad (1.21)$$

Recalling that magnitudes are defined in terms of the flux F via $m = -2.5 \log_{10} F +$ const, one finds that Earth-like planets are expected to be 24–25 magnitudes fainter than their host stars. This faintness means that, quite apart from the very unfavorable contrast ratio between planet and star, moderately deep exposures are needed to have even a chance of directly imaging planets in reflected light. Alternatively, one may contemplate imaging extrasolar planets in their thermal emission. If we crudely approximate the emission at frequency ν from a planet with surface temperature T as a blackbody, then the spectrum is described by the Planck function:

$$B_\nu(T) = \frac{2h\nu^3}{c^2} \frac{1}{\exp(h\nu/k_B T) - 1}, \tag{1.22}$$

where h is Planck's constant, c the speed of light, and k_B the Boltzmann constant. The peak of the spectrum falls at $h\nu_{max} = 2.8 k_B T$, which lies in the mid-IR for typical terrestrial planets (the wavelength corresponding to ν_{max} is $\lambda \approx 20$ μm for the Earth with $T = 290$ K). If the star, with radius R_*, also radiates as a blackbody at temperature T_*, the flux ratio at frequency ν is

$$f = \left(\frac{R_p}{R_*}\right)^2 \frac{\exp(h\nu/k_B T_*) - 1}{\exp(h\nu/k_B T) - 1}. \tag{1.23}$$

For an Earth-analog around a solar-type star the contrast ratio for observations at a wavelength of 20 μm is $f \sim 10^{-6}$, which is some four orders of magnitude more favorable than the corresponding ratio in reflected light. This advantage of working in the infrared is, however, offset by the need for a larger telescope in order to spatially resolve the planet. The spatial resolution of a telescope of diameter D working at a wavelength λ is

$$\theta \sim 1.22 \frac{\lambda}{D}. \tag{1.24}$$

A spatial resolution of 0.5 AU at a distance of 5 pc corresponds to an angular resolution of 0.1 arcsec, which is theoretically achievable in the visible ($\lambda = 550$ nm) with a telescope of diameter $D \approx 1.5$ m. At 20 μm, on the other hand, the required diameter balloons to $D \approx 50$ m, which is unfeasibly large to contemplate constructing as a monolithic structure in space.

These elementary considerations show that the difficulty of directly imaging planets depends strongly on their orbital radii. Giant planets orbiting at large or very large radii (tens of AU) are relatively easy to image, with known examples including the multiple planets of the HR 8799 system (Marois *et al.*, 2008). State-of-the-art direct imaging surveys employ a combination of telescope hardware and novel modes of observation to overcome the unfavorable contrast ratio between star and planet. Adaptive optics can recover close to diffraction-limited performance of large ground-based telescopes in the near-IR, and a coronograph can be used to suppress the dominant stellar contribution. Techniques such as angular differential imaging (Marois *et al.*, 2005), which combines observations that are rotated in the sky plane to reduce imaging artifacts caused by the point-spread-function of the optics, further improve the available dynamic range for planet discovery.

The challenge of imaging terrestrial planets orbiting within the habitable zone is altogether more formidable that that of detecting giant planets, but carries the payoff of being able to characterize their atmospheres through spectroscopy. The composition of the Earth's atmosphere is strongly modified by the presence of life, and the overarching goal of proposals to image terrestrial planets is to search for biomarkers such as oxygen, ozone, or methane. In principle, the depth, resolution, and contrast required to first image and then obtain low resolution ($R \sim 10^2$) spectra of potentially habitable planets could be realized in three different ways:

- A highly optimized coronagraph on a space telescope of at least 4 m class, operating with near diffraction-limited performance in the ultraviolet (UV) to near-IR spectral range (0.1 μm $\lesssim \lambda \lesssim$ 2 μm). Larger apertures offer greater throughput and higher resolution, but become more challenging due to the high degree of wavefront stability that is required for planetary detection.
- A similar sized telescope that operates in tandem with a starshade. An optimally shaped starshade of 50–100 m diameter, flying $\sim 10^5$ km away from a space telescope, creates a narrow patch of extremely deep shadow that is highly effective at suppressing starlight and allowing planet detection and characterization.
- An interferometer, with optical elements on multiple free-flying spacecraft, operating in the mid-IR.

These architectures are all optically possible but technically very challenging, and which approach is selected for implementation first will depend more on engineering considerations of cost and technical risk than on any simple physical principle. Over the longer term it is likely that observations in both the visible and mid-IR wavebands will be needed, since detailed characterization of the possible existence of life on any nearby habitable planets will require spectroscopy over as wide a band as possible.

1.7.2 Radial Velocity Searches

A star hosting a planet describes a small orbit about the center of mass of the star–planet system. The radial velocity method for finding extrasolar planets works by measuring this stellar orbit via detection of periodic variations in the radial velocity of the star. The radial velocity is determined from the Doppler shift of the stellar spectrum, which in favorable cases can be measured at better than meter per second precision.

Circular Orbits

The principles of the radial velocity technique can be illustrated with the simple case of circular orbits. Consider a planet of mass M_p orbiting a star of mass M_* in a circular orbit of semi-major axis a. For $M_\mathrm{p} \ll M_*$ the Keplerian orbital velocity of the planet is

$$v_\mathrm{K} = \sqrt{\frac{GM_*}{a}}. \tag{1.25}$$

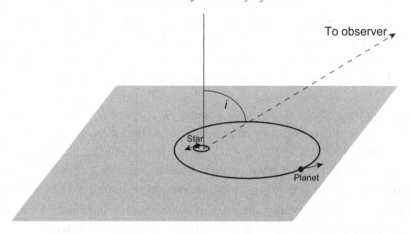

Figure 1.5 The reflex motion (greatly exaggerated) of a star orbited by a planet on an eccentric orbit. The inclination of the system is the angle i between the normal to the orbital plane and the observer's line of sight.

Conservation of linear momentum implies that the orbital velocity v_* of the star around the center of mass is determined by $M_* v_* = M_p v_K$. If the planetary system is observed at an inclination angle i, as shown in Fig. 1.5, the radial velocity varies sinusoidally with a semi-amplitude

$$K = v_* \sin i = \left(\frac{M_p}{M_*}\right) \sqrt{\frac{GM_*}{a}} \sin i. \qquad (1.26)$$

K is a directly observable quantity, as is the period of the orbit

$$P = 2\pi \sqrt{\frac{a^3}{GM_*}}. \qquad (1.27)$$

If the stellar mass can be estimated independently but the inclination is unknown, as is typically the case, then we have two equations in three unknowns and the best that we can do is to determine a lower limit to the planet mass via the product $M_p \sin i$. Since the average value of $\sin i$ for randomly inclined orbits is $\pi/4$ the statistical correction between the minimum and true masses is not large. The correction for individual systems is not normally known, however, and can be an important uncertainty, for example when analyzing the dynamics and stability of multiple planet systems.

Consideration of the solar radial velocity that is induced by the planets in the Solar System gives an idea of the typical magnitude of the signal. For Jupiter $v_* = 12.5$ m s^{-1} while for the Earth $v_* = 0.09$ m s^{-1}. For planets of a given mass there is a selection bias in favor of finding planets with small a. In an idealized survey in which the noise per observation is constant from star to star, Eq. (1.26) implies that the selection limit scales as

$$M_p \sin i|_{\text{minimum}} = C a^{1/2}, \qquad (1.28)$$

with C a constant. Planets with masses below this threshold are undetectable, as are planets with orbital periods that exceed the duration of the survey (since orbital solutions are generally poorly constrained when only a fraction of an orbit has been observed).

Eccentric Orbits

For real applications it is necessary to consider the radial velocity signature produced by planets on eccentric orbits. The derivation of most of the required results is worked through in Appendix B, so here we just state the main results that are required.

Consider a planet on an orbit of eccentricity e, semi-major axis a, and period P. The orbital radius varies between $a(1 + e)$ (apocenter) and $a(1 - e)$ (pericenter). Suppose that passage of the planet through pericenter occurs at time t_{peri}. In terms of these quantities, the *eccentric anomaly*[3] E is defined implicitly via Kepler's equation,

$$\frac{2\pi}{P} \left(t - t_{peri}\right) = E - e \sin E. \tag{1.29}$$

Kepler's equation is transcendental and cannot be solved for E in terms of simple functions. However, it can readily be solved numerically. Once E is known, the *true anomaly* f is given by Eq. (B.16),

$$\tan \frac{f}{2} = \sqrt{\frac{1 + e}{1 - e}} \tan \frac{E}{2}. \tag{1.30}$$

The true anomaly is the angle between the vector joining the bodies and the pericenter direction. Finally, in terms of these quantities, the radial velocity of the star is

$$v_*(t) = K \left[\cos(f + \varpi) + e \cos \varpi\right], \tag{1.31}$$

where the longitude of pericenter ϖ is the angle in the orbital plane between pericenter and the line of sight to the system. The eccentric generalization of Eq. (1.26) in the same limit in which $M_p \ll M_*$ is

$$K = \frac{1}{\sqrt{1 - e^2}} \left(\frac{M_p}{M_*}\right) \sqrt{\frac{GM_*}{a}} \sin i. \tag{1.32}$$

For a planet of given mass and semi-major axis, the amplitude of the radial velocity signature therefore increases with increasing e, due to the rapid motion of the planet (and star) close to pericenter passages.

Figure 1.6 illustrates the form of the radial velocity curves as a function of the eccentricity of the orbit and longitude of pericenter. Both e and ϖ can be measured given measurements of v_* as a function of time. Compared to the circular orbit

---

[3] The eccentric anomaly has a rather complex geometrical interpretation, but for our purposes all that matters is that it is a monotonically increasing function of t that specifies the location of the body around the orbit.

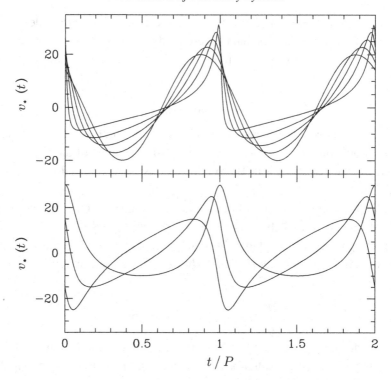

Figure 1.6 Time dependence of the radial velocity of a star hosting a single planet. The upper panel shows the stellar radial velocity when the planet has a circular orbit (the symmetric sinusoidal curve), and when the planet has an eccentric orbit with $e = 0.2, e = 0.4, e = 0.6$, and $e = 0.8$. In all cases the longitude of pericenter $\varpi = \pi/4$. The lower panel shows, for a planet with $e = 0.5$, how the stellar radial velocity varies with the longitude of pericenter.

of equal period, a planet on an eccentric orbit produces a stellar radial velocity signal of greater amplitude, but there are also long periods near apocenter where the gradient of v_* is rather small. These two properties of eccentric orbits mean that, depending upon the observing strategy employed, a radial velocity survey can be biased either in favor of or against finding eccentric planets. Such bias, however, is not a major concern for current samples, and the most important selection effects are those already discussed for circular orbits.

Noise Sources

The amplitude of the radial velocity signal produced by Jovian analogs in extra-solar planetary systems is of the order of 10 m s^{-1}. High resolution astronomical spectrographs operating in the visible part of the spectrum have resolving powers $R \sim 10^5$, which correspond to a velocity resolution $\Delta v \approx c/R$ of a few km s^{-1}. The Doppler shift in the stellar spectrum due to orbiting planets therefore results in a periodic translation of the spectrum on the detector by a few *thousandths* of a

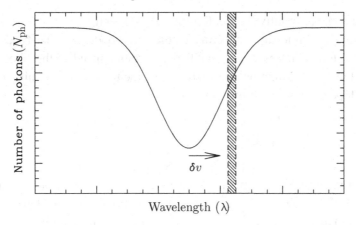

Figure 1.7 Schematic spectrum in the vicinity of a single spectral line of the host star. The wavelength range that corresponds to a single pixel in the observed spectrum is shown as the vertical shaded band. If the spectrum shifts by a velocity δv the number of photons detected at that pixel will vary by an amount that depends upon the local slope of the spectrum.

pixel.[4] Detecting such small shifts reliably requires both exceptional instrumental stability (extending, for long period planets, over periods of many years) and careful consideration of the potential sources of noise in the measurement of the radial velocity signal.

Shot noise (i.e. uncertainty in the number of photons due purely to counting statistics) defines the ultimate radial velocity precision that can be attained from an observation of specified duration. An estimate of the shot noise limit can be derived by starting from a very simple problem: how accurately can the velocity shift of a spectrum be estimated given measurement of the flux in a single pixel on the detector? To do this, we follow the basic approach of Butler *et al.* (1996) and consider the spectrum in the vicinity of a spectral line, as shown in Fig. 1.7. Assume that in an observation of some given duration, N_{ph} photons are detected in the wavelength interval corresponding to the shaded vertical band. If we now imagine displacing the spectrum by an amount (expressed in velocity units) δv the change in the mean number of photons is

$$\delta N_{ph} = \frac{\mathrm{d}N_{ph}}{\mathrm{d}v}\delta v. \tag{1.33}$$

Since a 1σ detection of the shift requires that $\delta N_{ph} \approx N_{ph}^{1/2}$, the minimum velocity displacement that is detectable is

$$\delta v_{min} \approx \frac{N_{ph}^{1/2}}{\mathrm{d}N_{ph}/\mathrm{d}v}. \tag{1.34}$$

---

[4] For simplicity, we assume that one spectral resolution element corresponds to one pixel on the detector. A real instrument may over-sample the spectrum, but this practical point does not alter any of the basic results.

This formula makes intuitive sense. Regions of the spectrum that are flat are useless for measuring δv while sharp spectral features are good. For solar-type stars with photospheric temperatures $T_{eff} \approx 6000$ K the sound speed at the photosphere is around 10 km s^{-1}. Taking this as an estimate of the thermal broadening of spectral lines, the slope of the spectrum is at most

$$\frac{1}{N_{ph}} \frac{dN_{ph}}{dv} \sim \frac{1}{10 \text{ km s}^{-1}} \sim 10^{-4} \text{ m}^{-1} \text{ s}. \tag{1.35}$$

Combining Eqs. (1.34) and (1.35) with knowledge of the number of photons detected per pixel yields an estimate of the photon-limited radial velocity precision. For example, if the spectrum has a signal to noise ratio of 100 (and there are no other noise sources) then each pixel receives $N_{ph} \sim 10^4$ photons and $\delta v_{min} \sim 100$ m s^{-1}. If the spectrum contains N_{pix} such pixels the combined limit to the radial velocity precision is

$$\delta v_{shot} = \frac{\delta v_{min}}{N_{pix}^{1/2}} \sim \frac{100 \text{ m s}^{-1}}{N_{pix}^{1/2}}. \tag{1.36}$$

Although this discussion ignores many aspects that are practically important in searching for planets from radial velocity data, it suffices to reveal several key features. Given a high signal to noise spectrum and stable wavelength calibration, photon noise is small enough that a radial velocity measurement with the meters per second (m s^{-1}) precision needed to detect extrasolar planets is feasible. The resolution of the spectrograph needs to be high enough to resolve the widths of spectral lines, but does not need to approach the magnitude of the planetary signal. In fact the intrinsic precision of the method depends first and foremost on the amount of structure that is present within the stellar spectrum, and the measurement precision will be degraded for stars whose lines are additionally broadened, for example by rotation.

Once precisions of the order of m s^{-1} have been attained, intrinsic radial velocity jitter due to motions at the stellar photosphere presents a second limit. The vertical velocity at the photosphere of a star is not zero, due to the presence of both convection and p-mode oscillations (acoustic modes trapped in the star). Although p-mode oscillations are of great intrinsic interest for helio- and asteroseismology, when under-sampled they represent noise for radial velocity searches for extrasolar planets. The effects of resolved stellar oscillations can be removed using appropriate observing strategies (Dumusque *et al.*, 2011), allowing radial velocity measurements with a precision that approaches the 10 cm s^{-1} needed to find Earth analogs.

1.7.3 Astrometry

Astrometric measurement of the stellar reflex motion in the plane of the sky provides a complementary method for detecting planets. From the definition of the

center of mass, the physical size of the stellar orbit is related to the planetary
semi-major axis via $a_* = (M_p/M_*)a$. For a star at distance d from the Earth the
angular displacement of the stellar photo-center during the course of an orbit has a
characteristic scale,

$$\theta = \left(\frac{M_p}{M_*}\right)\frac{a}{d}. \tag{1.37}$$

Unlike radial velocity searches, which are biased toward detecting short period
planets, astrometry favors large semi-major axes. A further difference is that
astrometry measures two independent components of the stellar motion (versus
a single component via radial velocity measurements), and this yields more
constraints on the orbit. As a result there is no $\sin i$ ambiguity and all of the
important planetary properties can be directly measured. Numerically the size of
the signal is

$$\theta = 5 \times 10^{-4} \left(\frac{M_p}{M_J}\right)\left(\frac{M_*}{M_\odot}\right)^{-1}\left(\frac{a}{5\ \text{AU}}\right)\left(\frac{d}{10\ \text{pc}}\right)^{-1} \quad \text{arcsec}. \tag{1.38}$$

Even though the parameters adopted here are rather optimistic this is still a very
small displacement, and none of the planets found in the first decade of discovery
were identified this way. In principle, however, there are no fundamental obstacles
to achieving astrometric accuracies of 1–10 μarcsec, which is good enough to detect
a wide range of hypothetical planets. A $10M_\oplus$ planet at 1 AU, for example, yields
an astrometric signature of 3 μarcsec at $d = 10$ pc.

1.7.4 Transits

A planet whose orbit causes it to transit the stellar disk can be detected by monitoring the stellar flux for periodic dips. At the most basic level transit events
can be characterized in terms of their depth (the fraction of the stellar flux that
is obscured), duration, and probability of being observed by a randomly oriented
external observer. These quantities are readily estimated. The depth or amplitude of
the transit signal is independent of the distance between the planet and the star, and
provides a measure of the relative size of the two bodies. If a planet of radius R_p
occults a star of radius R_*, with the geometry shown in Fig. 1.8, the fractional
decrement in the stellar flux during the transit (assuming a uniform brightness
stellar disk) is just

$$f = \left(\frac{R_p}{R_*}\right)^2. \tag{1.39}$$

Gas giant planets have a rather flat mass–radius relation, so it is reasonable to use
Jupiter's radius $R_J = 7.142 \times 10^9$ cm as a proxy for all giant planet transits. The

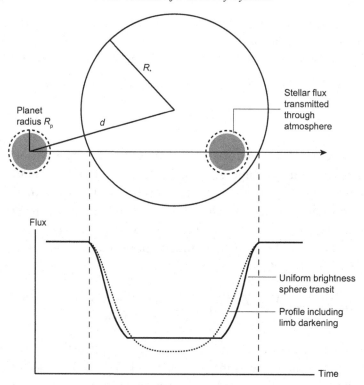

Figure 1.8 The geometry and light curve of a stellar transit by a planet (not to scale). The stellar and planetary radii are R_* and R_p respectively, and the distance between the projected centers of the bodies is d. The simplest model for the transit light curve corresponds to a perfectly opaque disk moving across a circle of uniform brightness. Deviations from this light curve occur due to the phenomenon of limb darkening. The photosphere is physically deeper looking toward the center of the disk, so we see hotter gas and greater intensity there than close to the limb. A small fraction of the starlight passes *through* the atmosphere of the planet, allowing measurement of the atmospheric constituents via transit spectroscopy.

amplitude of the signal for a gas giant transiting a solar-radius star is $f = 0.01$. For rocky planets with the same mean density as the Earth,

$$f = 8.4 \times 10^{-5} \left(\frac{M_p}{M_\oplus} \right)^{2/3}. \qquad (1.40)$$

The probability that a planet will be observed to transit follows from elementary geometric arguments. For a planet on a circular orbit with semi-major axis a, the condition that some part of the planet will be seen to graze the stellar disk is that the inclination angle i satisfies $\cos i \leq (R_* + R_p)/a$. If an ensemble of such systems has random inclinations then the probability of observing transits is

$$P_{\text{transit}} = \frac{R_* + R_p}{a}. \qquad (1.41)$$

For a planet similar to the Earth, the probability is $P_{transit} \approx 5 \times 10^{-3}$. If the geometry is favorable for observing transits, their maximum duration is roughly $2R_*/v_K$. More accurately (Quirrenbach, 2006),

$$t_{transit} = \frac{P}{\pi} \sin^{-1} \left(\frac{\sqrt{(R_* + R_p)^2 - a^2 \cos^2 i}}{a} \right), \tag{1.42}$$

where P is the orbital period. A planet similar to the Earth whose orbit is seen edge-on ($i = 90°$) transits its host star for 13 hours. Combining this result with the transit probability, one finds that if every star were to host an Earth-like planet, about 1 in 10^5 stars would be being transited at any given time.

The predicted light curve for an idealized (noise-free) planetary transit can be computed using various approximate or empirical models for the brightness distribution of the stellar disk. The simplest model treats the planet as a perfectly occulting disk passing across the face of a star whose brightness or specific intensity is everywhere constant. Following Mandel and Agol (2002) we set up the problem as shown in Fig. 1.8. The planet and star have radii R_p and R_* respectively, and the instantaneous separation of the centers of the two bodies is d. We define auxiliary variables $z \equiv d/R_*$ and $p = R_p/R_*$. In terms of these variables the fractional decrement f is given as

$$f(p,z) = \begin{cases} 0 & 1 + p < z, \\ \frac{1}{\pi} \left[p^2 \kappa_0 + \kappa_1 - \sqrt{\frac{4z^2 - (1+z^2-p^2)^2}{4}} \right] & |1 - p| < z \le 1 + p, \\ p^2 & z \le 1 - p, \\ 1 & z \le p - 1. \end{cases} \tag{1.43}$$

In these expressions,

$$\kappa_0 = \cos^{-1} \left[\frac{p^2 + z^2 - 1}{2pz} \right], \tag{1.44}$$

$$\kappa_1 = \cos^{-1} \left[\frac{1 - p^2 + z^2}{2z} \right]. \tag{1.45}$$

Unfortunately this model, which is simple enough to be written entirely in terms of elementary functions, is not very realistic. Real transit light curves do not show the perfectly flat bottoms with $f = p^2$ that would be predicted for a black disk crossing a star of uniform brightness. The problem is limb darkening. The line of sight to the star is perpendicular to the stellar surface at the center of the disk, but increasingly tangential as we move toward the limb. As a consequence, photons coming from the center of the star last scattered, on average, at a greater physical depth than those coming from near the limb. Greater physical depth corresponds to a higher temperature, so the central brightness is greatest and the limb regions are relatively darkened. There is no exact first principles description of limb darkening, but a variety of empirical laws fit observational data well. Mandel and Agol (2002) provide more complex but still analytic light curves for some of these laws.

Noise sources for transit searches include photon shot noise, stellar variability (including a contribution from the asteroseismic modes), and, for ground-based experiments, atmospheric fluctuations. For stars of roughly solar radii ground-based experiments are an efficient means of finding planets with $R_p \simeq R_J$, while smaller planets with $R_p \simeq R_\oplus$ are only detectable given the lower noise levels attainable from space. Ground-based surveys are sensitive to roughly Earth-radius planets around M-dwarfs, which are physically much smaller. An example is GJ 1132b, a 1.2 $R_\oplus$ planet orbiting a 0.18 $M_\odot$ star that was discovered from the ground (Berta-Thompson *et al.*, 2015). In addition to these true noise sources all transit surveys have to contend with false positives that are caused by planetary transit-like signals of alternate astronomical origin. A stellar eclipsing binary, for example, may source a signal that looks like a planet if its light is diluted by the presence of a third unresolved star within the photometric aperture. A combination of follow-up observations, yielding radial velocity or higher spatial resolution adaptive optics data, and statistical arguments are used to discriminate between such imposters and true planets.

Detection of repeated planetary transit signals provides an immediate measure of the orbital period, and of the ratio between the physical size of the planet relative to the star. In cases where multiple planets are observed to transit the same star, gravitational perturbations between the planets lead to small changes in the time at which transits occur. When detectable these *Transit-Timing Variations* (TTVs) provide constraints and in some cases measurements of the planetary masses (Agol *et al.*, 2005; Holman & Murray, 2005; Holczer *et al.*, 2016). Planetary masses can also be determined for systems in which transit data are supplemented by radial velocity measurements. In this case knowledge of the inclination removes the usual $\sin i$ uncertainty in the mass. Time-resolved measurement of the radial velocity signal *during* the transit also opens up the possibility of detecting the Rossiter–McLaughlin effect (originally discussed in the context of eclipsing stellar binaries; Rossiter, 1924; McLaughlin, 1924). The Rossiter–McLaughlin effect is a perturbation to the apparent stellar radial velocity that is caused by the fact that when a planet transits across the face of a rotating star, it obscures portions of the stellar photosphere that are either redshifted or blueshifted relative to the whole-disk average. Figure 1.9 illustrates the principle. Detection of the effect allows for a measurement of the projected obliquity, on the plane of the sky, between the stellar and orbital angular momentum vectors. This, in turn, informs and constrains theoretical models for the origin of planets in short-period orbits.

Transiting systems are also a rich source of informative data on planetary atmospheres. In some cases one can detect the secondary eclipse when the planet moves behind the star, and in others it is possible to measure the out-of-eclipse phase modulation as the planet orbits. Depending upon the wavelength of the observations,

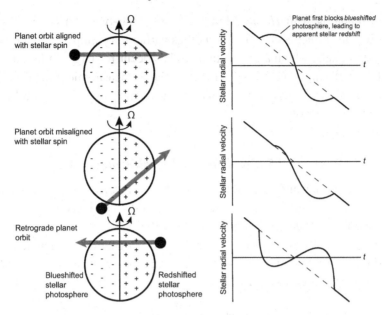

Figure 1.9 Illustration of the Rossiter–McLaughlin effect. Selective obscuration of the photosphere of a rotating star introduces a time-dependent perturbation to the measured radial velocity signal as a planet transits the star. The form of the perturbation depends upon whether the planet first blocks a redshifted or blueshifted piece of photosphere, and hence probes for misalignment between the stellar and orbital angular momentum vectors.

data of this type can constrain the albedo of the planet or the size of the difference between the temperature of the day side and that of the night side. Even more powerful are observations of the apparent size of the planet at different wavelengths, due to light that passes through the atmosphere at the planet's limb. This is a small effect. For a planet whose atmosphere has temperature T, mean molecular weight μ, and surface gravity g, the exponential scale-height is

$$H_{\mathrm{p}} = \frac{k_{\mathrm{B}} T}{\mu g}, \tag{1.46}$$

where k_{B} is the Boltzmann constant. Suppose that we observe a planet at two wavelengths, one where the atmosphere is entirely transparent (so that the apparent planetary size is given by that of the solid surface) and one where an atmospheric absorber is optically thick up to n scale-heights above the surface. The transit depth measured at the two wavelengths will differ by an amount

$$\begin{aligned}
\delta f &= \frac{(R_{\mathrm{p}} + n H_{\mathrm{p}})^2}{R_*^2} - \frac{R_{\mathrm{p}}^2}{R_*^2} \\
&\simeq \frac{2 n R_{\mathrm{p}} H_{\mathrm{p}}}{R_*^2}.
\end{aligned} \tag{1.47}$$

For an Earth analog with $H_p = 8$ km orbiting a solar-type star $\delta f \sim 2 \times 10^{-7}$, and detecting this effect is close to 1000 times harder than simply seeing the transit itself. Larger signals, which are detectable today, occur for larger or hotter planets, and for those orbiting smaller stars. In these cases measurement of the apparent planetary radius as a function of observing wavelength provides a handle on the atmospheric composition.

1.7.5 Gravitational Microlensing

Gravitational microlensing is a powerful method for detecting extrasolar planets that is based upon indirectly detecting the gravitational deflection of light that passes through the planetary system from a background source. The foundations of the method were laid by Einstein, who derived the general relativistic result for light bending by gravitating objects. A photon that passes a star of mass M_* with impact parameter b is deflected by an angle

$$\alpha = \frac{4GM_*}{bc^2}. \tag{1.48}$$

If two stars lie at different distances along the same line of sight, consideration of the geometry illustrated in Fig. 1.10 implies that the image of the background star (the "source") is distorted by the deflection introduced by the foreground star (the "lens") into a ring. Writing the distance between the observer and the lens as d_L, the observer–source distance as d_S, and the lens–source distance as d_{LS}, the angular radius of the so-called *Einstein ring* is

$$\theta_E = \frac{2}{c}\sqrt{\frac{GM_* d_{LS}}{d_L d_S}}. \tag{1.49}$$

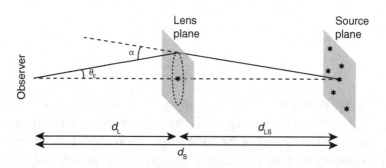

Figure 1.10 Microlensing geometry. Light from a background star is deflected by a small angle α when it passes a foreground star of mass M_*. If the alignment between the foreground and background star is perfect, then by symmetry the observer sees the image of the background star distorted into a circular Einstein ring around the foreground star. The ring has an angular radius θ_E. When lensing is used as a tool to detect planets, the distances involved are such that the ring is unresolved. The observable is the change in brightness of the background star as the two stars first align, and then move apart.

If the alignment between the source and the lens is nearly but not quite perfect, the axial symmetry is broken and the source is lensed into multiple images whose angular separation is also of the order of θ_E.

For planet detection, a typical configuration of interest observationally is that where the source star lies in the Galactic bulge ($d_S = 8\,\mathrm{kpc}$) and the lens star is in the disk ($d_L \approx 4\,\mathrm{kpc}$). The Einstein ring radius for a solar mass lens with these parameters is $\theta_E \sim 10^{-3}$ arcsec, so the multiple images cannot be spatially resolved. A property of lensing, however, is that it conserves surface brightness. The area of the lensed image is greater than it would have been in the absence of lensing, and hence the flux from the lensed images exceeds that of the unlensed star. This means that if the alignment between the source and the lens varies with time, monitoring of the light curve of the source star can detect unresolved lensing via the time-variable magnification caused by lensing. Light curves in which a single disk star lenses a background star are smooth, symmetric about the peak, and magnify the source achromatically. If the proper motion of the lens relative to that of the source is μ, the characteristic time scale of the lensing event is $t_E = \theta_E/\mu$. For events toward the Galactic bulge this time scale is about a month, scaling with the lens mass as $\sqrt{M_*}$.

Microlensing light curves are altered when the lens star is part of a binary. For star–planet systems in which the mass ratio $q = M_p/M_* \ll 1$ the conceptually simplest case to consider is when the planet orbits close to the physical radius of the Einstein ring at the distance of the lens. If one of the multiple images of the source passes close to the planet during the lensing event the planet's gravity introduces an additional perturbation to the light curve. To an order of magnitude, the time scale of the perturbation is just $t_p \sim q^{1/2}t_E$, while the probability that the geometry will be such that a perturbation occurs is $P \sim Aq^{1/2}$, where A is the magnification of the source at the moment when the image passes near the planet. A second channel for planet detection occurs in rare high magnification events, during which planets close to the Einstein ring modify the light curve near peak regardless of their orbital position. This channel is valuable observationally since high magnification events can be anticipated based on photometric observations made well before the peak, allowing for detailed monitoring of those events that are most favorable for detecting planets.

Determining the properties of planets from the analysis of microlensing light curves is a difficult inverse problem, but the above discussion is enough to explain the attractive aspects of the method. First, the physical radius of the Einstein ring for disk stars lensing background stars in the Galactic bulge corresponds to a few AU. Lensing is therefore well suited to detect planets at relatively large orbital radii (roughly corresponding to the location of the snow line in the Solar System) which are challenging to detect via radial velocity or transit methods. Second, the weak $\sqrt{M_P}$ dependence of the time scale on planet mass allows for a wide range of planets to be detected. In particular, Jupiter mass planets yield a perturbation time scale of around a day, while Earth mass planets have a characteristic time scale of about an hour. Both time scales are quite accessible observationally. Moreover, *when* a planetary perturbation occurs its magnitude can be significant, even for Earth mass planets. Low mass planets can therefore be detected from the ground via gravitational microlensing, though the superior precision and uninterrupted viewing possible from space affords large advantages.

1.8 Properties of Extrasolar Planets

A central result of exoplanet studies is the extraordinary diversity of observed extrasolar planetary systems. Figure 1.11 illustrates this property by plotting the orbital radii of planets in some celebrated systems on a logarithmic scale. Solar System planets are restricted to about two orders of magnitude in semi-major axis. Extrasolar planets have been found across four orders of magnitude in orbital radius (and are particularly abundant close-in), though whether this full range is populated in any individual system remains unknown. Planets around low mass stars are common, as are planets and planetary systems that violate the clear dichotomy between terrestrial and giant planets that is a feature of the Solar System.

It is sometimes implied that the discovery of extrasolar planetary systems that do not resemble the Solar System came as a complete surprise to observers and

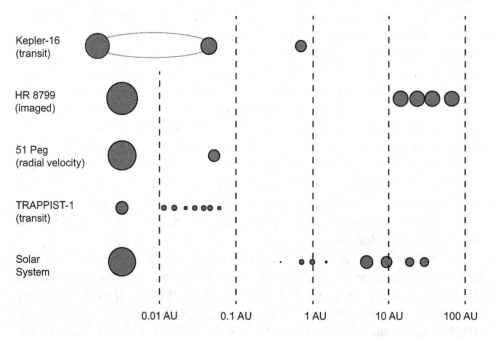

Figure 1.11 Observed planetary systems display diverse orbital architectures. Systems of moderately low mass planets around low mass stars – TRAPPIST-1 is a spectacular example (Gillon *et al.*, 2017) – are common, but there are also systems with circumbinary planets (illustrated here is the Kepler 16 system; Doyle *et al.*, 2011), with massive planets at orbital radii substantially larger than in the Solar System (HR 8799; Marois *et al.*, 2008), and with hot Jupiters (51 Peg b; Mayor & Queloz, 1995).

theorists alike. This is an over-statement. Otto Struve, in proposing a program of high-precision radial velocity measurements, suggested that it is "not unreasonable that a planet might exist at a distance of 0.02 AU" (Struve, 1952), and theorists expected that planets ought to be common (e.g. Wetherill, 1991) and identified in advance some of the mechanisms that lead to orbital migration (Goldreich & Tremaine, 1980). It is the extent of the differences between the Solar System and many extrasolar planetary systems that constitutes a genuine surprise. It has prompted both a re-evaluation of the effects of previously known planet formation processes, and a search for new ones.

1.8.1 Parameter Space of Detections

Figure 1.12 shows the distribution of semi-major axes for a sample of extrasolar planets with known masses. This plot, and most like it, is neither complete nor unbiased, and it omits most transit-detected planets as these often lack mass estimates. It suffices, however, to illustrate the main populations of known extrasolar planets.

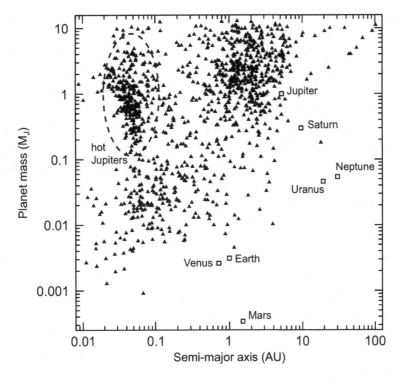

Figure 1.12 The distribution of a sample of known extrasolar planets in semi-major axis and planet mass (either M_p or $M_p \sin i$, depending upon the available observations). Selected Solar System planets are shown as the open squares. Data from the *NASA Exoplanet Archive*.

In the realm of giant planets with masses equal to or larger than Saturn, a population of "warm" and "cold" Jupiters extends from a few tenths of an AU out to at least Jupiter's orbital radius. Knowledge of the true outer extent of this population is currently limited by selection effects. The masses of extrasolar giant planets range up to the deuterium burning limit (an order of magnitude larger than in the Solar System), but there are relatively few brown dwarfs with AU-scale orbits around solar mass stars, a property known as the "brown dwarf desert." Statistical studies show that giant planets with orbital periods $P <$ few yr orbit approximately 10% of Sun-like stars. Over the observed range of orbital radii the probability density is roughly uniform in $\log P$. Low mass giants are more common than high mass ones. Cumming *et al.* (2008) estimated that $dN_p/d \ln M \propto M^\alpha$ with $\alpha \simeq -0.3 \pm 0.2$ for this population.

Closer in the "hot Jupiters" have clearly distinct physical and dynamical properties, and possibly a different formation history, than their colder brethren. These planets orbit within about 0.1 AU of their hosts, and have correspondingly high effective temperatures. They have near circular orbits and are expected to be in tidally synchronized rotation states. Hot Jupiters are present around approximately 1% of solar-type stars, and are less common around M dwarfs (Dawson & Johnson, 2018).

Lower mass planets, although under-represented in Fig. 1.12, are substantially more common. Statistics of this population were derived from the results of the *Kepler* mission, which monitored $\approx$150 000 stars of spectral type FGKM for transits for 4 years. Roughly half of Sun-like stars have planets with radii $1-4$ $R_\oplus$ with orbital periods $P \leq 1$ yr. Closely spaced systems with multiple planets in these relatively short-period orbits are common (Winn & Fabrycky, 2015).

A small number of stars are known to host massive planets with orbital radii substantially larger than any of the Solar System's giant planets. An example is HR 8799, a 1.5 $M_\oplus$ star that is still young enough to have a debris disk. HR 8799 has at least four planets, with masses of $\sim$5–10 M_J and orbital radii between 15 and 70 AU (Marois *et al.*, 2008).

Important parts of the parameter space of possible planetary systems remain poorly surveyed. Knowledge of the abundance of planets with masses or radii comparable to the Earth is extremely limited beyond $\approx$1 AU, while beyond 10 AU information even on giant planets is scant. A full picture of the architecture of typical extrasolar planetary systems is also lacking. At the time of writing most known close-in low mass planets were discovered using *Kepler* data, and orbit relatively faint stars that are hard to monitor for long-period radial velocity signals. As a consequence, knowledge of the relationship between low mass and giant planets in extrasolar planetary systems is more limited than one might expect given the amount of good data on each individual population. Basic questions, such as the average number of planets that orbit stars of different masses, remain open.

1.8.2 Orbital Properties

The eccentricity of giant extrasolar planets is a function of their orbital radius. The closest-in planets, including most of the hot Jupiters with $a \leq 0.1$ AU, have circular or low eccentricity orbits. A qualitatively similar result is found for binary stars, and in both regimes it is attributed to the circularizing influence of tides. At orbital radii where the effects of tides can be neglected, giant extrasolar planets have a broad eccentricity distribution which includes highly eccentric planets that have no Solar System analogs. HD 80606b, for example, has an eccentricity $e \simeq 0.93$ which in the Solar System would be more characteristic of comets than any larger bodies. Figure 1.13 shows the semi-major axes and eccentricity for a sample of planets discovered as a result of radial velocity searches. Excluding the hot Jupiters we see that small to moderate eccentricities $e = 0-0.3$ are common, and that there is a tail that extends to high values. Restricting the sample to about 400 planets with $0.3\,M_J \leq M_p < 10\,M_J$ and 0.1 AU $< a < 5$ AU the mean eccentricity is $\langle e \rangle \simeq 0.25$, while the median $\langle e \rangle \simeq 0.19$.

The detailed interpretation of exoplanet eccentricities is complex, and various models have been proposed to explain the data. Most invoke gravitational dynamics,

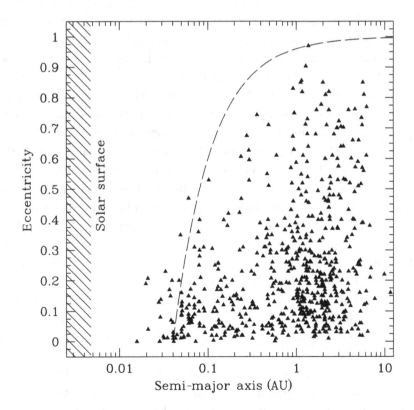

Figure 1.13 The distribution of a sample of extrasolar planets discovered via radial velocity searches in semi-major axis and eccentricity. The dashed line shows the eccentricity of a planet whose pericenter distance is 0.04 AU.

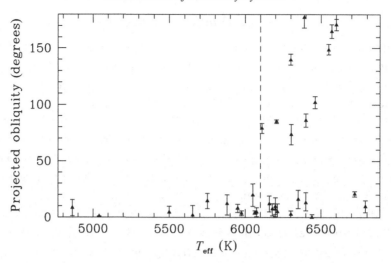

Figure 1.14 The sky-projected angle between the stellar spin axis and the orbital plane of close-in massive planets. The sample includes hot Jupiters with orbital period $P < 7$ days, masses between 0.3 and 3 M_J, and quoted measurement errors of less than 10° (data from Wright *et al.*, 2011a).

taking place after the main phase of planet formation, to excite the eccentricity of initially near-circular giant planets. The observed distribution may also be affected by interactions (also mediated by gravity) between planets and gas that occur before or during the dispersal of the protoplanetary disk.

Measurements of the Rossiter–McLaughlin effect (Fig. 1.9) provide a constraint on how well planetary orbits align with the equators of their host stars. The observable quantity is not the misalignment itself but rather the *projected obliquity*, that is the angle between the vectors describing the orbit and the stellar spin projected on the sky plane. Figure 1.14 shows measurements of the projected obliquity for a sample of hot Jupiters orbiting stars of varying effective temperature T_{eff}. A subset of hot Jupiters have high projected obliquities, which implies that some giant planets orbit their stars in highly tilted, polar, or even retrograde orbits. The observed distribution has a striking dependence on the stellar effective temperature. Low projected obliquities are the norm for $T_{eff} \lesssim 6100$ K, while a wide range of obliquities characterize hot Jupiters around hotter stars (Winn *et al.*, 2010).

A small number of dynamical processes, including the Kozai–Lidov effect which we will discuss in Chapter 7, are able to tilt the orbits of planets that form initially in the plane defined by the stellar spin. They were fingered as prime suspects as soon as the first examples of misaligned hot Jupiters were identified. Alternatively, the disk plane itself could be tilted away from the equatorial one in some fraction of systems, leading to a primordial origin for the misalignments (Bate *et al.*, 2010; Batygin, 2012). In either case the fact that the threshold in effective temperature coincides with known structural differences in stars suggests that stellar rotation and tidal dissipation play a role in shaping the late-time distribution.

Definitive determination of whether specific extrasolar planetary systems with multiple planets are resonant or not is often not possible. In many cases where a pair of planets are close to a commensurability the available data are ambiguous as to whether a true resonance exists. Nonetheless, some clear examples of resonant systems are known. Among massive extrasolar planetary systems three of the planets in the Gliese 876 system occupy a Laplace-type resonance in which the orbital periods are in the ratio of 4:2:1 (Rivera *et al.*, 2010; Millholland *et al.*, 2018). A number of resonances and resonant chains are also observed among lower mass planetary systems, including the TRAPPIST-1 system (Gillon *et al.* 2017).

As in the Solar System the existence of resonant configurations among extrasolar planets is strong evidence for the importance of dissipative processes at an earlier time in the systems' history. Statistically, however, resonant configurations are not the most common state, though there is a marked dependence on planet mass. Wright *et al.* (2011b) analyzed a sample of 43 well-characterized multiple planet systems that were discovered from radial velocity searches. Among this sample – dominated by massive planets – roughly a third showed evidence for low-order period commensurability. This fraction is substantially greater than would be expected by chance. In contrast the distribution of period ratios for the mostly lower-mass planets discovered by the *Kepler* mission, shown in Fig. 1.15, shows no preference for resonant configurations. Instead there is a modest enhancement in the number of systems with period ratios just *larger* than would be expected for low-order mean-motion resonances (Fabrycky *et al.*, 2014).

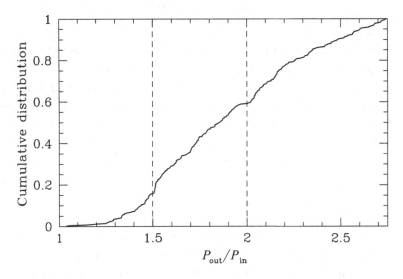

Figure 1.15 The cumulative distribution of period ratios for pairs of extrasolar planets detected via transit. Period ratios indicative of low-order mean-motion resonances are uncommon, but there is a modest excess of pairs with period ratios slightly larger than commensurabilities such as 3:2 and 2:1 (data from Fabrycky *et al.*, 2014).

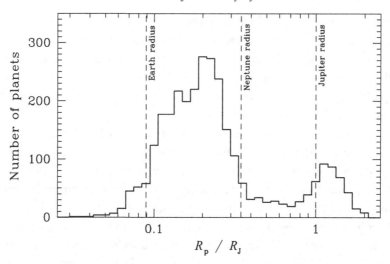

Figure 1.16 The distribution of radii for approximately 3000 extrasolar planets that were discovered via the transit method (using data from the *NASA Exoplanet Archive*). The data have not been corrected for completeness, and becomes increasingly incomplete for small planets. "Small" planets – those with radii intermediate between the Earth and Neptune – are substantially more abundant in transit surveys than gas giants. Many stars host detectable planets with radii that are intermediate between the Solar System's terrestrial and ice giant planets.

1.8.3 Mass–Radius Relation

The distribution of radii for planets discovered via the transit method is shown in Fig. 1.16. The data used to make this plot were not corrected for bias and incompleteness, but it is clear that the distribution is bimodal, and that "small" planets with radii of $R_p = 0.1$–0.4 R_J are substantially more abundant than gas giants with $R_p \simeq R_J$. What is *not* seen is a clear break between planets with $R_p \approx R_\oplus$, which Solar System experience would suggest are rocky terrestrial worlds, and planets with $R_p \approx R_{Nep}$, which in the Solar System are ice giants. Many stars host planets with radii that are midway between the Solar System's terrestrial and ice giant classes. The basic possibilities are that these are scaled-up versions of terrestrial planets ("super-Earths") or scaled-down versions of ice giants ("mini-Neptunes").

The physical nature of extrasolar planets of different sizes can be determined with greater confidence for the subset of planets whose masses (and hence densities) are also known. Unfortunately knowing the density does not yield a unique answer as to what the planet is made of. Planets with entirely different compositions can have identical masses and radii, making classification necessarily ambiguous. Several structures are well motivated by the existence of Solar System analogs and/or formation considerations.

- **Gas giants** with masses roughly in the range between M_{Sat} and 13 M_J. Theoretical models of gas giant structure show that the predicted radius is a weak function of mass, so to leading order gas giants ought to have $R_p \simeq R_J$. At the next order of approximation the radius is a decreasing function of the total mass of heavy elements within the planet, independent of whether that material is concentrated in a core or dispersed throughout the envelope.

- **Rocky planets.** There is no clear lower limit to the mass of rocky (or icy) worlds, though bodies smaller than a few hundred km in size are typically irregular in shape and fall into the class of minor bodies. There is also no absolute upper limit, though once we reach masses of 5–10 $M_\oplus$ capture of an envelope dense enough to increase the radius measured via transit becomes increasingly likely. Zeng *et al.* (2016) provide a semi-empirical model for the mass-radius relation of broadly Earth-like planets,

$$R_p = (1.07 - 0.21 f_{core}) \left(\frac{M_p}{M_\oplus} \right)^{1/3.7} R_\oplus, \tag{1.50}$$

valid for the range between 1 and 8 Earth masses. In this expression f_{core} is the core mass fraction (for the Earth $f_{core} \simeq 0.33$). Smaller radii are possible for iron-rich planets (Mercury-like and super-Mercury worlds), with the minimum physically plausible radii being attained for objects made up of 100% Fe.

- **Low mass core/envelope planets.** A variety of structures are possible within the generic class of a rocky or icy core (typically of a few to about ten $M_\oplus$) surrounded by a sub-dominant gaseous envelope. End members are "gas dwarfs," where the envelope is made up of hydrogen and helium, and objects where the envelope is heavily enriched with water and other species such as NH_3 and CH_4. Planets with these structural features would be seen as low density objects with masses of a few to about ten Earth masses.

- **Water worlds.** The high predicted abundance of water beyond the snow line in protoplanetary disks opens up the possibility of planets with a water dominated bulk composition. Such planets have a mass–radius relation that is similar to rocky planets, but 30–40% larger in size.

Figure 1.17 shows the observed masses and radii of a sample of extrasolar planets for which both quantities have been measured with moderate to good precision. Among giant planets the weak predicted dependence of radius on mass is clearly evident. A small number of giant planets have anomalously small radii, indicative of a large bulk heavy element abundance. Among the hot Jupiters, many planets have anomalously *large* radii, exceeding in some cases the prediction for a planet that lacks a significant mass of heavy elements. These radius anomalies correlate with the incident stellar irradiation, giving support to the hypothesis that some fraction of the incident stellar energy finds its way into the planets' convective zones. Energy input into the convection zone is known to stall contraction or inflate

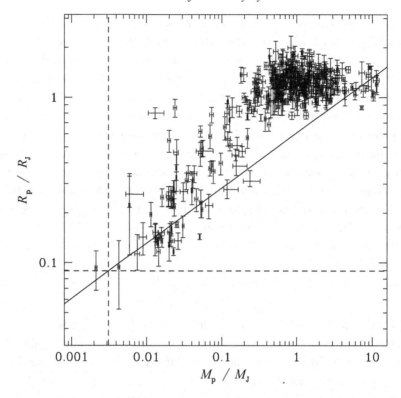

Figure 1.17 The distribution of a sample of extrasolar planets with well-determined mass and radius estimates.

giant planets, though the mechanism that injects stellar energy deep into the planet remains unclear.

Mass and radius data for smaller planets are sparser, especially for planets with masses similar to or smaller than that of the Earth. The evidence is consistent with the idea that many planets with $R_p \lesssim 1.5\ R_\oplus$ are rocky super-Earths, while larger planets have a range of radii that reflect variations in the amount of gas within their envelopes (Rogers, 2015). Not very much is known about the detailed composition of envelopes in the mini-Neptune size range, nor can it be excluded that some observed planets have water dominated compositions.

1.8.4 Host Properties

Important constraints on planet formation models come from the observation that the abundance of planets correlates with properties of the host star. A key observation was the discovery by Debra Fischer and Jeff Valenti that the fraction of stars with observed giant planets increases with the metal abundance measured spectrosopically in the stellar photosphere (Fischer & Valenti, 2005). Figure 1.18 shows an updated version of this correlation (Sousa *et al.*, 2011). For planets (and

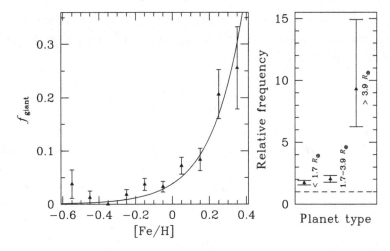

Figure 1.18 Left panel: the fraction of a sample of stars that host known giant extrasolar planets (using data from Sousa *et al.*, 2011). Right panel: the enhancement to the observed planet frequency for planets of different radii, corresponding approximately to super-Earths, mini-Neptunes, and gas giants. Note that astronomical convention is to measure the abundance of heavy elements relative to hydrogen, with the resulting ratio being normalized to the ratio found in the Sun. A logarithmic scale is used, denoted by square brackets. Thus, a star with "[Fe/H] = 0" has the same fractional abundance of iron as the Sun, whereas one with [Fe/H] = 0.5 is enriched in iron by a factor $\sim$3 as compared to the Sun.

a small number of more massive objects) with masses between 0.1 M_J and 25 M_J the detected planet frequency is found to rise steeply with metallicity, following a scaling law $P_{\text{planet}} \propto Z^{2.58}$.

The planet–metallicity correlation was discovered from, and is strongest for, giant planets. Using a sample of around 400 *Kepler*-discovered planets, Wang and Fischer (2015) determined the extent to which being metal rich ([Fe/H] > 0.05) increased the planet occurrence rate as opposed to being metal poor ([Fe/H] < −0.05), for different planet sizes. The boost that a high metal content gives to planet abundance, shown as the right panel in Fig. 1.18, is substantially larger for planets with radii characteristic of gas giants than for smaller gas dwarf or terrestrial planets.

Because most of the heavy elements in a collapsing molecular cloud core end up in the star rather than in planets, the stellar photospheric metal abundance is reasonably indicative of the bulk enrichment of the protoplanetary disk that star once had. Under the assumption that the initial gas mass of disks is not a strong function of metallicity, the current stellar metallicity is then a marker for the mass of solid material that was once available to form planets. It is then certainly not surprising that higher stellar metallicities lead to a greater probability of planet formation, or to a greater number of planets in systems where more than one planet forms, as is observed. As we will discuss later, the fact that the planet–metallicity correlation is strongest for giant planets appears to be consistent with the idea that giant planet formation involves a time-limited threshold step – to form a giant planet a massive *core* needs to assemble in a limited period of time before the gas in the protoplanetary disk disperses.

1.9 Habitability

Several factors appear to contribute to the hospitable environment that the Earth offers for life. The surface temperature and pressure allow for the presence of liquid water across much of the surface, and these conditions are maintained over time due to geological processes (volcanism and plate tectonics) that stabilize the climate. Convective motions within the Earth's core sustain a magnetic field, which reduces the rate at which water is lost from the upper atmosphere due to collisions with high energy solar wind particles. Together with a moderate global abundance of water, established at early times, the low loss rate means that the current surface conditions – with both major oceans and large landmasses – have remained roughly constant over billions of years. Dynamical effects may also be important. Our large Moon, for example, stabilizes the Earth's obliquity against perturbations from the other planets, and the Solar System's architecture does not lead to large excursions in the Earth's eccentricity.

A full accounting of the planetary conditions that are necessary and sufficient for life to survive is not known, and it is a more difficult task still – bordering on guesswork – to enumerate the characteristics of planets where life might form in the first place. It is therefore customary to adopt a deliberately limited definition of the "habitable zone" as the range of orbital distances where a planet of given mass, orbiting a star of specified mass or spectral type, can sustain liquid water on its surface. To leading order this requires that the incident stellar flux be similar to the $1360 \ \mathrm{W \ m^{-2}}$ received by the Earth, and the steep scaling of stellar luminosity with stellar mass implies that the habitable zone has a strong dependence on host star mass. As shown in Fig. 1.19 a planet around a $0.2 \ M_\odot$ star needs to orbit at roughly $a \simeq 0.1$ AU to experience similar insolation as the Earth, while for the lowest mass hydrogen-burning objects the habitable zone is at just a few hundredths of an AU.

A crude estimate of the width of the habitable zone can be obtained by ignoring atmospheric physics and assuming that a planet reradiates intercepted stellar radiation as a single temperature blackbody. For a planet with radius R_p, albedo A, and an orbital distance a from a star of luminosity L, the temperature given these assumptions would be determined by

$$4\pi R_p^2 \sigma T^4 = \frac{L(1-A)}{4\pi a^2} \pi R_p^2, \qquad (1.51)$$

where σ is the Stefan–Boltzmann constant. Identifying the outer and inner edges of the habitable zone with the freezing and boiling points of water at the Earth's atmospheric pressure we estimate the relative width of the habitable zone to be

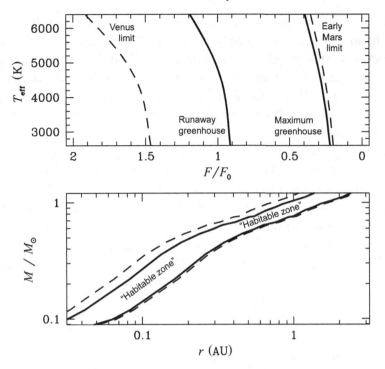

Figure 1.19 Estimates of the extent of the habitable zone, where liquid water can stably exist on the surface of a roughly Earth mass terrestrial planet. The upper panel, based on calculations by Kopparapu *et al.* (2013), shows theoretical estimates based on limiting cases of the greenhouse effect (solid lines) along with the extensions that follow if we assume that Mars and/or Venus were once but are no longer habitable (dashed lines). The width of the zone is expressed for stars of different effective temperature T_{eff} in terms of the incident planetary flux normalized to that of the Earth. The lower panels shows the same estimates converted into orbital radii for main-sequence stars of given mass.

$$\frac{a_{\text{out}}}{a_{\text{in}}} = \left(\frac{273 \text{ K}}{373 \text{ K}}\right)^{-2} \approx 1.9. \tag{1.52}$$

Although the neglect of the greenhouse effect means that the temperatures predicted by Eq. (1.51) are inaccurate, this estimate of the width of the zone is not too far off from actual calculations.

The use of climate models to map out the extent of the habitable zone was pioneered by Kasting *et al.* (1993). A climate model predicts the surface temperature of a planet given a specified atmospheric composition and stellar radiation field.[5] Geophysical assumptions are needed to define the limits of plausible

---

[5] In practice, calculations of the habitable zone often use an "inverse" climate model in which the surface temperature is specified and the model is used to compute the stellar flux that would yield that temperature.

atmospheric compositions, which change over time. On the Earth the amount of CO_2 in the atmosphere is regulated by the carbonate–silicate cycle. Carbonate minerals react with SiO_2 under the high temperature conditions found in subduction zones to liberate CO_2, which is vented into the atmosphere via volcanic activity. Burton *et al.* (2013) estimate the magnitude of this source term to be about 0.5 MT yr^{-1}. Although dwarfed by current anthropogenic emissions this natural source of CO_2, if unchecked, would significantly increase the atmospheric budget of greenhouse gases and raise the global temperature on a time scale of the order of a Myr. The counter-balancing sink is provided by weathering. Atmospheric CO_2 reacts with rainwater to form carbonic acid, H_2CO_3, which dissolves silicate rocks on land. The weathering products flow into the oceans, where they ultimately reform carbonate minerals. Crucially, the weathering rate is an increasing function of the temperature, so it is possible (though not guaranteed) to reach an equilibrium state in which volcanic sources of CO_2 balance the weathering sink. In the extreme case where a planet becomes entirely ice covered (a "snowball Earth" state) the cessation of weathering and resulting build-up of CO_2 provides a mechanism to return the surface to habitability on a relatively short time scale.

The formation of calcium carbonate by living organisms in the oceans is an important part of the Earth's carbonate–silicate cycle, whose quantitative details would certainly be different on other planets. It seems reasonable, however, to assume that a qualitatively similar negative feedback loop acts to stabilize the climate of terrestrial planets that, like the Earth, have active plate tectonics, a large reservoir of carbon, and substantial but not dominant amounts of water. Within this class of planets the limits to habitability can be identified with catastrophic failures of climate stabilization mechanisms. As the stellar flux decreases for planets at larger orbital distance, the expected drop in the surface temperature is partially offset by the development of successively thicker CO_2 atmospheres which provide warming via the greenhouse effect. This stabilization fails once the atmosphere reaches a "maximum greenhouse" limit, beyond which CO_2 starts to condense out and is unable to provide additional warming. Calculations by Kopparapu *et al.* (2013) for a solar-type star place the maximum greenhouse limit at $F \simeq 0.35 F_0$ (where F_0 is the flux received by the Earth), corresponding to an orbital radius of 1.7 AU. Moving in the other direction, a closer-in analog of the Earth can maintain a stable but hotter climate until it reaches the "moist greenhouse" limit, when the stratosphere becomes dominated by water. Water can then be lost from the planet via a combination of dissociation of water molecules and escape of the light H atoms. The time scale for water loss becomes less than the age of the Earth for a surface temperature of $T \approx 340$ K, so this limit is more stringent than the simple requirement that the temperature remain below the boiling point of water. The oceans are expected to evaporate in their entirety at the "maximum greenhouse" limit, beyond which CO_2 can accumulate further in the absence of efficient sink processes. The nominal moist and runaway greenhouse limits for the Solar System are at 0.99 AU and 0.97 AU (Kopparapu *et al.*, 2013), although the cooling effect of clouds (which are not included in these calculations) means that the Earth is not as close to the inner edge of the habitable zone as these estimates would suggest.

Figure 1.19 shows how the inner and outer boundaries of the habitable zone are predicted to scale with stellar mass. The leading order effect, which dominates the plot of $a_{\mathrm{out}}(M_*)$ and $a_{\mathrm{in}}(M_*)$ in the lower panel, is the scaling of stellar luminosity with mass. Smaller but still significant effects occur because of the shift of the peak of the stellar spectrum with mass. Cooler stars emit a greater fraction of their bolometric luminosity in the near-IR as compared to the visible, and this near-IR flux is both scattered less (Rayleigh scattering is proportional to λ^{-4}) and absorbed better (by species such as H_2O and CO_2) by planetary atmospheres. An Earth analog around a lower mass star would therefore have a lower albedo, and both the inner and outer edges of the habitable zone would be pushed outward. The magnitude of the shift, expressed in terms of the threshold flux F/F_0, is about 20–25% for stars ranging in mass between 0.1 $M_\odot$ and 1 $M_\odot$.

Opportunities to assess the validity of theoretical estimates of the habitable zone are for now extremely limited. Mars cannot sustain liquid water on its surface today, and hence is not within the habitable zone by our definition. The presence of river valleys and canyons on the Martian surface, however, strongly suggests that liquid water flowed on Mars in the early history of the planet, and may have been stable until atmospheric loss under the low gravity conditions led to cooling. This line of reasoning is consistent with the theoretical expectation that Mars' orbit lies close to the outer edge of the habitable zone. Much less can be said about the inner edge of the habitable zone. Venus is assuredly not habitable today, but there is no direct way of knowing whether the planet had liquid water at an early time when the solar luminosity was smaller. Empirically, then, the inner edge of the habitable zone lies beyond the orbit of Venus, but could be significantly closer to the Sun than the theoretical estimate shown in Fig. 1.19.

It is worth emphasizing again that the conventional definition of the habitable zone does not necessarily match up to the plain English meaning of habitability. Planets within the habitable zone may be inhospitable to life, for example because they possess vanishing quantities of water, while bodies far outside the zone could in principle harbor habitable sub-surface oceans. (There is strong evidence for such reservoirs being present on both Jupiter's moon Europa and Saturn's moon Enceladus.) For these reasons, the habitable zone is best viewed as defining a region of exoplanet parameter space that can reasonably be prioritized in follow-up atmospheric characterization studies focused on the identification of biomarkers such as oxygen.

1.10 Further Reading

The properties of the planets, moons, and minor bodies of the Solar System are discussed in depth in any planetary science text. A good example is *Planetary Sciences* by I. de Pater and J. J. Lissauer (updated second edition 2015, Cambridge: Cambridge University Press).

The *Handbook of Exoplanets* (editors H. J. Deeg and J. A. Belmonte, 2018, Cham: Springer) includes review articles on different exoplanet detection methods along with analyses on the properties of the exoplanet population.

2

Protoplanetary Disk Structure

Planets form from protoplanetary disks of gas and dust that are observed to surround young stars for the first few million years of their evolution. Disks form because stars are born from relatively diffuse gas (with particle number density $n \sim 10^5 \, \mathrm{cm}^{-3}$) that has too much angular momentum to collapse directly to stellar densities ($n \sim 10^{24} \, \mathrm{cm}^{-3}$). Disks survive as well-defined quasi-equilibrium structures because once gas settles into a disk around a young star its specific angular momentum *increases* with radius. To accrete, angular momentum must be lost from, or redistributed within, the disk gas, and this process turns out to require time scales that are much longer than the orbital or dynamical time scale.

In this chapter we discuss the structure of protoplanetary disks. Anticipating the fact that angular momentum transport is slow, we assume here that the disk is a static structure. This approximation suffices for a first study of the temperature, density, and composition profiles of protoplanetary disks, which are critical inputs for models of planet formation. It also permits investigation of the predicted emission from disks which can be compared to a large body of astronomical observations. We defer for Chapter 3 the tougher question of how the gas and solids within the disk evolve with time.

2.1 Disks in the Context of Star Formation

Stars form in the Galaxy today from the small fraction of gas that exists in molecular form within relatively dense, cool, molecular clouds. Observationally, most star formation results in the formation of stellar clusters (which may subsequently disperse), within which most stars are part of binary or small multiple systems (Duquennoy & Mayor, 1991). Protoplanetary disks inherit their initial mass, size, magnetic flux, and chemical composition from this broader star formation environment, while their subsequent evolution may be influenced by environmental effects such as stellar radiation, ongoing gas accretion, or stellar flybys. The importance of environmental effects varies markedly depending upon the star formation environment. The Trapezium cluster at the core of the Orion Nebula, for example, has a stellar density in excess of 10^4 stars per pc^3, while the small cluster around the young massive star MWC 297 has a much lower density of $\rho_* \sim 10^2 \, pc^{-3}$ (Lada & Lada, 2003). Evidently the common approximation that disks evolve in isolation will be better in the low density case. Surveys suggest that most stars form within rich clusters containing 100 or more stars, though even within such clusters the local environment can vary substantially (much of the Orion star forming region, for example, is much less dense than the Trapezium). Once disks have formed, the dominant environmental effects are normally due to close binary companions (if present) and external radiation produced by other stars within the cluster, especially massive ones that produce strong UV fluxes.

Molecular clouds are not homogeneous structures. Rather, their density and velocity fields exhibit structure across a wide range of spatial scales (Larson, 1981) which is interpreted as a combination of turbulence and self-gravitating infall. Any collapsing region will therefore possess nonzero angular momentum, leading to the formation of disks (and perhaps even binary companions) around newly formed stars. Within molecular clouds, stars form from dense knots of gas called molecular cloud cores. These cores have typical scales of ~ 0.1 pc and densities of $n \sim 10^5 \, cm^{-3}$. On these scales the *dynamical* importance of rotation is rather modest. An observational analysis by Goodman *et al.* (1993), for example, estimated the typical ratio of rotational to gravitational energy in dense cores to be

$$\beta \equiv \frac{E_{\text{rot}}}{|E_{\text{grav}}|} \sim 0.02. \tag{2.1}$$

Even these small values, however, correspond to very substantial reservoirs of angular momentum. If we crudely model a core as a uniform density sphere in solid body rotation, then we find that a $\beta = 0.02$ core with a mass of $M_\odot$ and a radius of 0.05 pc has an angular momentum

$$J_{\text{core}} \simeq 10^{54} \, \text{g cm}^2 \, \text{s}^{-1}, \tag{2.2}$$

which is three to four orders of magnitude in excess of the total angular momentum in the Solar System. Understanding how this angular momentum is lost or redistributed either during collapse, or subsequently, is at the heart of the *angular momentum problem* of star formation. Finessing this problem for now, we simply note that if a solar mass star forms from such a cloud the mean specific angular momentum $l_{\text{core}} = J_{\text{core}}/M_\odot \approx 4 \times 10^{20} \, cm^2 \, s^{-1}$. Gas with this much angular momentum will circularize around the newly formed star at a radius r_{circ} where the specific angular momentum of a Keplerian orbit equals that of the core,

$$l_{\text{core}} = \sqrt{GM_\odot r_{\text{circ}}}. \tag{2.3}$$

For the parameters discussed above $r_{\text{circ}} \sim 100$ AU. The formation of disks with sizes comparable to or larger than the Solar System is thus an inevitable consequence of the collapse of molecular cloud cores.

2.2 Observations of Protostellar Disks

Young Stellar Objects (YSOs) frequently show more emission in the infra-red than would be expected from a pre-main-sequence star's photosphere. The IR excess is attributed to the presence of dust in the vicinity of the star, and its strength forms the basis for an empirical classification scheme for YSOs (Lada & Wilking, 1984). We define the slope of the spectral energy distribution (SED) between near-IR and mid-IR wavelengths,

$$\alpha_{IR} \equiv \frac{d \log \nu F_\nu}{d \log \nu} \equiv \frac{d \log \lambda F_\lambda}{d \log \lambda}. \tag{2.4}$$

The anchor points in the near- and mid-IR for the determination of α_{IR} vary from study to study, but are typically something like 2 μm and 25 μm. Based on α_{IR} four or five classes of YSO are recognized (Williams & Cieza 2011):

- **Class 0**: heavily obscured sources with no optical or near-IR emission (α_{IR} is therefore undefined).
- **Class I**: $\alpha_{IR} > 0.3$.
- **Flat spectrum sources**: $-0.3 < \alpha_{IR} < 0.3$.
- **Class II**: $-1.6 < \alpha_{IR} < -0.3$.
- **Class III**: $\alpha_{IR} < -1.6$. These sources have at most weak IR-excess emission, and have SEDs that resemble isolated pre-main-sequence stellar photospheres.

Figure 2.1 shows these empirically derived classes, and their theoretical interpretation in terms of the expected evolution of circumstellar material over time (Adams *et al.*, 1987). The youngest (Class 0 and Class I) YSOs are surrounded by disks that are fed by infall from envelopes. Older YSOs (Class II) have lost their envelopes but retain relatively massive and often actively accreting gas disks. Finally the gas disk dissipates, leaving behind a Class III YSO.

For the subset of YSOs that are optically visible a separate classification scheme is based on the equivalent width of the Hα line (Joy, 1945). Hα is a diagnostic of gas accreting onto the star, so Class II sources with disks are largely equivalent to "classical T Tauri stars" with high Hα equivalent widths. "Weak-lined T Tauri stars" are likewise essentially the same objects as Class III sources.

Determining the absolute ages of young stars is a difficult exercise that leads to uncertainty in estimates of the ages of individual star–disk systems. It is easier to estimate the *relative* durations of the different YSO phases. Given a large sample of regions of recent and ongoing star formation, it can be assumed that the proportions of YSOs in the different classes reflect the amount of time that a typical YSO spends in each class. Figure 2.1 shows the distribution of YSOs among the classes, as determined by the "Cores to Disks" legacy project which used the *Spitzer* space telescope (Evans *et al.*, 2009). Class I sources with envelopes are much less numerous than Flat Spectrum and Class II which that have disks but little or no

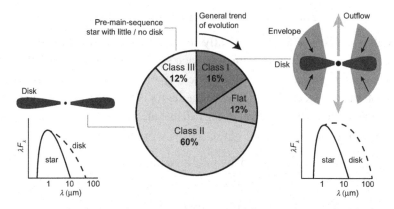

Figure 2.1 The physical picture underlying the classification of Young Stellar Objects together with statistics from the Spitzer c2d Legacy survey for the fraction of sources falling into each class.

envelope. Most of the circumstellar disk lifetime therefore consists of relatively isolated star–disk systems that have largely completed accreting their envelopes.

Many studies of planet formation assume that the disk environment of interest corresponds to Class II YSOs. Important phases of planet formation, however, have estimated time scales that are shorter than the likely lifetime of the Class 0 and Class I phases of YSO evolution. Planet formation could therefore be well advanced prior to young stars becoming optically visible.

2.2.1 Accretion Rates and Lifetimes

The accretion rate $\dot{M}$ of gas through the disk is an important quantity that we would like to determine. It can only be measured with any confidence at small radii, where the gas from the disk is flowing onto the star. As we will discuss in Section 3.9, for typical classical T Tauri stars the stellar magnetic field disrupts the inner disk at a magnetospheric radius R_m. Gas interior to R_m follows trajectories that are tied to magnetic field lines, before crashing onto the stellar surface at close to the free-fall speed. For a star of mass M_* and radius R_* the accretion luminosity will be

$$L_{\text{acc}} \simeq GM_*\dot{M}\left(\frac{1}{R_*} - \frac{1}{R_m}\right)$$

$$\sim 0.2\left(\frac{M_*}{M_\odot}\right)\left(\frac{R_*}{1.5\,R_\odot}\right)^{-1}\left(\frac{\dot{M}}{10^{-8}\,M_\odot\,\text{yr}^{-1}}\right)L_\odot, \qquad (2.5)$$

where for the numerical estimate we have taken $R_m \gg R_*$ and adopted typical numbers for T Tauri stars. The accretion luminosity will be radiated from shocks or hotspots on the stellar surface. $\dot{M}$ can then be estimated from UV data provided that we know the basic stellar parameters and can distinguish the emission arising from accretion from the stellar photospheric emission (Gullbring *et al.*, 1998).

The inferred accretion rates in Class II YSOs are found to vary strongly with the stellar mass M_*. In the Chamaeleon I star forming region, for example, Manara *et al.* (2016) determine a stellar mass/accretion rate relation of the form

$$\log\left(\frac{\dot{M}}{M_\odot \text{ yr}^{-1}}\right) = (1.83 \pm 0.35) \log\left(\frac{M_*}{M_\odot}\right) - (8.13 \pm 0.16). \tag{2.6}$$

This is a representative result. Mean accretion rates through protoplanetary disks around solar mass stars are of the order of $10^{-8}\ M_\odot \text{ yr}^{-1}$, with a stellar mass scaling that is significantly steeper than linear.

2.2.2 Inferences from the Dust Continuum

At disk temperatures $T < 1500$ K the opacity in protoplanetary disks is dominated by the contribution from rocky or icy grains (generically "dust"). Observations of continuum radiation from thermal dust emission provide information about the radial temperature distribution in the disk, the disk mass, and the size of dust particles. To illustrate how this works, we start with a toy model of a thin disk in which dust emits thermal radiation with a single temperature at each radius. We assume that the gas surface density Σ and dust temperature T_{dust} are power laws in radius,

$$\Sigma \propto r^{-p},$$
$$T_{\text{dust}} \propto r^{-q}, \tag{2.7}$$

and adopt a frequency-dependent opacity (defined per unit mass of gas),

$$\kappa_\nu = \kappa_0 \nu^\beta. \tag{2.8}$$

The vertical optical depth through the disk is then $\tau_\nu = \Sigma \kappa_\nu$. For a face-on disk the solution of the radiative transfer equation (Rybicki & Lightman, 1979) gives the flux density F_ν as

$$F_\nu = \frac{1}{D^2} \int_{r_{\text{in}}}^{r_{\text{out}}} B_\nu(T_{\text{dust}})(1 - e^{-\tau_\nu})2\pi r dr, \tag{2.9}$$

where D is the distance to the source, r_{in} and r_{out} are the inner and outer radii of the disk, and B_ν is the Planck function,

$$B_\nu = \frac{2h\nu^3}{c^2} \frac{1}{\exp[h\nu/k_B T_{\text{dust}}] - 1}. \tag{2.10}$$

Important inferences can be made from F_ν in the limit where the disk is either optically thick ($\tau_\nu > 1$) or optically thin.

Taking the $\tau_\nu \gg 1$ limit first, let us assume that we observe the disk at wavelengths where the entire disk is optically thick (typically in the near- and mid-IR). Setting $e^{-\tau_\nu} = 0$ in Eq. (2.9) we obtain,[1]

$$\nu F_\nu \propto \nu^{4-2/q}. \tag{2.11}$$

The slope of the infrared spectral energy distribution thus provides a constraint on the radial variation of dust temperature within the disk.

In the mm/sub-mm region of the spectrum (i.e. $\lambda \sim 1$ mm) the bulk of the emission comes instead from optically thin regions of the disk. Taking the limit where the entire disk is optically thin at the frequencies of interest, we have that $(1 - e^{-\tau_\nu}) \approx \tau_\nu = \Sigma \kappa_\nu$. Equation (2.9) then takes the form

$$F_\nu = \frac{B_\nu(\bar{T}_{\text{dust}})\kappa_\nu}{D^2} \int_{r_{\text{in}}}^{r_{\text{out}}} 2\pi r \Sigma \, dr, \tag{2.12}$$

where $\bar{T}_{\text{dust}}$ is a weighted average of the temperature of the emitting material that we can determine from the infrared part of the SED. From this we can infer two disk properties. First, we note that the integral in the above expression is just the disk mass, M_{disk}. If we know the distance to the source, the opacity, and the disk temperature, a measurement of the flux density at optically thin wavelengths determines the disk mass. Second, at sufficiently long wavelengths we will be on the Rayleigh–Jeans tail of the Planck function. In this limit $B_\nu \propto \nu^2$ and we find

$$\nu F_\nu \propto \nu^{\beta+3}. \tag{2.13}$$

A measurement of the spectral slope at optically thin wavelengths in the mm/sub-mm thus determines the frequency dependence of the opacity. The opacity, in turn, is determined by the size distribution, structure, and composition of the solid particles within the disk. For many disks, the integrated emission implies a value for $\beta \approx 1 \pm 0.5$ which is significantly smaller than the value ($\beta \approx 2$) found for dust in either the interstellar medium or in molecular clouds (Beckwith & Sargent, 1991). This result provides evidence for the growth of solid particles up to sizes of at least mm within protoplanetary disks (Ricci *et al.*, 2010).

The interpretation of dust continuum observations is subject to a number of uncertainties. The most directly inferred quantity is the mass of *optically thin dust* with particle sizes roughly the same as the observing wavelengths. Additional mass could be hidden in similarly sized particles at radii where the emission is optically thick, and in particles that are large compared to the observing wavelength. Conversion to a *gas* mass involves additional complications, because several physical processes affect the solid and gas components in different ways. For example, radial drift changes the gas to dust ratio as a function of both radius and time,

---

[1] To see this, substitute $x \equiv h\nu/k_B T$ and approximate the limits as $r_{\text{in}} = 0$ and $r_{\text{out}} = \infty$.

while photoevaporation preferentially removes gas from the disk leaving all but the smallest solid particles behind.

2.2.3 Molecular Line Observations

In addition to the continuum emission in the IR and mm/sub-mm, a number of molecules and molecular ions have been detected in line radiation. In the mm/sub-mm the workhorse molecule is CO (and its isotopologues such as ^{13}CO and $C^{18}O$), but several other fairly simple and abundant species including CS, HCN, N_2H^+, and DCO^+ are observable. Some of the same molecules (albeit at different disk radii) are also observable in the near-IR, including CO. The far-IR provides access to additional molecules, including water and ammonia.

Molecular observations are powerful, but subject to a key limitation that comes from the fact that H_2 is a homonuclear diatomic molecule with no electric dipole moment. As a result H_2 cannot produce a rotational or vibrational spectrum in the dipole approximation, and the bulk of the mass in cold H_2 is observationally inaccessible. There are some loopholes. Hot H_2 in the atmosphere of the disk (with $T > 2000$ K) can be excited by stellar Lyα radiation, and detected via electronic transitions that lie in the far-UV. Emission from hydrogen deuteride HD (specifically the $J = 1 \to 0$ rotational transition at $\lambda \approx 112$ μm) is also detectable, and has been used to measure a small number of disk masses (McClure *et al.*, 2016).

Molecular line observations provide a wealth of kinematic information. The instrumental resolution of spectra in the mm/sub-mm is typically better than the expected thermal width of molecular lines, so for bright disks that are well resolved spatially an observation yields a "data cube" expressing the line emission as a function of sky position (x, y) and line of sight velocity v_{los}. At the lowest order, such data can be used to measure the rotation profile of the disk gas and hence the mass of the central star. Beyond that we may try to measure the contributions that thermal and turbulent broadening make to the line profile. In turbulent disk the total line width is

$$\Delta v = \frac{v}{c} \sqrt{\frac{2k_B T}{\mu m_H} + v_{turb}^2}, \qquad (2.14)$$

where μ is the molecular weight of the observed species in units of the mass of a hydrogen atom m_H, T is the temperature, and v_{turb} is a root mean square estimator of the turbulence. We generally expect turbulence in disks to be subsonic, but by observing relatively heavy molecules such as CO or CS it is possible to attain sensitivity to v_{turb} values that are significantly below the sound speed. Using molecules with differing optical depths (e.g. isotopologues of CO) opens up the possibility of mapping v_{turb} as a function of height above the disk mid-plane (Flaherty *et al.*, 2017).

2.2.4 Transition Disks

The term transition disk (Espaillat *et al.*, 2014) is an umbrella for a subset of disks that do not fit neatly into the SED-based classification scheme for YSOs. A significant number of sources show little-to-no near-IR excess (resembling Class III sources) but robust mid- and far-IR emission (resembling Class II sources). The transitional disk class includes several well-studied systems, including TW Hya, GM Aur, IRS 48 and LkCa15.

The geometric interpretation of transition disk SEDs is straightforward. Near-IR excesses originate in the inner disk, which in a normal Class II source is expected to be the region with the highest optical depth. Seeing little or no near-IR emission, while mid-IR emission persists, implies a disk with a hole or cavity in the dust distribution. This inference is supported by imaging in the sub-mm, which directly reveals the presence of dust cavities within transition disks. Nonetheless, the cavities in transition disks are for the most part not empty. Most transitional sources are accreting, and although there may be some suppression the rate can be as high as normal classical T Tauri stars. Moreover, although the defining feature of transitional disks is a cavity in the dust distribution, some transition disks' spectra indicate the presence of low levels of dust close to the star. The overall picture is thus one in which a relatively unexceptional outer disk is truncated. Inside the cavity the column density of dust is severely depleted, but gas persists and continues to accrete onto the star.

2.2.5 Disk Large-Scale Structure

Long baseline observations with *ALMA* at mm and sub-mm wavelengths have enabled imaging of protoplanetary disks with an angular resolution as good as 0.025 arcseconds. Surprisingly, it is found that relatively high contrast axisymmetric structures are common features of protoplanetary disks. In one of the prototypical cases, HL Tau, continuum imaging at wavelengths between 0.87 mm and 2.9 mm detects seven pairs of bright/dark ring-like structures at orbital radii between 20 AU and 100 AU (ALMA Partnership *et al.*, 2015). The Disk Substructures at High Angular Resolution Project (DSHARP) results suggest that similar ring-like features are the most common type of large-scale structure in protoplanetary disks (Andrews *et al.*, 2018). Rings are observed in dust continuum images at a wide variety of orbital radii, and with a broad distribution of strengths.

Nonaxisymmetric structure is also evident in high resolution disk images. One class of nonaxisymmetric structure is horseshoe-shaped emission in the mm/sub-mm. A particularly vivid example is the transition disk system IRS 48. Van der Marel *et al.* (2013) found that the mm-sized particles surrounding the cavity at $r \approx 60$ AU, traced by 0.44 mm continuum emission, were strongly concentrated in a crescent-shaped feature on one side of the star. No such asymmetry was evident in either the gas or in micrometer-sized dust traced by mid-IR imaging.

Spiral structure represents a separate class of nonaxisymmetric features. Spirals are seen in both scattered light and dust continuum images, for example in the disk around the young star Elias 2-27 (Perez *et al.*, 2016).

2.3 Vertical Structure

The equilibrium structure of gas orbiting a star is determined in general by solving for a steady-state solution to the hydrodynamic equations and the Poisson equation for the gravitational potential. Such an exercise is nontrivial, and even when a solution can be found its dynamical stability is not guaranteed (Papaloizou & Pringle, 1984).

For protoplanetary disks two simplifications make the problem a great deal more straightforward. First, it is usually justified to assume that the total disk mass $M_{\text{disk}} \ll M_*$, and this allows us to neglect the gravitational potential of the disk and consider only stellar gravity. This approximation is accurate for disks with masses comparable to the minimum mass Solar Nebula ($M_{\text{disk}} \simeq 0.01 M_\odot$), becomes marginal for some of the most massive observed disks ($M_{\text{disk}} \simeq 0.1 M_\odot$), and fails at sufficiently early epochs when the disk mass may be comparable to that of the protostar. We will discuss some of the dynamical effects that are important for massive disks later on, but for now we arbitrarily define a protoplanetary disk as one whose mass is small enough ($M_{\text{disk}} \leq 0.1 M_\odot$) that the star dominates the potential.[2] Second, the vertical thickness of the disk h is invariably a small fraction of the orbital radius. This follows from the fact that a disk has a large surface area and can cool via radiative losses rather efficiently. Efficient cooling implies relatively low disk temperatures and pressures, which are unable to support the gas against gravity except in a geometrically thin disk configuration with $h/r \ll 1$.

The structure of a geometrically thin protoplanetary disk follows from considering the vertical force balance at height z above the mid-plane in a disk orbiting at cylindrical radius r around a star of mass M_*. The vertical component of the gravitational acceleration (Fig. 2.2),

$$g_z = g \sin \theta = \frac{GM_*}{(r^2 + z^2)} \frac{z}{(r^2 + z^2)^{1/2}}, \qquad (2.15)$$

must balance the acceleration due to the vertical pressure gradient in the gas $(1/\rho)(\mathrm{d}P/\mathrm{d}z)$. If we assume that the disk is vertically isothermal (a reasonable first guess if the temperature of the gas is set by stellar irradiation) and write the pressure as $P = \rho c_{\text{s}}^2$, with c_{s} being the sound speed, we have

---

[2] More massive disks are sometimes described as protostellar rather than protoplanetary disks, but there is no consistent usage of these terms in the literature.

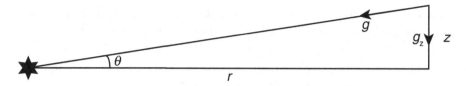

Figure 2.2 The vertical structure of geometrically thin disks is set by a balance between the vertical component of the star's gravity g_z and a pressure gradient.

$$c_s^2 \frac{d\rho}{dz} = -\frac{GM_*z}{(r^2 + z^2)^{3/2}}\rho. \tag{2.16}$$

The solution is

$$\rho = C \exp\left[\frac{GM_*}{c_s^2(r^2 + z^2)^{1/2}}\right], \tag{2.17}$$

where the constant of integration C is set by the mid-plane density. This expression is rarely used. Rather, we note that for a thin disk $z \ll r$ and $g_z \simeq \Omega^2 z$, where $\Omega = \sqrt{GM_*/r^3}$ is the Keplerian angular velocity. In this limit the vertical density profile has the simple form

$$\rho = \rho_0 e^{-z^2/2h^2}, \tag{2.18}$$

where the mid-plane density ρ_0 can be written in terms of the full-plane surface density Σ as

$$\rho_0 = \frac{1}{\sqrt{2\pi}} \frac{\Sigma}{h}, \tag{2.19}$$

and h, the vertical disk scale-height, is given by

$$h \equiv \frac{c_s}{\Omega}. \tag{2.20}$$

Note that this means that the geometric thickness of the disk $h/r = \mathcal{M}^{-1}$, where the Mach number of the flow $\mathcal{M} = v_K/c_s$.

The assumptions that went into this calculation are generally self-consistent. Detailed disk models computed by Bell *et al.* (1997) show that for a disk accreting at a low rate onto a solar mass star the mid-plane temperature at 1 AU is around $T \simeq 100$ K. The corresponding isothermal sound speed is

$$c_s^2 = \frac{k_B T}{\mu m_p}, \tag{2.21}$$

where k_B is the Boltzmann constant and μ is the mean molecular weight in units of the proton mass m_p. Taking $\mu = 2.3$ for a fully molecular gas of cosmic composition we have that $c_s \approx 0.6$ km s^{-1} and (at 1 AU around a solar mass star) $h/r \approx 0.02$. The condition that the disk is geometrically thin is adequately satisfied. We should also verify that the vertical structure is not modified by the gravitational force from the disk itself. Representing the disk as a thin infinite sheet of mass with surface density Σ, a straightforward application of Gauss' theorem shows that the acceleration outside the sheet is constant with distance,

$$g_{z,\text{disk}} = 2\pi G\Sigma. \tag{2.22}$$

Equating this disk contribution to the acceleration due to the vertical component of stellar gravity at $z = h$, the condition that the self-gravity of the disk can be neglected when computing the vertical structure becomes

$$\Sigma < \frac{M_*h}{2\pi r^3}. \tag{2.23}$$

For the minimum mass Solar Nebula at 1 AU this condition is satisfied by more than an order of magnitude. More generally, if we write the enclosed disk mass as $M_{\text{disk}}(r) \sim \pi r^2 \Sigma$, self-gravity is ignorable provided that

$$\frac{M_{\text{disk}}}{M_*} < \frac{1}{2}\left(\frac{h}{r}\right). \tag{2.24}$$

At large radii $h/r \sim 0.1$, so for a massive disk with $M_{\text{disk}}/M_* = 0.1$ the neglect of disk self-gravity in the calculation of the vertical structure is not justified. Such disks require additional consideration, however, since the condition $M_{\text{disk}}/M_* \sim h/r$ also describes the mass above which disk self-gravity results in the disk becoming unstable to the development of nonaxisymmetric structure in the form of spiral waves. We discuss this separate (and more important) effect of self-gravity in Chapter 3.

The shape of the disk depends upon $h(r)/r$. If we parameterize the radial variation of the sound speed via

$$c_s \propto r^{-\beta}, \tag{2.25}$$

then the aspect ratio varies as

$$\frac{h}{r} \propto r^{-\beta+1/2}. \tag{2.26}$$

The disk will flare – i.e. h/r will increase with radius giving the disk a bowl-like shape – if $\beta < 1/2$. This requires a temperature profile $T(r) \propto r^{-1}$ or shallower. In an unwarped flaring disk the star is visible from all points on the surface of the disk.

2.4 Radial Force Balance

The density profile of the disk in the radial direction cannot be derived without either considering the nature of angular momentum transport, or by appealing to observational constraints (such as the minimum mass Solar Nebula). The orbital velocity of disk gas, however, can be determined given a surface density and temperature profile. We start from the momentum equation for an unmagnetized and inviscid fluid,

$$\frac{\partial \mathbf{v}}{\partial t} + (\mathbf{v} \cdot \nabla)\mathbf{v} = -\frac{1}{\rho}\nabla P - \nabla \Phi, \tag{2.27}$$

where $\mathbf{v}$ is the velocity, ρ the density, P the pressure, and Φ the gravitational potential. Specializing to a stationary axisymmetric flow in which the potential is dominated by that of the star, the radial component of the momentum equation implies that the orbital velocity of the gas $v_{\phi,\mathrm{gas}}$ is given by

$$\frac{v_{\phi,\mathrm{gas}}^2}{r} = \frac{GM_*}{r^2} + \frac{1}{\rho}\frac{dP}{dr}. \tag{2.28}$$

Since the pressure near the disk mid-plane normally decreases outward, the second term on the right-hand-side is negative and the azimuthal velocity of the gas is slightly less than the Keplerian velocity (Eq. 1.25) of a point mass particle orbiting at the same radius. To quantify the difference we write the variation of the mid-plane pressure as a power-law near some fiducial radius r_0,

$$P = P_0 \left(\frac{r}{r_0}\right)^{-n}, \tag{2.29}$$

where $P_0 = \rho_0 c_\mathrm{s}^2$. Substituting we find that

$$v_{\phi,\mathrm{gas}} = v_\mathrm{K}\left(1 - n\frac{c_\mathrm{s}^2}{v_\mathrm{K}^2}\right)^{1/2}. \tag{2.30}$$

Recalling the definition of the vertical scale-height (Eq. 2.20) we note that the deviation from Keplerian velocity is $\mathcal{O}(h/r)^2$, and hence small for geometrically thin disks. For example, if the disk has a constant value of $h(r)/r = 0.05$ and a surface density profile $\Sigma \propto r^{-1}$ we obtain $n = 3$ and

$$v_{\phi,\mathrm{gas}} \simeq 0.996 v_\mathrm{K}. \tag{2.31}$$

When considering the motion of the gas alone this difference is utterly negligible and we can safely assume the gas moves at the Keplerian velocity. The slightly lower gas velocity is however important for the evolution of solid bodies within the disk, since it results in aerodynamic drag and resultant orbital decay.

2.5 Radial Temperature Profile of Passive Disks

The time scale for the disk to attain thermal equilibrium is generally much less than the time scale for evolution of either the disk or the star. The temperature profile of the disk is then set by the balance between cooling and heating, for which there are two main sources:

- Intercepted stellar radiation, which is absorbed (usually by dust) and subsequently reradiated at longer wavelengths. Disks for which this is the main heating process are described as passive disks.
- Dissipation of gravitational potential energy, as matter in the disk spirals in toward the star.

The heating per unit area from both of these sources drops off with increasing distance from the star. If the disk extends out to large enough radii, heating due to the ambient radiation field provided by nearby stars can become significant and prevent the disk temperature from dropping below some floor level. For normal star forming environments this floor temperature might be 10–30 K.

The relative importance of accretional heating versus stellar irradiation depends upon the accretion rate and the amount of intercepted stellar radiation. Globally, inspiralling matter will radiate an amount of energy per unit mass that is approximately given by the gravitational potential at the stellar surface, GM_*/R_*, while the disk will intercept a fraction $f < 1$ of the stellar radiation. Taking $f = 1/4$ (the correct result for a razor-thin disk that extends all the way to the stellar surface) the accretion rate $\dot{M}$ above which accretional heating dominates can be crudely estimated as

$$\frac{GM_*\dot{M}}{R_*} = \frac{1}{4}L_*. \tag{2.32}$$

For a young solar mass star with a luminosity $L_* = L_\odot = 3.9 \times 10^{33}$ erg s^{-1} and a radius $R_* = 2R_\odot$,

$$\dot{M} \approx 2 \times 10^{-8} \, M_\odot \, \text{yr}^{-1}. \tag{2.33}$$

Measured accretion rates for classical T Tauri stars range from an order of magnitude above this critical rate to two orders of magnitude below, so it is oversimplifying to treat protoplanetary disks as being either always passive or always active. Rather, the thermal structure of disks at early epochs (when accretion is strongest) is likely dominated by internal heating due to the accretion, whereas at later times reprocessing of stellar radiation dominates. At a single epoch, due to the flaring effect discussed below, accretion heating is relatively more significant close to the star.

2.5.1 Razor-Thin Disks

The temperature profile and spectral energy distribution of a passive protoplanetary disk are determined by the shape of the disk (whether it is flat, flared, or warped) and by the mechanism by which the absorbed stellar radiation is re-emitted. We begin by considering the simplest model: a flat razor-thin disk in the equatorial

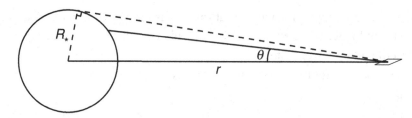

Figure 2.3 Geometry for calculation of the radial temperature profile of a razor-thin protoplanetary disk. The flux impinging on the surface of the disk at distance r from the star is computed by integrating over the visible stellar surface.

plane that absorbs all incident stellar radiation and re-emits it locally as a single temperature blackbody. The back-warming of the star by the disk is neglected.

We consider a surface in the plane of the disk at distance r from a star of radius R_*. The star is assumed to be a sphere of constant brightness I_*. Setting up spherical polar coordinates such that the axis of the coordinate system points to the center of the star, as shown in Fig. 2.3, the stellar flux passing through this surface is

$$F = \int I_* \sin\theta \cos\phi \, d\Omega, \tag{2.34}$$

where $d\Omega$ represents the element of solid angle. We count the flux coming from the top half of the star only (and to be consistent equate that to radiation from only the top surface of the disk), so the limits on the integral are

$$-\pi/2 < \phi \le \pi/2,$$
$$0 < \theta < \sin^{-1}\left(\frac{R_*}{r}\right). \tag{2.35}$$

Substituting $d\Omega = \sin\theta \, d\theta \, d\phi$, the integral for the flux is

$$F = I_* \int_{-\pi/2}^{\pi/2} \cos\phi \, d\phi \int_0^{\sin^{-1}(R_*/r)} \sin^2\theta \, d\theta, \tag{2.36}$$

which evaluates to

$$F = I_* \left[\sin^{-1}\left(\frac{R_*}{r}\right) - \left(\frac{R_*}{r}\right)\sqrt{1 - \left(\frac{R_*}{r}\right)^2} \right]. \tag{2.37}$$

For a star with effective temperature T_*, the brightness $I_* = (1/\pi)\sigma T_*^4$, with σ the Stefan–Boltzmann constant (e.g. Rybicki & Lightman, 1979). Equating F to the one-sided disk emission σT_{disk}^4 we obtain a radial temperature profile

$$\left(\frac{T_{\text{disk}}}{T_*}\right)^4 = \frac{1}{\pi}\left[\sin^{-1}\left(\frac{R_*}{r}\right) - \left(\frac{R_*}{r}\right)\sqrt{1 - \left(\frac{R_*}{r}\right)^2} \right]. \tag{2.38}$$

Integrating over radii, we obtain the total disk luminosity,

$$L_{\text{disk}} = 2 \times \int_{R_*}^{\infty} 2\pi r\sigma T_{\text{disk}}^4 \, dr$$

$$= \frac{1}{4} L_*. \tag{2.39}$$

We conclude that a flat passive disk extending all the way to the stellar equator intercepts a quarter of the stellar flux. The ratio of the observed bolometric luminosity of such a disk to the stellar luminosity will vary with viewing angle, but clearly a flat passive disk is predicted to be less luminous than the star.

The form of the temperature profile given by Eq. (2.38) is not very transparent. Expanding the right-hand-side in a Taylor series in the limit that $(R_*/r) \ll 1$ (i.e. far from the stellar surface), we obtain

$$T_{\text{disk}} \propto r^{-3/4}, \tag{2.40}$$

as the limiting temperature profile of a thin, flat, passive disk. For fixed molecular weight μ this in turn implies a sound speed profile

$$c_{\text{s}} \propto r^{-3/8}. \tag{2.41}$$

Assuming vertical isothermality, the aspect ratio given by Eq. (2.26) is

$$\frac{h}{r} \propto r^{1/8}, \tag{2.42}$$

and we predict that the disk ought to flare modestly to larger radii. If the disk does flare then the outer regions intercept a larger fraction of stellar photons, leading to a higher temperature. As a consequence, a temperature profile $T_{\text{disk}} \propto r^{-3/4}$ is probably the steepest profile we would expect to obtain for a passive disk.

2.5.2 Flared Disks

The next step in sophistication is to consider a flared disk. If at cylindrical distance r from the star the disk absorbs stellar radiation at a height h_{p} above the mid-plane, the disk is described as flared if the ratio h_{p}/r is an increasing function of radius. Features of flared disks are, first, that all points on the surface of the disk have a clear line of sight to the star and, second, that as seen from the star the disk subtends a greater solid angle than a razor-thin disk. Flared disks absorb a greater fraction of the stellar radiation than flat disks, and thus produce stronger IR excesses. The temperature profile is also modified and this changes the shape of the resulting SED.

The temperature profile of a flared disk can be computed in the same way as for a razor-thin disk, namely by evaluating the flux (Eq. 2.34) by integrating over the part of the stellar surface visible from the disk surface at radius r. The exact calculation can be found in the appendix to Kenyon and Hartmann (1987), and

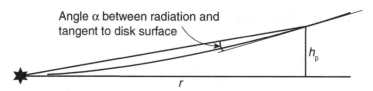

Figure 2.4 Geometry for calculation of the radial temperature profile of a flared protoplanetary disk. At distance $r \gg R_*$, radiation from the star is absorbed by the disk at height h_p above the mid-plane. The angle between the tangent to the disk surface and the radiation is α.

while it is conceptually simple some messy geometry is required. Here we adopt an approximate treatment valid for $r \gg R_*$, and consider the star to be a point source of radiation. At cylindrical distance r, stellar radiation is absorbed by the disk at height h_p above the mid-plane. Note that h_p is *not* the same as the disk scale-height h, since the absorption of stellar radiation depends not just on the density but also on the opacity of the disk material to stellar photons (the height of the disk's photosphere – i.e. the surface at which the optical depth to the disk's own thermal radiation is $\tau = 2/3$ – is yet a third different height). From consideration of the geometry in Fig. 2.4 the angle between the incident radiation and the local disk surface is given by[3]

$$\alpha = \frac{dh_p}{dr} - \frac{h_p}{r}.$$ (2.43)

The rate of heating per unit disk area at distance r is

$$Q_+ = 2\alpha \left(\frac{L_*}{4\pi r^2} \right),$$ (2.44)

where the factor of two comes from the fact that the disk has two sides and we have assumed that *all* of the stellar surface is visible from the surface of the disk (i.e. that the optically thick disk does not extend all the way to the stellar surface). Equating the heating rate to the rate of cooling by blackbody radiation,

$$Q_- = 2\sigma T_{disk}^4,$$ (2.45)

the temperature profile becomes

$$T_{disk} = \left(\frac{L_*}{4\pi\sigma} \right)^{1/4} \alpha^{1/4} r^{-1/2}.$$ (2.46)

Since $L_* = 4\pi R_*^2 \sigma T_*^4$ an equivalent expression is

$$\frac{T_{disk}}{T_*} = \left(\frac{R_*}{r} \right)^{1/2} \alpha^{1/4}.$$ (2.47)

---

[3] A more accurate approximation including a correction for the finite size of the star is $\alpha \simeq 0.4 R_*/r + r d(h_p/r)/dr$.

The interior of an irradiated protoplanetary disk can reasonably be assumed to be isothermal, so this equation specifies the central sound speed and hence the vertical scale-height h via Eq. (2.20). If we additionally specify a relation between h and h_p – for example by assuming that $h_p \propto h$, which may be plausible if the disk is very optically thick – then Eqs. (2.20), (2.21), (2.43), and (2.47) form a closed system of equations whose solution determines α. At large radii, Kenyon and Hartmann (1987) find that the surface temperature approaches

$$T_{\text{disk}}(r) \propto r^{-1/2}, \qquad (2.48)$$

which is as expected substantially flatter than the $T_{\text{disk}} \propto r^{-3/4}$ profile of a flat disk (Eq. 2.40).

2.5.3 Radiative Equilibrium Disks

Thus far we have assumed that the stellar radiation intercepted by the disk is reradiated as a single temperature blackbody. Additional complexity arises for real disks because the dominant opacity is provided by dust, which absorbs relatively short wavelength starlight (at around 1 μm) more efficiently than it emits longer wavelength thermal radiation. Figure 2.5 depicts the absorption and reradiation of stellar radiation for an optically thick disk in which dust is mixed evenly with the gas in the vertical direction. Short wavelength stellar radiation is absorbed by dust within a relatively thin surface layer for which the optical depth to grazing starlight $\tau \lesssim 1$. The layer is optically thin ($\tau \ll 1$) to the longer wavelength thermal

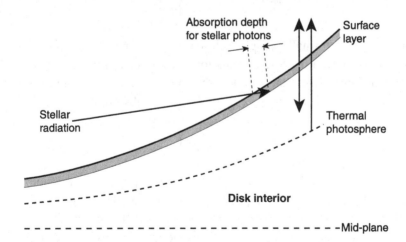

Figure 2.5 The physics underlying radiative equilibrium models for protoplanetary disks. Stellar radiation is absorbed in a thin surface layer by dust, which reradiates in the infrared both upward into space and downward where it is absorbed once more and acts to heat the disk interior. The local emission from the disk is a superposition of radiation from the hot surface layer and the cooler interior.

radiation emitted by the hot dust. Approximately half of the incoming stellar flux is then reradiated directly to space, while the downward directed half is absorbed by and heats the interior of the disk. The local emission from the disk is then a superposition of a cool blackbody component from the disk interior with a warmer dilute blackbody generated by the surface layer.

The temperature of the dust in the surface layer can be calculated given knowledge of the radiative properties of the dust. We define the emissivity ϵ as the ratio of the efficiency with which the dust emits or absorbs radiation relative to a blackbody surface (by definition a perfect absorber and emitter with $\epsilon = 1$). Dust is generally a good absorber of radiation at wavelengths that are small compared to the particle size, with the emissivity dropping at longer wavelengths. For spherical dust particles of radius s we have, approximately,

$$
\begin{aligned}
\epsilon &= 1, & \lambda \leq 2\pi s, \\
\epsilon &= \left(\frac{\lambda}{2\pi s}\right)^{-1}, & \lambda > 2\pi s.
\end{aligned}
\tag{2.49}
$$

This is equivalent to assuming that the wavelength dependence of the monochromatic opacity $\kappa_\nu \propto \lambda^{-1}$. If a dust particle at radius r is exposed to a stellar flux $F_* = L_*/(4\pi r^2)$, its equilibrium temperature T_{dust} is determined by balancing heating at a rate $\pi s^2 \epsilon_* F_*$, where ϵ_* is the weighted emissivity for absorption of the stellar radiation spectrum, with cooling at a rate $4\pi s^2 \sigma T_{\text{dust}}^4 \epsilon_{\text{dust}}$, where ϵ_{dust} is the weighted emissivity for the dust particle's thermal emission. Equating,

$$
\frac{T_{\text{dust}}}{T_*} = \left(\frac{\epsilon_*}{\epsilon_{\text{dust}}}\right)^{1/4} \left(\frac{R_*}{2r}\right)^{1/2}.
\tag{2.50}
$$

For particles that are small enough such that $\epsilon < 1$ for both absorption of stellar radiation and emission of thermal radiation, we can estimate the ratio $\epsilon_*/\epsilon_{\text{dust}}$ by evaluating the emissivity at the wavelength where thermal radiation peaks. We then have that $\epsilon \propto \lambda^{-1} \propto T$, so that $\epsilon_*/\epsilon_{\text{dust}} = T_*/T_{\text{dust}}$. Substituting in Eq. (2.50),

$$
\frac{T_{\text{dust}}}{T_*} = \left(\frac{R_*}{2r}\right)^{2/5}.
\tag{2.51}
$$

Since the mismatch between the emissivity for absorption and emission becomes more severe at larger radii, the temperature of dust exposed to stellar radiation at the disk surface drops even more slowly with increasing distance than the temperature of a single temperature flared disk. Inserting fiducial values for the stellar radius $R_* = 2R_\odot$ and stellar effective temperature $T_* = 4000$ K we find that

$$
T_{\text{dust}} = 470 \left(\frac{r}{1 \text{ AU}}\right)^{-2/5} \text{ K}.
\tag{2.52}
$$

This scaling was derived by Chiang and Goldreich (1997).

This analysis assumes that the dust particles can be considered in isolation, whereas in reality they are embedded within gas whose temperature will not in general be equal to the dust temperature given by Eq. (2.50). If $T_{gas} < T_{dust}$, collisions between molecules and the dust particles will result in nonradiative cooling of the dust. If this cooling process is efficient enough it will invalidate the radiative equilibrium that was assumed in deriving Eq. (2.50).

To evaluate the conditions for which it is reasonable to assume that the dust and gas are thermally decoupled, we note that collisions between molecules and dust particles occur at a rate (per unit area of dust) $n_{gas} v_{th}$, where n_{gas} is the number density of molecules, and v_{th}, the thermal velocity of the molecules, is given by

$$v_{th} = \left(\frac{8k_B T_{gas}}{\pi \mu m_H} \right)^{1/2}, \tag{2.53}$$

where μm_H is the mean particle mass. Each collision carries away of the order of $k_B T_{dust}$ worth of energy. Collisions will be ignorable for thermal equilibrium of dust particles if

$$k T_{dust} n_{gas} v_{th} < \epsilon_{dust} \sigma T_{dust}^4. \tag{2.54}$$

Evaluating the thermal velocity of the gas assuming that its temperature is comparable to that of the dust, we find that collisions can be neglected provided that

$$n_{gas} < \left(\frac{\pi}{8} \right)^{1/2} \frac{(\mu m_H)^{1/2}}{k_B^{3/2}} \epsilon_{dust} \sigma T_{dust}^{5/2}. \tag{2.55}$$

The two-temperature disk model of Fig. 2.5 will therefore persist provided that the gas density in the region where dust is marginally optically thick to stellar radiation is low enough. To give a specific example, we can estimate whether a model in which the gas and dust temperatures are unequal is self-consistent at 1 AU, where we would predict that $T_{dust} = 470$ K. For dust particles whose size is such that $\epsilon_{dust} \simeq 0.1$ the above expression suggests that collisions with hydrogen molecules are ignorable for gas densities $n_{gas} \lesssim 2 \times 10^{13}$ cm^{-3}. Taking a minimum mass Solar Nebula surface density at 1 AU of 1.7×10^3 g cm^{-2} with $h/r = 0.05$, the *mid-plane* hydrogen number density is a few $\times 10^{14}$ cm^{-3}. The dust and the gas will therefore be thermally coupled if the dust has settled to the mid-plane and absorbs stellar radiation there. Conversely, n_{gas} will be below the threshold for thermal coupling if the dust is well mixed with the gas and absorption occurs at a height $z \gtrsim 2.5h$. Evidently the extent of dust–gas energy exchange at 1 AU is rather subtle, and strictly it cannot be considered in isolation from questions of dust settling and agglomeration. At larger radii, where the surface density is smaller and the scale-height larger, the effect of collisions on the thermal state of irradiated dust can more safely be neglected.

For studies of disk chemistry, and for modeling of molecular line observations, the gas temperature is of more importance than that of the dust. If the density is high enough, the thermal accommodation process discussed above will establish a common temperature for both, and there is no need to explicitly consider the thermal physics of the gas. At lower optical depth, however, $T_{gas} \neq T_{dust}$. In particular, gas that is exposed to stellar UV photons is strongly heated by the photoelectric effect. The work function of graphite grains (the minimum energy required to free an electron from them) is around 5 eV, so 10 eV far ultraviolet (FUV) photons can eject electrons from uncharged grains with 5 eV of kinetic energy that is ultimately thermalized and acts to heat the gas. The ejection probability is of the order of 0.1, so the overall efficiency (the fraction of incident FUV energy that goes into heating the gas) can be rather high, around 5%.

Thermal models of gaseous protoplanetary disks include photoelectric heating, cooling via line radiation, and energy exchange with dust, among other processes. The qualitative picture of gas disk structure that emerges includes three layers.

- A cool mid-plane region, where $T_{gas} \approx T_{dust}$ and dust cooling is dominant.
- A warm surface layer in which both dust and gas have temperatures that exceed the mid-plane value. The gas in the warm layer can be substantially hotter than the dust ($T \sim 10^3$ K at 1 AU), and cools both by dust–gas collisions and by CO rotational–vibrational transitions.
- A hot, low density atmosphere, where Lyα radiation and other atomic lines cool the gas.

2.5.4 The Chiang–Goldreich Model

Although computation of disk models that include all of the aforementioned physics generally requires numerical techniques, useful analytic approximations are available. The most widely used is that developed by Chiang and Goldreich (1997). They considered a disk surrounding a 0.5 $M_\odot$ star with radius $R_* = 2.5 \, R_\odot$ and effective temperature $T_* = 4000$ K. The disk has a surface density profile $\Sigma(r) \propto r^{-3/2}$, a surface density at 1 AU of 10^3 g cm^{-2}, and a dust to gas ratio of 10^{-2}. The resulting temperature of dust at the surface of the disk is

$$T_{dust} = 550 \left(\frac{r}{1 \text{ AU}}\right)^{-2/5} \text{ K.} \tag{2.56}$$

The temperature of the disk interior T_i and the height of the visible photosphere h_p are approximated by piecewise polynomials. At 0.4 AU $< r <$ 84 AU,

$$T_i = 150 \left(\frac{r}{1 \text{ AU}}\right)^{-3/7} \text{ K,}$$
$$\frac{h_p}{r} = 0.17 \left(\frac{r}{1 \text{ AU}}\right)^{2/7}. \tag{2.57}$$

At 84 AU $< r <$ 209 AU,

$$T_i = 21 \text{ K},$$
$$\frac{h_p}{r} = 0.064 \left(\frac{r}{1 \text{ AU}}\right)^{1/2}. \tag{2.58}$$

At 209 AU $< r <$ 270 AU,

$$T_i = 200 \left(\frac{r}{1 \text{ AU}}\right)^{-19/45} \text{ K},$$
$$\frac{h_p}{r} = 0.20 \left(\frac{r}{1 \text{ AU}}\right)^{13/45}. \tag{2.59}$$

These expressions, which can be generalized without difficulty to other stellar or disk properties, are quite useful in practice for giving an idea of the physical conditions within weakly accreting protoplanetary disks. Their main limitation arises from the fact that in ignoring heating due to accretion they tend to underestimate the temperature at the disk mid-plane. This deficiency can be remedied (e.g. Garaud & Lin, 2007), but there remains considerable uncertainty in *all* disk models due to the uncertain distribution (both radially and vertically) of dust and its evolution over time.

2.5.5 Spectral Energy Distributions

The spectral energy distribution of the disk can be computed by summing up the local disk emission at each radius, weighted by the area. The simplest case is where each annulus in the disk radiates as a blackbody at the local temperature $T_{disk}(r)$. If the disk extends from r_{in} to r_{out} the disk spectrum is

$$F_\lambda \propto \int_{r_{in}}^{r_{out}} 2\pi r \, B_\lambda[T_{disk}(r)] dr, \tag{2.60}$$

where B_λ is the Planck function,

$$B_\lambda(T) = \frac{2hc^2}{\lambda^5} \frac{1}{\exp[hc/\lambda k_B T] - 1}. \tag{2.61}$$

The behavior of the spectrum implied by Eq. (2.60) is easy to derive. At long wavelengths $\lambda \gg hc/k_B T_{disk}(r_{out})$ we recover the Rayleigh–Jeans form

$$\lambda F_\lambda \propto \lambda^{-3}, \tag{2.62}$$

while at short wavelengths $\lambda \ll hc/k_B T_{disk}(r_{in})$ there is an exponential cut-off that matches that of the hottest annulus in the disk,

$$\lambda F_\lambda \propto \lambda^{-4} \exp\left[\frac{-hc}{\lambda k_B T_{disk}(r_{in})}\right]. \tag{2.63}$$

For intermediate wavelengths,

$$\frac{hc}{k_B T_{\text{disk}}(r_{\text{in}})} \ll \lambda \ll \frac{hc}{k_B T_{\text{disk}}(r_{\text{out}})}, \tag{2.64}$$

the form of the spectrum depends upon the radial profile of the disk temperature distribution. For the razor-thin disk model with $T_{\text{disk}}(r) \propto r^{-3/4}$ we substitute

$$x \equiv \frac{hc}{\lambda k_B T_{\text{disk}}(r_{\text{in}})} \left(\frac{r}{r_{\text{in}}}\right)^{3/4}, \tag{2.65}$$

into Eq. (2.60). We then have, approximately,

$$F_\lambda \propto \lambda^{-7/3} \int_0^\infty \frac{x^{5/3}\mathrm{d}x}{e^x - 1} \propto \lambda^{-7/3}, \tag{2.66}$$

and so

$$\lambda F_\lambda \propto \lambda^{-4/3}. \tag{2.67}$$

The overall spectrum, shown schematically in Fig. 2.6, is that of a "stretched" blackbody (Lynden-Bell & Pringle, 1974). This simple spectrum is too steep in the infrared to account for the observed properties of Class I and many Class II sources. The more realistic flared disk models, or those that incorporate a warm surface layer, yield flatter spectral energy distributions in the infrared which are able to reproduce the features of many YSO SEDs.

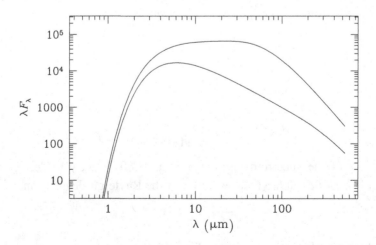

Figure 2.6 The spectral energy distributions (λF_λ, plotted in arbitrary units on the vertical axis) of disks radiating as multicolor blackbodies with temperature profiles $T_{\text{disk}}(r) \propto r^{-1/2}$ (upper curve) and $T_{\text{disk}}(r) \propto r^{-3/4}$ (lower curve). For these toy models the disk was assumed to have a temperature of 1000 K at the inner edge at 0.1 AU, and to extend out to 30 AU.

2.6 Opacity

Dust is the dominant opacity source within protoplanetary disks everywhere except in the very innermost regions, where the temperature is high enough ($T \simeq 1500$ K) for dust particles to be destroyed leaving only molecular opacity. For an individual dust particle the cross-section to radiation of a given wavelength will depend upon the particle size, structure (for example the particle may have a spherical or more complex geometry), and composition. At a particular location within the disk, the total opacity due to dust will depend primarily upon the temperature and chemical composition of the disk (which determine which types of dust or ice are present) and the size distribution of the particles, which may change as coagulation proceeds. To a lesser extent, the gas density also influences the opacity by modifying the temperatures at which different species of dust particle are predicted to evaporate. Given a model for each of the above effects, it is possible to compute the frequency-dependent opacity κ_ν of the gas and dust within the disk as a function of disk temperature and density.

Within the inner disk (where "inner" encompasses most of the radii of greatest interest for planet formation) it is typically true that the mean-free path of thermal radiation from the disk interior is small compared to the disk scale-height. In this regime the radiation field is approximately isotropic and blackbody and the flux is proportional to the gradient of the energy density of the radiation $\mathbf{F}_{rad} = -c\nabla(aT^4)/(3\kappa_R\rho)$. The relevant opacity that enters into the equation for the flux is the Rosseland mean,

$$\frac{1}{\kappa_R} = \frac{\int_0^\infty (1/\kappa_\nu)(\partial B_\nu/\partial T)d\nu}{\int_0^\infty (\partial B_\nu/\partial T)d\nu}, \tag{2.68}$$

where B_ν is the Planck function.

Figure 2.7 shows the temperature dependence of the Rosseland mean opacity for dusty gas of density 10^{-9} g cm^{-3}, as calculated by Semenov *et al.* (2003). The elemental abundances represent best estimates of the solar values. The assumed size distribution of particles follows that proposed by Pollack *et al.* (1985):

$$\begin{aligned} n(s) &= 1, & s &< 0.005 \text{ }\mu\text{m}, \\ n(s) &= \left(\frac{s}{0.005 \text{ }\mu\text{m}}\right)^{-3.5}, & 0.005 \text{ }\mu\text{m} &\le s < 1 \text{ }\mu\text{m}, \\ n(s) &= 4 \times 10^4 \left(\frac{s}{0.005 \text{ }\mu\text{m}}\right)^{-5.5}, & 1 \text{ }\mu\text{m} &\le s < 5 \text{ }\mu\text{m}, \\ n(s) &= 0, & s &\ge 5 \text{ }\mu\text{m}. \end{aligned} \tag{2.69}$$

This distribution is itself a modified version of the Mathis *et al.* (1977) distribution which was derived based on observations of the wavelength dependence of stellar extinction. The specific model plotted assumes that the dust is composed of homogeneous spherical particles.

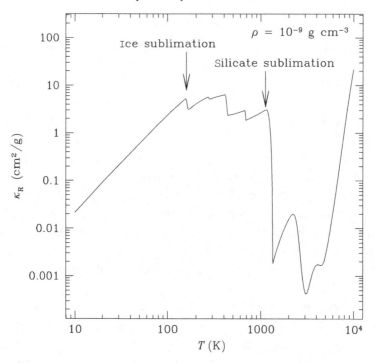

Figure 2.7 The Rosseland mean opacity for dusty gas in the protoplanetary disk, as calculated by Semenov *et al.* (2003) for a gas density of 10^{-9} g cm^{-3}. The dust is assumed to be composed of homogeneous spherical particles whose size distribution follows a modified Mathis *et al.* (1977) law.

At low temperatures ($T \lesssim 150$ K), corresponding to regions of the disk beyond the snow line, water ice and volatile organic materials are the dominant particulates and the Rosseland mean opacity rises as T^2. At somewhat higher temperatures the opacity displays a sawtooth pattern that mirrors the temperatures at which successively hardier materials evaporate: first water ice, then in succession organics, troilite (FeS), iron, and silicates. Above the dust destruction temperature for silicates ($T \simeq 1500$ K) the opacity (which in this regime is the summed contribution of millions of molecular lines) plummets by at least two orders of magnitude. At still higher temperatures – which are occasionally of interest when modeling high accretion rate disks such as those encountered in FU Orionis objects – the opacity rises due to H$^-$ scattering, enters a regime where bound–free and free–free absorption dominate, and finally plateaus at the electron scattering value.

It is frequently useful to have analytic expressions for the opacity as a function of temperature and density. For specified grain properties the opacity can normally be well approximated as piece-wise series of power-laws. Bell and Lin (1994) and Zhu *et al.* (2009) provide such fits, which are widely used in situations where detailed opacity computations are superfluous.

2.7 The Condensation Sequence

Since stars and their disks form from the collapse of the same material it is reasonable to assume that the elemental abundances in the disk (i.e. the ratio of the abundance of each element to the abundance of hydrogen) are very similar to those in the star. For the Solar System, the measured solar photospheric abundances, supplemented by abundances derived from a small number of primitive chondritic meteorites, provide a proxy for the primordial Solar Nebula composition. Even once the raw abundances are known, however, there remains the task of determining the chemical composition of the disk – i.e. how the elements are distributed among the possible chemical compounds and states – as a function of radius and time. This is not an easy task. Oxygen, for example, may be present in many forms, including carbon monoxide (CO), water (H_2O) in the form of vapor or ice, oxides such as Fe_3O_4, and silicates such as Mg_2SiO_4. The distribution of oxygen between these (and many other) forms will affect the opacity in the disk, the total surface density of solid materials, and the composition of larger bodies that form at a given radius.

The most general formulation of the problem of determining the composition of the disk is as an initial value problem. If the initial composition of the disk with radius is known, a chemical network describing the rates of possible reactions can be integrated forward in time to determine the composition at subsequent epochs. This approach allows for the disk's chemical composition to be out of equilibrium (for example in the outer disk, where the low temperatures mean that the rates of energetically favorable reactions can be very low), and can be readily extended to accommodate radial flow as the disk evolves.

In practice, solving for the evolution of a disk coupled to a fully consistent treatment of the chemistry is challenging. It is therefore useful to pose a simpler question: for a given temperature and pressure what composition is the most thermodynamically stable? Conceptually, we can imagine starting with a high temperature gas in which all of the elements are in the vapor phase and cooling it very slowly so that the elements successively condense out into various solid phases. Provided that the cooling is slow enough all reactions will have sufficient time to reach equilibrium, and the resulting condensation sequence will depend only on the abundances, the temperature, and the pressure. Under conditions of constant temperature and pressure, the thermodynamically preferred state is the one that minimizes the Gibbs free energy,

$$G \equiv H - TS, \tag{2.70}$$

where H is the enthalpy and S the entropy. If the chemical system comprises a number of homogeneous (or ideal) phases, $i = 1, 2, \ldots,$

$$G = \Sigma \left(\mu_i f_i \right), \tag{2.71}$$

where f_i is the fraction in phase i with chemical potential μ_i. The composition can then be determined by finding the f_i that minimize G, subject to the constraint imposed by the specified overall abundance of the elements. In practice, matters are complicated by the existence of *non-ideal* solid solutions – solid phase mixtures of two or more components in which there is both a change of entropy and enthalpy on mixing. The Gibbs free energy is then written as

$$G = \Sigma \left(\mu_i a_i \right) \tag{2.72}$$

where the a_i, known as the *thermodynamic activities*, are the quantities that need to be determined. Although schematically straightforward, calculation of the condensation sequence is difficult in practice due both to the need for accurate thermodynamic data and because a very large number of different species need to be considered simultaneously. The calculation of the condensation temperatures of the elements by Lodders (2003) includes 2000 gaseous species and 1600 different condensates.

Table 2.1 gives the values of a small number of condensation temperatures, as calculated by Lodders (2003) at a fiducial pressure of $P = 10^{-4}$ bar. This pressure is somewhat higher than the typical mid-plane pressure in protoplanetary disks except in the innermost AU, and this slightly modifies the condensation temperatures (in particular, water ice is only stable below $T \simeq 150$–170 K under realistic disk conditions near the snow line). Nevertheless, the condensation sequence is primarily a function of temperature, and hence the sequence can be approximately mapped into a predicted radial distribution of different solid materials. There is a transition between purely rocky condensates in the hot inner disk and a combination of rocky and icy materials at greater radii. Depending upon the adopted abundances, Lodders (2003) finds a total condensate fraction between 1.3% and 1.9%, of which between 0.44% and 0.49% is in the form of rocky materials. These results imply a jump in the surface density of solids at the snow line that is substantial, but smaller than the factor of $\simeq 4$ implied by the minimum-mass Solar Nebula model of Hayashi (1981).

Table 2.1 *The condensation temperatures for a handful of important species, as computed by Lodders (2003) for a pressure of 10^{-4} bar.*

Species	Composition	Condensation temperature
Methane	CH_4	41 K
Argon hydrate	$Ar.6H_2O$	48 K
Methane hydrate	$CH_4.7H_2O$	78 K
Ammonia hydrate	$NH_3.H_2O$	131 K
Water ice	H_2O	182 K
Magnetite	Fe_3O_4	371 K
Troilite	FeS	704 K
Forsterite	Mg_2SiO_4	1354 K
Perovskite	$CaTiO_3$	1441 K
Aluminum oxide	Al_2O_3	1677 K

2.8 Ionization State of Protoplanetary Disks

The interior temperatures of typical protoplanetary disks range from a few thousand K in the immediate vicinity of the star to just tens of K in the outer regions. These temperatures are not high enough to ionize the gas fully. As a rule of thumb, an element with an ionization potential χ is ionized once the temperature exceeds

$$T \sim \frac{\chi}{10 k_B}. \tag{2.73}$$

The ionization temperature for hydrogen is about 10^4 K and even potassium – one of the most easily ionized species with $\chi = 4.34$ eV – requires $T \gtrsim 5000$ K before it becomes ionized. As a first and excellent approximation we can therefore assume that protoplanetary disks are neutral. However, as we will see later when we discuss disk evolution, even apparently negligible ionization fractions suffice to couple magnetic fields dynamically to the gas, and this coupling in turn may be a critical element that permits angular momentum transport and accretion. An accurate understanding of the ionization fraction of protoplanetary disks as a function of radius and height above the mid-plane is thus important. This requires balancing ionization sources, which may be thermal or nonthermal (radioactive decay of short-lived nuclides, stellar X-rays, or cosmic rays), against recombinations in the gas phase or on the surfaces of dust grains. In more complex models it may be necessary to go beyond such equilibrium considerations and allow for the possibility that (relatively) well-ionized gas may be mixed by turbulent processes into neutral regions on a time scale shorter than the recombination time (Turner *et al.*, 2007; Ilgner & Nelson, 2008).

2.8.1 Thermal Ionization

Thermal ionization of the alkali metals dominates the ionization balance in the very innermost regions of the disk, usually well inside 1 AU. This regime can be treated straightforwardly. In thermal equilibrium the ionization state of a single species with ionization potential χ is described by the Saha equation (e.g. Rybicki & Lightman, 1979):

$$\frac{n^{ion} n_e}{n} = \frac{2 U^{ion}}{U} \left(\frac{2\pi m_e k_B T}{h^2} \right)^{3/2} \exp[-\chi / k_B T]. \tag{2.74}$$

In this equation, n^{ion} and n are the number densities of the ionized and neutral species, and n_e ($= n^{ion}$) is the electron number density. The partition functions for the ions and neutrals are U^{ion} and U respectively. The electron mass is m_e. The slightly modified temperature dependence (an extra factor of $T^{3/2}$) as compared to the Boltzmann factor, which governs the occupancy of atomic energy levels, arises because the ionized state is favored on entropy grounds over the neutral state.

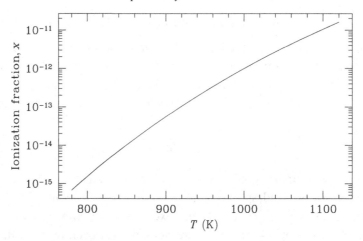

Figure 2.8 The Saha equation prediction for the ionization state of the disk due to the thermal ionization of potassium (ionization potential $\chi = 4.34\,\text{eV}$). The fractional abundance of potassium has been taken to be $f = 10^{-7}$, and the total number density of neutrals $n_n = 10^{15}\,\text{cm}^{-3}$.

In protoplanetary disks the first significant source of thermal ionization arises when the temperature becomes high enough to start ionizing the alkali metals. For potassium, the ionization potential $\chi = 4.34$ eV. We write the fractional abundance of potassium relative to all other neutral species as $f = n_K/n_n$, and define the ionization fraction x as

$$x \equiv \frac{n_e}{n_n}. \tag{2.75}$$

While potassium remains weakly ionized, the Saha equation yields

$$x \simeq 10^{-12} \left(\frac{f}{10^{-7}}\right)^{1/2} \left(\frac{n_n}{10^{15}\,\text{cm}^{-3}}\right)^{-1/2} \left(\frac{T}{10^3\,\text{K}}\right)^{3/4}$$
$$\times \frac{\exp[-2.52 \times 10^4/T]}{1.14 \times 10^{-11}}, \tag{2.76}$$

where the final numerical factor in the denominator is the value of the exponent at the fiducial temperature of 10^3 K. The predicted ionization fraction is shown as a function of temperature in Fig. 2.8. Ionization fractions that are large enough to be interesting for studies of magnetic field coupling are attained at temperatures of $T \sim 10^3$ K, although the numbers are still *extremely* small – of the order of $x \sim 10^{-12}$ for our assumed parameters.

2.8.2 Nonthermal Ionization

Outside the inner thermally ionized region, disk irradiation by stellar X-rays, cosmic rays, and radioactive decay of short-lived nuclides can all be significant sources of nonthermal ionization. Although the basic physics of each of these mechanisms is reasonably well understood there is at least an order of magnitude uncertainty in the absolute value of the resulting ionization rates. To calculate the ionization *fraction*, which is the quantity of interest for most purposes, requires knowledge of both the ionization and recombination rates, and poor understanding of the latter introduces large additional uncertainties.

T Tauri stars are extremely luminous X-ray sources as compared to main-sequence stars of the same mass. Many disk-bearing classical T Tauri stars are detected in X-ray surveys of star forming regions, often with a soft X-ray luminosity in the range of $10^{28.5}$–$10^{30.5}$ erg s^{-1} (Feigelson & Montmerle, 1999; Feigelson *et al.*, 2007). Flares are observed with substantially higher luminosities. The spectra in the X-ray band are complex and include both line and continuum emission, which can be represented as a superposition of optically thin thermal bremsstrahlung components. The hardest spectral components have $T_X \sim$ few keV, which implies a tail of emission extending out to energies of 10 keV or more. Analysis of the X-ray properties supports an interpretation of the emission as originating in a magnetically powered corona that is a scaled-up version (both in power and in the size of the X-ray emitting magnetic field loops) of the Sun's. The existence of many other diagnostics of stellar magnetic activity, including Zeeman splitting of optical spectral lines and nonthermal radio emission, support this conclusion.

Some fraction of the stellar X-ray luminosity will be intercepted by the disk, either directly or following scattering in the disk atmosphere or in a stellar wind. If the gas is depleted of heavy elements (i.e. grains have condensed) the photoionization cross-section can be fit with a power-law (Igea & Glassgold, 1999),

$$\sigma(E) = 8.5 \times 10^{-23} \left(\frac{E}{\text{keV}} \right)^{-2.81} \text{cm}^2. \tag{2.77}$$

The steep power-law dependence of the cross-section with energy means that the penetrating power of X-rays is strongly energy dependent. X-rays with $E = 1$ keV are absorbed by a column $\Delta\Sigma \ll 1\,\text{g\,cm}^{-2}$, while at 5 keV the stopping depth is a few g cm^{-2}. The outermost skin of the disk will therefore be ionized by soft X-rays, while the hard tail of the spectrum will penetrate further toward the interior. Expressions for computing the attenuation are given in Krolik and Kallman (1983) and in Fromang *et al.* (2002). For our purposes, it suffices to consider a rough fit to numerical results published by Igea and Glassgold (1999). For an X-ray luminosity of $L_X = 2 \times 10^{30}$ erg s^{-1} and a 5 keV thermal spectrum, Turner and Sano (2008) fit an ionization profile of the form

$$\zeta_X = 2.6 \times 10^{-15} \left(\frac{r}{1\,\text{AU}} \right)^{-2} \exp[-\Delta\Sigma/\Sigma_{\text{stop}}]\ \text{s}^{-1}, \tag{2.78}$$

where $\Delta\Sigma$ is the column density measured downward from the disk surface and $\Sigma_{\text{stop}} = 8\,\text{g\,cm}^{-2}$ is an approximate stopping depth.

The high column density of gas in the inner regions of protoplanetary disks means that even the hardest stellar X-rays measured to date will be unable to penetrate to the disk mid-plane. Interstellar cosmic rays have a spectrum that extends up to much higher energies, with correspondingly greater penetrating power. Adopting the fiducial value for the interstellar cosmic ray flux (Spitzer & Tomasko, 1968) the resulting ionization rate is

$$\zeta_{CR} = 10^{-17} \exp[-\Delta\Sigma/\Sigma_{stop}] \; s^{-1}, \qquad (2.79)$$

where in this case $\Sigma_{stop} = 10^2 \, g\,cm^{-2}$ (Umebayashi & Nakano, 1981). The main difficulty in applying this expression to protoplanetary disks comes from the uncertain effect of stellar winds on the cosmic ray flux. The solar wind – which is probably much less powerful than typical T Tauri winds – is able to modulate the flux of Galactic cosmic rays measured at the Earth, and it is quite conceivable that T Tauri winds (from either the star or the disk) may be able to exclude the bulk of the interstellar cosmic ray flux. Equation (2.79) is therefore realistically an upper limit to the ionization expected from cosmic rays, and the actual cosmic ray ionization rate could be negligibly small.

Radioactive decay provides a final source of ionization. It takes about $\Delta\epsilon = 36\,eV$ to produce an ion pair in molecular hydrogen, so a decay that yields an amount of energy E will create about $E/\Delta\epsilon$ ions. If the nuclide under consideration has a fractional abundance (relative to hydrogen) f and a decay constant λ,[4] the ionization rate is (Stepinski, 1992)

$$\zeta_R = \frac{\lambda f E}{\Delta\epsilon}. \qquad (2.80)$$

Abundant short-lived nuclides yield the highest ionization rates. Provided that it is originally present and has not yet had time to decay, ^{26}Al is predicted to dominate. Note that ^{26}Al has a very short half-life of 0.72 Myr ($\lambda \simeq 3 \times 10^{-14} \, s^{-1}$), an energy per decay of $E = 3.16$ MeV, and an estimated Solar System abundance of $f \sim 10^{-10}$. With these parameters $\zeta_R \simeq 2.6 \times 10^{-19} \, s^{-1}$, though this value will drop substantially over the disk lifetime due to the ongoing exponential decay. The rate may be further suppressed if the radioactive material is locked up within small pebble-sized solid bodies, and may be zero if the disk in question was not polluted with material from a relatively recent supernova.

Figure 2.9 shows the ionization rate as a function of column density for the simple models of X-ray, cosmic ray, and radioactive ionization discussed above. Despite the numerous uncertainties it is safe to conclude that X-ray ionization will dominate all other sources close to the disk surface, down to a column of 10–$100 \, g\,cm^{-2}$. Close to the mid-plane in the inner disk – where $\Sigma > 10^3 \, g\,cm^{-2}$ – either cosmic rays or ^{26}Al decays can potentially furnish a lower level of ionization.

---

[4] Related to the half-life via $t_{1/2} = \ln 2/\lambda$.

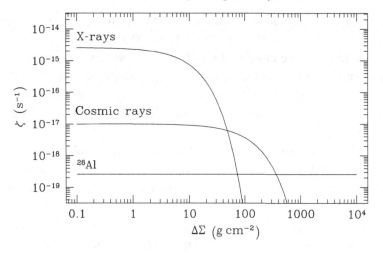

Figure 2.9 The dependence of the ionization rate ζ on column density $\Delta\Sigma$, measured from the disk surface. Approximate contributions from the three main sources of nonthermal ionization are plotted, based on the simplified model described in the text: (a) cosmic rays, assuming that the interstellar flux is *not* shielded by a stellar wind; (b) stellar X-rays, evaluated at 1 AU using a single exponential fit (Turner & Sano, 2008) to numerical results by Igea & Glassgold (1999); and (c) radioactive decay of ^{26}Al present at a fractional abundance (relative to hydrogen) of $f = 10^{-10}$ (Stepinski, 1992).

The equilibrium ionization fraction x (Eq. 2.75) is obtained by balancing ionization against recombination. If we ignore for the moment both metal ions and dust particles, the dominant channel for absorbing free electrons would be dissociative recombination reactions with molecular ions. In the case of HCO$^+$, for example, the reaction is

$$e^- + HCO^+ \rightarrow H + CO. \tag{2.81}$$

Denoting the number density of molecular ions as n_{m^+}, ionization equilibrium requires that

$$\frac{dn_e}{dt} = \zeta n_n - \beta n_e n_{m^+} = 0. \tag{2.82}$$

A generic expression for the temperature dependence of the rate coefficient β is

$$\beta = 3 \times 10^{-6} T^{-1/2} \text{ cm}^3 \text{ s}^{-1}, \tag{2.83}$$

which yields an equilibrium electron abundance

$$x = \sqrt{\frac{\zeta}{\beta n_n}}. \tag{2.84}$$

We can evaluate the predicted electron abundance for conditions typical of protoplanetary disks. In the inner disk, the disk mid-plane is shielded from cosmic rays and a representative ionization rate due to radioactive decay might be $\zeta \sim 10^{-19}\,\mathrm{s}^{-1}$. For $n_n = 10^{15}\,\mathrm{cm}^{-3}$ and $T = 300\,\mathrm{K}$ we obtain $x \simeq 2.4 \times 10^{-14}$. For the same temperature close to the disk surface we may have $n_n = 10^{10}\,\mathrm{cm}^{-3}$ and a substantially enhanced ionization rate due to stellar X-rays, $\zeta_X \sim 10^{-15}\,\mathrm{s}^{-1}$. These parameters yield $x \simeq 7.6 \times 10^{-10}$. Although both values may appear negligibly small, in fact the difference is sufficient to imply substantially different coupling between the gas and magnetic fields.

The actual ionization equilibrium of disks is very probably a great deal more complex than this toy model suggests. In more realistic disk models, H_2^+ can charge exchange either to other molecules such as HCO^+, or to metal ions such as Mg^+. Radiative recombination with metal ions is much slower than dissociative recombination, so the presence of such charge exchange reactions can substantially alter the resulting electron fraction (Fromang *et al.*, 2002). A chemical network is required to assess the importance of these effects (e.g. Sano *et al.*, 2000; Ilgner & Nelson, 2006). Dust is also important: as a reservoir for metal atoms, as a charge carrier in its own right, and as a surface on which electrons and ions can recombine. Although all of these processes have been studied in some detail, it is unclear whether our knowledge of the composition of disks and of the distribution of solid particles is yet good enough to facilitate a comprehensive model for the ionization state.

2.9 Disk Large-Scale Structure

Imaging of protoplanetary disks, discussed in Section 2.2.5, shows that the dust continuum emission is frequently more structured than an axisymmetric model with a radial power-law profile. Ring-like structures are common, and some disks display large-amplitude deviations from axisymmetry in the form of "horseshoe" shaped crescents of dust emission. Rings have also been detected in molecular line observations.

The observation of large-scale structure in disks suggests that the formerly common assumption that they could be adequately described by simple power-laws may be false. Disks can support long-lived radial perturbations, and departures from axisymmetry that are persistent on time scales much longer than the dynamical time. Hydrodynamic structures, known as *zonal flows* (Johansen *et al.*, 2009a) and *vortices* (Barge & Sommeria, 1995), are candidates for the background gas flow that could sustain the observed structure. There may also be observable structure associated with ice lines, including the water snow line and ice lines of species such as CO.

2.9.1 Zonal Flows

Radial force balance in the disk (Eq. 2.27) can be satisfied in a simple model where the mid-plane pressure is a monotonically decreasing power-law function of radius.

However, one can equally well set up an equilibrium in which radial forces from a complex pressure profile (which may include local pressure maxima) are balanced by radial variations in azimuthal velocity. In a local frame co-rotating with the disk at angular velocity Ω_0 the balance is between the pressure gradient $\nabla P/\rho$ and the Coriolis force $2\Omega_0 \times \mathbf{v}$. When these terms balance, such that

$$\frac{dP}{dx} = 2\Omega_0 \rho v_y, \tag{2.85}$$

the system is said to be in *geostrophic balance* (here x is the radial direction, and y the azimuthal). The pressure gradient is compensated by variations in the orbital velocity, creating *zonal flows* analogous to the banded structure of winds in giant planet atmospheres.

A disk zonal flow is an equilibrium solution to the fluid equations, but that equilibrium may not be stable. Too pronounced a deviation from Keplerian rotation results in a shear profile that is unstable to a process loosely analogous to Kelvin–Helmholtz instability, known as Rossby wave instability (Lovelace *et al.*, 1999). Even a dynamically stable perturbation will be erased on a longer time scale if angular momentum transport processes in the gas disk have the character of a diffusive viscosity. Persistent zonal flows are thus not expected in classical disk theory. Zonal flows have been seen, however, in numerical simulations of turbulent magnetized disks, starting with the work of Johansen *et al.* (2009a), and hence it is possible that the effects of turbulence or magnetic field within real disks support their spontaneous formation. Axisymmetric pressure maxima with the same sort of properties can also be formed due to the response of a gas disk to the gravitational perturbation of a neighboring planet.

2.9.2 Vortices

In a Keplerian disk two points that are initially separated radially by Δr move apart by the same distance azimuthally on a time scale of only $\sim \Omega_K^{-1}$. Generically, this strong shear makes the persistence of nonaxisymmetric structure in disks surprising. One of the few ways in which it can be done is if a localized region of vorticity is superimposed on the background shear flow, forming a vortex that is analogous to terrestrial weather systems. In disks, however, only anticyclonic vortices are stable. Anticyclonic vortices have central pressures that exceed those in the background disk, and this property enables them to trap aerodynamically coupled solid particles (Barge & Sommeria, 1995).

The magnitude of the vorticity $\omega = \nabla \times \mathbf{v}$ in a strictly Keplerian disk is $\omega_K = -(3/2)\Omega_K$. A Keplerian disk containing a vortex can be modeled as a spatially localized elliptical perturbation with vorticity $\omega = \omega_K + \omega_v$, with ω_v a constant. Other types of vortex are possible, but this type has the useful property of being described by an exact nonlinear solution (Kida, 1981) that is useful for both analytic and numerical studies.

The Kida solution describes a vortex within a shearing-sheet approximation to disk flow (for a derivation, see Chavanis, 2000). Following Lesur and Papaloizou (2009) we define a Cartesian coordinate system (x, y) which is centered at radius r_0 and which co-rotates with the background disk flow at angular velocity $\Omega_K = \Omega_K(r_0)$,

$$
\begin{aligned}
x &= r_0\phi, \\
y &= -(r - r_0).
\end{aligned}
\tag{2.86}
$$

Time-independent vortex solutions are possible if the semi-major axis of the vortex is aligned with the azimuthal direction (x in the local shearing sheet model), and the vorticity perturbation satisfies

$$
\frac{\omega_v}{\omega_K} = \frac{1}{\chi}\left(\frac{\chi + 1}{\chi - 1}\right).
\tag{2.87}
$$

Here $\chi = a/b$ is the aspect ratio of the vortex, which forms an elliptical patch with semi-major axis a and semi-minor axis b. The right-hand-side is positive, which implies that the only steady Kida vortices in Keplerian disks are anticyclonic (with ω_v having the opposite sign to Ω_K).

The complete Kida solution is written in terms of a streamfunction ψ in an elliptic coordinate systems (μ, ν), where

$$
\begin{aligned}
x &= f\cosh(\mu)\cos(\nu), \\
y &= f\sinh(\mu)\sin(\nu),
\end{aligned}
\tag{2.88}
$$

and $f = a\sqrt{(\chi^2 - 1)/\chi^2}$. The solution can be split into a core and an exterior part,

$$
\begin{aligned}
\psi_{\text{core}} &= -\frac{3\Omega_K f^2}{4(\chi - 1)}\left[\chi^{-1}\cosh^2(\mu)\cos^2(\nu) + \chi\sinh^2(\mu)\sin^2(\nu)\right], \\
\psi_{\text{ext}} &= -\frac{3\Omega_K f^2}{8(\chi - 1)^2}\bigg[1 + 2(\mu - \mu_0) + 2(\chi - 1)^2\sinh^2(\mu)\sin^2(\nu) \\
&\qquad + \frac{\chi - 1}{\chi + 1}\exp[-2(\mu - \mu_0)]\cos(2\nu)\bigg],
\end{aligned}
\tag{2.89}
$$

which match at $\mu = \mu_0 = \tanh^{-1}(\chi^{-1})$. The Cartesian velocity field is then given by $v_x = -\partial\psi/\partial y$, $v_y = \partial\psi/\partial x$. In general the Cartesian representation of the velocity has no simple form, but within the core it is

$$
\begin{aligned}
v_{x,\text{core}} &= \frac{3\Omega_K\chi}{2(\chi - 1)}y, \\
v_{y,\text{core}} &= -\frac{3\Omega_K}{2\chi(\chi - 1)}x,
\end{aligned}
\tag{2.90}
$$

describing simple elliptical motion. Figure 2.10 shows the contours (logarithmically spaced) of the full streamfunction for Kida vortices with two different aspect ratios.

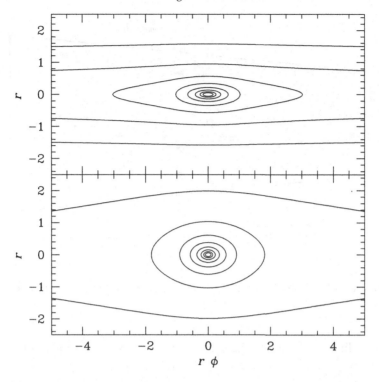

Figure 2.10 Contours showing the steady-state Kida vortex streamfunction for different values of the vortex aspect ratio (top, $a/b = 3$, bottom, $a/b = 1.5$). The solution has an elliptical core that merges smoothly into the shear flow of the disk further out.

Vortices are extremely stable structures in two-dimensional fluid flows with simple equations of state (for example, barotropic disks in which the pressure is solely a function of density) and little dissipation. They exhibit dynamics characteristic of an inverse cascade of turbulent energy, in which multiple small vortices at initially similar radii eventually merge to form a single large vortex. In realistic disk models, dissipative processes and the effects of the third dimension mean that vortices are expected to be quasi-stable structures with a long but finite lifetime. Observed structure in disks, if it is to be associated with the trapping of solid particles in the high pressure cores of vortices, therefore likely requires a mechanism that is able to generate vortices *in addition* to the vorticity perturbations that would have been present during disk formation.

2.9.3 Ice Lines

Zonal flows and vortices are examples of hydrodynamic structures that can, in principle, form and persist within disks across a wide range of radii. A different type of observable structure may be associated to changes in the chemical makeup or size distribution of small solid particles in the vicinity of the water snow line

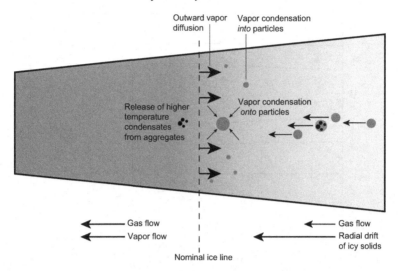

Figure 2.11 Illustration of physical processes that can occur near an ice line in the protoplanetary disk.

and other ice lines. Potential ice line structure is tied to specific radial locations that satisfy the temperature/pressure conditions needed for condensation/sublimation. Unlike zonal flows or vortices, which are fundamentally *gas* disk structures that change the dust distribution via aerodynamic effects, ice line physics changes the solid content directly and affects the gas, if at all, only via secondary processes.

Figure 2.11 shows some of the significant physical processes that may occur near an ice line. A partial list includes:

- A change in the total solid surface density corresponding to the total mass in condensible species interior and exterior to the ice line. This is the most basic expectation for an ice line, already discussed in Section 2.7.
- Vapor diffusion/condensation effects. The physics of sublimation and condensation is exponentially sensitive to temperature, and as a result low temperature condensates that drift inward through an ice line sublimate entirely into vapor within a radially narrow region. This leads to a strong radial gradient of vapor concentration, which, in the presence of turbulence, sources a diffusive outward flux of vapor beyond the nominal ice line location. The vapor will condense, either into new solid particles or onto existing ones, growing them in size.
- The size distribution of the solids may differ across the ice line (even absent the effects of vapor condensation). For example, relatively large aggregates of icy and rocky particles outside the water snow line may break up inside to release smaller purely rocky particles. Because the rate of radial drift depends upon the size, changes in the size distribution translate into changes in the surface density that differ from the equilibrium expectation.

Models incorporating some of the above effects have been developed (e.g. Stevenson & Lunine, 1988; Ros & Johansen, 2013; Pinilla *et al.*, 2017). Strong effects, which could have a substantial impact on multiple phases of planet formation, are likely to occur around the water snow line (and potentially also at the silicate "ice" line). Observable effects, of uncertain magnitude, are possible around the lower temperature ice lines further out.

2.10 Further Reading

"Protoplanetary Disks and Their Evolution," J. P. Williams & L. A. Cieza (2011), *Annual Review of Astronomy and Astrophysics*, **49**, 67.

"Models of the Structure and Evolution of Protoplanetary Disks," C. P. Dullemond, D. Hollenbach, I. Kamp, & P. D'Alessio in *Protostars and Planets V* (2007), B. Reipurth, D. Jewitt, & K. Keil (eds.), Tucson: University of Arizona Press.

3

Protoplanetary Disk Evolution

Observationally it is clear that protoplanetary disks are not static structures, but rather evolve slowly over time. The observational manifestations of disks are almost always associated with stars in or near star forming regions, and are absent in (still youthful) clusters such as the Pleiades. Explaining theoretically why disks should evolve is not, however, an easy task. For a geometrically thin disk the angular velocity is essentially that of a Keplerian orbit (Eq. 2.30), and the specific angular momentum,

$$l = r v_{\phi, \text{gas}} = \sqrt{GM_* r}, \qquad (3.1)$$

is an *increasing* function of radius. For gas in the disk to flow inward and be accreted by the star, it therefore needs to lose angular momentum. Understanding the mechanisms that can result in angular momentum loss is the central problem in the theory of *accretion disks*, not just in the protoplanetary context but also for disks around black holes and other compact objects. It is a difficult problem because the effect of interest is subtle. Protoplanetary disks have an observed lifetime of several million years, which equates to $\sim 10^4$ dynamical times at the outer edge of the disk. In other words they are almost, but not quite, stable. In this chapter we discuss the evolution of disks, the origin of angular momentum transport, and the processes that disperse the gas at the end of the disk lifetime. The focus is on the evolution of the gas, which by mass is the dominant disk component. The evolution of the solid component, which is partially coupled to the gas but which also involves distinct physical processes, is discussed in Chapter 4.

3.1 Observations of Disk Evolution

Evidence for disk evolution can be sought in any of the complementary tracers of the presence of circumstellar material or accretion. These include infrared excesses (above the stellar photospheric flux), indicative of warm dust close to the star, mm-continuum or line emission arising from cool dust or gas in the disk, and accretion signatures indicative of gas being accreted on to the star.

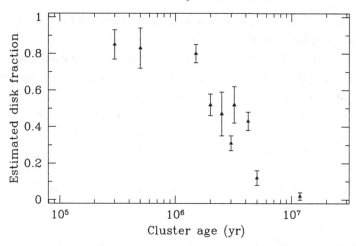

Figure 3.1 Compilation of estimates of the optically thick disk fraction in young clusters as a function of the estimated cluster age. The disk fraction is estimated from either ground-based infrared photometry (JHKL bands), or from *Spitzer* photometry. The plotted clusters are NGC 2024, the Trapezium, NGC 2264, NGC 2362 (all using data from Haisch *et al.*, 2001), NGC 1333 (disk fraction from Gutermuth *et al.*, 2008, plotted with an adopted age of 0.5 Myr), σ-Ori (Hernandez *et al.*, 2007), IC 348 (Lada *et al.*, 2006), Tr 37, NGC 7160 (Sicilia-Aguilar *et al.*, 2006), and Chamaeleon (Damjanov *et al.*, 2007). Note that there are substantial uncertainties in the absolute ages, especially for the younger clusters.

Figure 3.1 shows the fraction of young stars in different clusters that show evidence for protoplanetary disks in the form of IR excesses. This "disk fraction" is plotted against the estimated ages of the clusters, which are derived by comparing the locations of stars in the Hertzsprung–Russell diagram against theoretical stellar evolution tracks for pre-main-sequence stars.[1] The data sample includes the clusters compiled by Haisch *et al.* (2001), together with a selection of more recent results based on photometry obtained by the *Spitzer Space Telescope*. Where possible, the disk fraction refers to T Tauri stars (i.e. stars of around a solar mass) that host optically thick disks. This last qualifier is intended to distinguish so-called *primordial disks* – in which substantial amounts of dust are mixed in with gas around the youngest stars – from the generally weaker *debris disks* that are observed around older stars (the disk around β Pictoris being one well-known example). Although there is no unambiguous observational test to distinguish primordial from debris disks, most debris disks, when studied in detail, do not show evidence for any

---

[1] The derived ages need to be viewed with considerable caution for two reasons. First, for a very young cluster the spread in ages of the stars may be comparable to the mean age of the cluster. This makes the assignment of a single age to the system of dubious worth. Second, the evolutionary tracks used to assign ages return essentially arbitrary *absolute* ages for stars less than about 1 Myr old (Baraffe *et al.*, 2002), due to the neglect of accretion physics that matters at such early times. Statements about stars or clusters that rely on accurate determinations of ages of less than a Myr should generally be regarded skeptically.

significant gaseous component. The dust in these debris disks arises from erosive collisions between larger solid bodies.

The infrared excess measurements of disk frequency provide a robust measure of disk evolution. The disk frequency is close to 100% for clusters whose mean stellar age is less than about 1 Myr. For older clusters, the disk frequency drops steadily, reaching 50% at around 3 Myr and falling almost to zero by about 6 Myr. Evidence for primordial disks is essentially completely absent for stars of age 10 Myr or above. From these observations, one infers a disk lifetime of about 3–5 Myr.

Strictly speaking, the aforementioned constraint on the "disk lifetime" refers only to the survival time of small dust grains in the innermost (~ 1 AU) region of the disk. Does the disk as a whole evolve on the same time scale? Somewhat surprisingly, there is evidence that it may. First, the accretion rate of *gas* on to the star – measured entirely independently via spectroscopic observations of the hot continuum radiation produced when infalling gas impacts the stellar surface – decays with time on a similar time scale (Hartmann *et al.*, 1998). Second, the presence of one disk signature (such as an IR excess) usually, though not invariably, implies that the other signatures (such as a mm excess) are also present, *even when the different signatures arise at very different disk radii*. This observation implies that disks are dispersed across a range of radii on a relatively short time scale (Skrutskie *et al.*, 1990; Wolk & Walter, 1996; Andrews & Williams, 2005). We will discuss further the import of this observation for models of disk dispersal, but for now we note only that it makes the assignment of a single lifetime for the disk a meaningful concept.

3.2 Surface Density Evolution of a Thin Disk

Consider an axisymmetric protoplanetary disk whose gas surface density profile is given by $\Sigma(r,t)$. We assume that the radial velocity $v_r(r,t)$ of gas in the disk is small,[2] and note that the fact that the disk is geometrically thin ($h/r \ll 1$) implies that the predominant forces at work are rotational support and gravity (cf. Section 2.4). If the potential is time independent, local conservation of angular momentum implies that in the absence of angular momentum transport or loss $\Sigma(r,t)$ cannot change with time. Accretion and disk evolution will occur in the presence of angular momentum transport (often described as "viscosity," or, even more loosely, as "friction"), which allows local parcels of gas to reduce their angular momentum and spiral toward the star (global angular momentum conservation implies, of course, that gas elsewhere in the disk must *gain* angular momentum and move outward). This redistribution of angular momentum due to stresses within the disk is quite distinct from angular momentum loss – due for example to a

---

[2] The radial velocity is defined such that $v_r < 0$ corresponds to inflow.

magnetically driven outflow from the disk surface – and the evolution of disks under the action of winds is different from that due to internal redistribution.

The qualitative evolution of disks in the presence of dissipative processes was understood in the 1920s by, among others, the well-known geophysicist and astronomer Harold Jeffreys. The modern theory of thin disks was described in now-classic papers by Shakura and Sunyaev (1973) and Lynden-Bell and Pringle (1974). This theory is not fully predictive as it largely bypasses the central question of how efficiently angular momentum is transported within a disk flow, but it nonetheless forms the indispensable core to any discussion of disk evolution.

The evolution of $\Sigma(r,t)$ can be derived by considering the continuity equation (expressing the conservation of mass) and the azimuthal component of the momentum equation (expressing angular momentum conservation). The rate of change of the mass within an annulus in the disk extending between r and $r + \Delta r$ is given by

$$\frac{\partial}{\partial t}(2\pi r \Delta r \Sigma) = 2\pi r \Sigma(r)v_r(r) - 2\pi(r + \Delta r)\Sigma(r + \Delta r)v_r(r + \Delta r). \quad (3.2)$$

Writing for example $\Sigma(r + \Delta r) = \Sigma(r) + (\partial \Sigma/\partial r)\Delta r$, and taking the limit for small Δr, the continuity equation yields

$$r\frac{\partial \Sigma}{\partial t} + \frac{\partial}{\partial r}(r\Sigma v_r) = 0. \quad (3.3)$$

Following the same procedure (e.g. Pringle, 1981) conservation of angular momentum gives

$$r\frac{\partial}{\partial t}\left(r^2\Omega\Sigma\right) + \frac{\partial}{\partial r}\left(r^2\Omega \cdot r\Sigma v_r\right) = \frac{1}{2\pi}\frac{\partial G}{\partial r}, \quad (3.4)$$

where Ω, the angular velocity of the gas in the disk, is at this point unspecified and need not be the Keplerian angular velocity due to a point mass. The rate of change of angular momentum in the disk is determined by the change in surface density due to radial flows (the second term on the left-hand-side) and by the *difference* in the torque exerted on an annulus by stresses at the inner and outer edges. For a viscous fluid, the torque G can be written in the form

$$G = 2\pi r \cdot \nu\Sigma r\frac{d\Omega}{dr} \cdot r, \quad (3.5)$$

where ν is the kinematic viscosity. The torque on an annulus is the product of the circumference, the viscous force per unit length, and the lever arm r, and is proportional to the gradient of the angular velocity. Note that this dependence, which is characteristic of a viscous fluid, is only an assumption if the "viscosity" is not a true microscopic process but rather an effective viscosity resulting from turbulence. Proceeding, we eliminate v_r between Eq. (3.3) and Eq. (3.4) and specialize to a Keplerian potential for which $\Omega \propto r^{-3/2}$. We then obtain the evolution equation for the surface density of a geometrically thin disk under the action of internal angular momentum transport:

$$\frac{\partial \Sigma}{\partial t} = \frac{3}{r} \frac{\partial}{\partial r} \left[r^{1/2} \frac{\partial}{\partial r} \left(\nu \Sigma r^{1/2} \right) \right]. \tag{3.6}$$

The evolution equation is a diffusive partial differential equation for the surface density $\Sigma(r, t)$. It is linear if the viscosity ν is not itself a function of Σ. The equation can also be derived directly from the Navier–Stokes equations for a viscous fluid in cylindrical polar coordinates.

3.2.1 The Viscous Time Scale

The diffusive form of Eq. (3.6) can be made more transparent with a change of variables. Defining

$$X \equiv 2r^{1/2}, \tag{3.7}$$

$$f \equiv \frac{3}{2} \Sigma X, \tag{3.8}$$

and assuming that the viscosity ν is a constant, the evolution equation takes the prototypical form of a diffusion equation,

$$\frac{\partial f}{\partial t} = D \frac{\partial^2 f}{\partial X^2}, \tag{3.9}$$

with a diffusion coefficient D given by

$$D = \frac{12\nu}{X^2}. \tag{3.10}$$

Apart from its pedagogical value, this version of the evolution equation can be useful numerically, since even naive finite difference schemes preserve conserved quantities accurately when the equation is cast in this form. The diffusion time scale across a scale ΔX implied by Eq. (3.9) is just $(\Delta X)^2 / D$. Converting back to the physical variables, we then find that the time scale on which viscosity will smooth out surface density gradients on a radial scale Δr is

$$\tau_\nu \sim \frac{(\Delta r)^2}{\nu}. \tag{3.11}$$

If the disk has a characteristic size r, the surface density at all radii will evolve on a time scale

$$\tau_\nu \approx \frac{r^2}{\nu}. \tag{3.12}$$

This last time scale is described as the *viscous time scale* of the disk. It can be estimated observationally by measuring, for example, the rate at which accretion on to the star decays as a function of stellar age. For protoplanetary disks around solar-type stars it appears to be of the order of a million years.

3.2.2 Solutions to the Disk Evolution Equation

A steady-state solution to Eq. (3.6) can be derived by setting $\partial/\partial t = 0$ and integrating the resultant ordinary differential equation for the surface density. Applying the requisite boundary conditions is easiest if we start with the angular momentum conservation equation (Eq. 3.4) which does not assume Keplerian angular velocity. Setting the time derivative to zero and integrating, we have

$$2\pi r \Sigma v_r \cdot r^2 \Omega = 2\pi r^3 \nu \Sigma \frac{d\Omega}{dr} + \text{constant}. \tag{3.13}$$

Noting that the mass accretion rate $\dot{M} = -2\pi r \Sigma v_r$ we can write this equation in the form

$$-\dot{M} \cdot r^2 \Omega = 2\pi r^3 \nu \Sigma \frac{d\Omega}{dr} + \text{constant}, \tag{3.14}$$

where the constant of integration, which has the form of an angular momentum flux, remains to be determined. To specify the constant, we note that at a location in the disk where $d\Omega/dr = 0$ the viscous stress vanishes, and the constant is simply equal to the flux of angular momentum advected inward along with the mass,

$$\text{constant} = -\dot{M} \cdot r^2 \Omega. \tag{3.15}$$

A simple case to consider is that where the protoplanetary disk extends all the way down to the surface of a nonrotating (or slowly rotating) star. The disk and the star form a single fluid system, and the angular velocity (shown schematically in Fig. 3.2) must make a continuous transition between $\Omega = 0$ in the star and $\Omega \propto r^{-3/2}$ within the disk. The viscous stress vanishes at a radius $R_* + r_{bl}$, where r_{bl} is the width of the *boundary layer* that separates the star from the Keplerian part of the disk. Within the boundary layer the angular velocity increases with radius, and the sub-Keplerian rotational support cannot balance the inward gravitational force.

The hydrodynamics (and magnetohydrodynamics) of protoplanetary disk boundary layers is quite complex, and the flow structure in this region remains rather uncertain. Elementary arguments, however, suggest that in most cases the radial extent of the boundary layer is a small fraction of the stellar radius. If magnetic fields and viscosity are negligible, the momentum equation (Eq. 2.27) in axisymmetry can be written as

$$\frac{v_{\phi,\text{gas}}^2}{r} = \frac{GM_*}{r^2} + \frac{1}{\rho}\frac{dP}{dr} + v_r \frac{dv_r}{dr}, \tag{3.16}$$

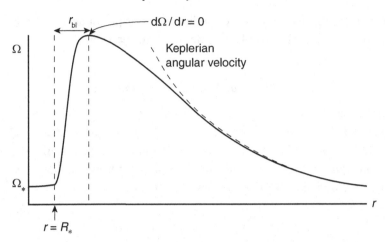

Figure 3.2 Schematic illustration of the angular velocity profile in a disk that extends to the equator of a slowly rotating star. The viscous stress vanishes at a radial location $r = R_* + r_{bl}$ where Ω has a maximum.

where P is the pressure. In the boundary layer – unlike in the disk itself – the rotational support (the term on the left-hand-side) cannot balance gravity. If, instead, radial pressure forces counteract the force of gravity,[3] we require that

$$\frac{1}{\rho}\frac{dP}{dr} \sim \frac{c_s^2}{r_{bl}} \sim \Omega_K^2 r, \tag{3.17}$$

where c_s is some characteristic sound speed in the boundary layer region and Ω_K is the Keplerian angular velocity. Recalling that the vertical scale-height $h = c_s/\Omega_K$ we find that

$$\frac{r_{bl}}{r} \sim \left(\frac{h}{r}\right)^2. \tag{3.18}$$

Provided that the boundary layer (like the disk) is geometrically thin, we conclude that force balance mandates that the *radial* extent of the boundary layer must also be narrow.

We are now in a position to evaluate the constant in Eq. (3.14). For a narrow boundary layer, $R_* + r_{bl} \simeq R_*$ and the maximum in Ω occurs close to the stellar surface. We have that

$$\text{constant} \simeq -\dot{M}R_*^2\sqrt{\frac{GM_*}{R_*^3}}, \tag{3.19}$$

and the steady-state solution for the disk (within which the angular velocity is Keplerian) simplifies to

---

[3] Similar arguments apply if we consider the influence of the $v_r dv_r/dr$ term rather than the pressure term, since causality requires that v_r be subsonic.

$$\nu\Sigma = \frac{\dot{M}}{3\pi}\left(1 - \sqrt{\frac{R_*}{r}}\right). \tag{3.20}$$

Once the viscosity is specified, this equation gives the steady-state surface density profile of a protoplanetary disk with a constant accretion rate $\dot{M}$. Away from the inner boundary one notes that $\Sigma(r) \propto \nu^{-1}$.

The solution given by Eq. (3.20) gives the surface density profile for a steady disk subject to a *zero-torque* boundary condition at the inner edge. Physically, this boundary condition is at least approximately realized for disks that extend to the equator of a slowly rotating star, and it is also the traditional choice in the more exotic circumstance of a disk of gas around a black hole. In classical T Tauri stars it is often the case that stellar magnetic fields truncate the disk before it reaches the stellar surface, and the resulting magnetic coupling between the star and the inner disk can violate the zero-torque assumption. More generally, one should note that the turnover in the surface density profile at small radii implied by Eq. (3.20) reflects the fact that for a Keplerian flow the only way in which the torque can vanish is if the surface density goes to zero. The turnover is therefore a purely formal result – in a real disk with a boundary layer the physical reason why the torque vanishes is because of the existence of a maximum in $\Omega(r)$ – and the inner boundary condition needs to be considered carefully if one needs the detailed form of the surface density very close to the inner edge of the disk.

Time-dependent analytic solutions to Eq. (3.6) can be derived for a number of simple forms for the viscosity and, although these forms are not particularly realistic for protoplanetary disks, the resulting solutions suffice to illustrate the essential behavior implied by the disk evolution equation. If $\nu = $ constant, a Green's function solution to the evolution equation is possible.[4] Suppose that at $t = 0$, all of the gas lies in a thin ring of mass m at radius r_0,

$$\Sigma(r, t = 0) = \frac{m}{2\pi r_0}\delta(r - r_0), \tag{3.21}$$

where $\delta(r - r_0)$ is a Dirac delta function. Given boundary conditions that impose zero torque at $r = 0$ and allow for free expansion toward $r = \infty$ the solution is (Lynden-Bell & Pringle, 1974)

$$\Sigma(x, \tau) = \frac{m}{\pi r_0^2}\frac{1}{\tau}x^{-1/4}\exp\left[-\frac{(1 + x^2)}{\tau}\right]I_{1/4}\left(\frac{2x}{\tau}\right), \tag{3.22}$$

where we have written the solution in terms of dimensionless variables $x \equiv r/r_0, \tau \equiv 12\nu r_0^{-2}t$, and $I_{1/4}$ is a modified Bessel function of the first kind. Since the evolution equation is linear for $\nu = f(r)$, the time-dependent solution for arbitrary initial conditions can formally be written as a superposition of these solutions. Although this approach is rarely illuminating, the solution

---

[4] Related solutions are known for the more general situation in which ν is a power-law in radius.

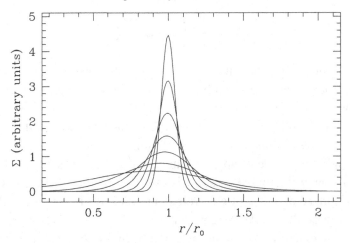

Figure 3.3 The Green's function solution to the disk evolution equation with $\nu = $ constant, showing the spreading of a ring of mass initially orbiting at $r = r_0$. From top down the curves show the behavior as a function of the scaled time variable $\tau = 12\nu r_0^{-2}t$, for $\tau = 0.004$, $\tau = 0.008$, $\tau = 0.016$, $\tau = 0.032$, $\tau = 0.064$, $\tau = 0.128$, and $\tau = 0.256$.

(Eq. 3.22), which is plotted in Fig. 3.3, illustrates several generic features of disk evolution. As t increases the initially narrow ring spreads diffusively, with the mass flowing toward $r = 0$ while simultaneously the angular momentum is carried by a negligible fraction of the mass toward $r = \infty$. This segregation of mass and angular momentum is a general feature of the evolution of a viscous disk, and is necessary if accretion is to proceed without overall angular momentum loss from the system.

Often of greater practical utility than the Green's function solution is the self-similar solution also derived by Lynden-Bell and Pringle (1974). Consider a disk in which the viscosity can be approximated as a power-law in radius,

$$\nu \propto r^\gamma. \tag{3.23}$$

Suppose that the disk at time $t = 0$ has the surface density profile corresponding to a steady-state solution (with this viscosity law) out to $r = r_1$, with an exponential cut-off at larger radii. Specifically, the initial surface density has the form

$$\Sigma(t = 0) = \frac{C}{3\pi\nu_1\tilde{r}^\gamma}\exp\left[-\tilde{r}^{(2-\gamma)}\right], \tag{3.24}$$

where C is a normalization constant, $\tilde{r} \equiv r/r_1$, and $\nu_1 \equiv \nu(r_1)$. The self-similar solution is then

$$\Sigma(\tilde{r},T) = \frac{C}{3\pi\nu_1\tilde{r}^\gamma}T^{-(5/2-\gamma)/(2-\gamma)}\exp\left[-\frac{\tilde{r}^{(2-\gamma)}}{T}\right], \tag{3.25}$$

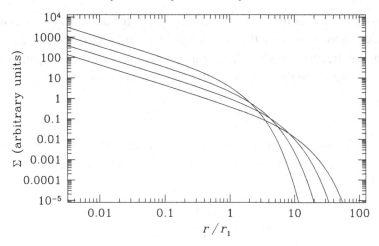

Figure 3.4 The self-similar solution to the disk evolution equation is plotted for a viscosity $\nu \propto r$. The initial surface density tracks the profile for a steady-state disk ($\Sigma \propto r^{-1}$, Eq. 3.20) at small radius, before cutting off exponentially beyond $r = r_1$. The curves show the surface density at the initial value of the scaled time $T = 1$, and at subsequent times $T = 2$, $T = 4$, and $T = 8$.

where

$$T \equiv \frac{t}{t_s} + 1, \qquad (3.26)$$

$$t_s \equiv \frac{1}{3(2-\gamma)^2} \frac{r_1^2}{\nu_1}. \qquad (3.27)$$

The evolution of related quantities such as the disk mass and accretion rate can readily be derived from the above expression for the surface density. The solution is plotted in Fig. 3.4. Over time, the disk mass decreases while the characteristic scale of the disk (initially r_1) expands to conserve angular momentum. This solution can be useful both for studying evolving disks analytically, and for comparing observations of disk masses, accretion rates, or radii with theory.

3.2.3 Temperature Profile of Accreting Disks

The radial dependence of the effective temperature of an actively accreting disk can be derived by considering the net torque on a ring of width Δr. This torque – $(\partial G/\partial r)\Delta r$ – does work at a rate

$$\Omega \frac{\partial G}{\partial r}\Delta r \equiv \left[\frac{\partial}{\partial r}(G\Omega) - G\Omega'\right]\Delta r, \qquad (3.28)$$

where $\Omega' = d\Omega/dr$. Written this way, we note that if we consider the whole disk (by integrating over r) the first term on the right-hand-side is determined solely by the boundary values of $G\Omega$. We therefore identify this term with the *transport of* energy, associated with the viscous torque, through the annulus. The second term, on the other hand, represents the rate of loss of energy to the gas. We assume that this is ultimately converted into heat and radiated, so that the dissipation rate per unit surface area of the disk (allowing that the disk has two sides) is

$$D(r) = \frac{G\Omega'}{4\pi r} = \frac{9}{8}\nu\Sigma\Omega^2, \tag{3.29}$$

where we have assumed a Keplerian angular velocity profile. For blackbody emission $D(r) = \sigma T_{\text{disk}}^4$. Substituting for Ω, and for $\nu\Sigma$ using the steady-state solution given by Eq. (3.20), we obtain

$$T_{\text{disk}}^4 = \frac{3GM_*\dot{M}}{8\pi\sigma r^3}\left(1 - \sqrt{\frac{R_*}{r}}\right). \tag{3.30}$$

We note that apart from near the inner boundary ($r \gg R_*$) the temperature profile of an actively accreting disk is $T_{\text{disk}} \propto r^{-3/4}$. This has the same form as for a razor-thin disk that reprocesses stellar radiation (Eq. 2.40). Moreover, the temperature profile does *not* depend upon the viscosity. This is an attractive feature of the theory given that there are uncertainties regarding the origin and efficiency of disk angular momentum transport, though it also eliminates many possible routes to learning about the physics underlying ν via observations of steady disks.

Substituting a representative value of $\dot{M} = 10^{-7}\ M_\odot\ \text{yr}^{-1}$ for the accretion rate, we obtain for a solar mass star at 1 AU an effective temperature $T_{\text{disk}} = 150\ \text{K}$. This is the *surface* temperature; as we will show shortly the central temperature is predicted to be substantially higher.

3.3 Vertical Structure of Protoplanetary Disks

In an actively accreting disk it is intuitively plausible that the dissipation of gravitational potential energy into heat will be concentrated close to the disk mid-plane where the density is highest. At most disk radii there is a large optical depth between the surface and the mid-plane, so the heated gas cannot simply radiate directly to space. Rather, energy will need to be transported vertically toward the photosphere via radiative diffusion or turbulent transport, and this requires that there exist a temperature gradient between $z = 0$ and the disk photosphere. Our goal here is to calculate this gradient, and its effect on the vertical structure of the disk. This is dynamically important because – in simple models – the magnitude of the viscosity that controls the evolution depends upon the *central* conditions in the disk rather than those at the surface. Given a model for the viscosity and knowledge of the vertical structure we can then derive a (in principle) self-consistent model for the disk structure and evolution.

The equations governing the vertical structure are essentially identical to those employed in stellar structure calculations except for a change in the geometry from spherically symmetric to plane parallel. If energy transport is due to radiative diffusion, the equations to be solved are (a) hydrostatic equilibrium,

$$\frac{dP}{dz} = -\rho g_z, \tag{3.31}$$

where g_z is the vertical gravity (Eq. 2.15); (b) an equation describing the vertical variation of the flux F_z,[5]

$$\frac{dF_z}{dz} = \frac{9}{4} \rho v \Omega^2; \tag{3.32}$$

and (c) the equation of radiative diffusion describing the relation between the flux and the temperature gradient in an optically thick medium,

$$\frac{dT}{dz} = -\frac{3\kappa_R \rho}{16\sigma T^3} F_z. \tag{3.33}$$

These equations must be supplemented with an equation of state relating the pressure P to the fundamental variables ρ and T, with expressions for the Rosseland mean opacity κ_R, and with appropriate boundary conditions. The dependence of the viscosity on the local physical conditions is also required. If the resulting temperature gradient turns out to be steeper than the adiabatic gradient the disk is convectively unstable, and the radiative flux needs to be supplemented with the convective flux. This can be done in the usual manner using mixing length theory, or more crudely by simply setting the temperature gradient in convectively unstable regions equal to the adiabatic gradient. Given these ingredients, calculation of the disk vertical structure is a straightforward numerical exercise.

3.3.1 The Central Temperature of Accreting Disks

An approximation to the vertical temperature structure can be derived under the assumption that the energy dissipation due to viscosity is strongly concentrated toward $z = 0$. To proceed, we define the optical depth to the disk mid-plane:

$$\tau = \frac{1}{2} \kappa_R \Sigma, \tag{3.34}$$

where κ_R is the Rosseland mean opacity and Σ is the disk surface density. The vertical density profile of the disk is $\rho(z)$. If the vertical energy transport occurs via radiative diffusion (in some regions convection may also be important), then for $\tau \gg 1$ the vertical energy flux $F(z)$ is given by the equation of radiative diffusion (Eq. 3.33 above):

$$F_z(z) = -\frac{16\sigma T^3}{3\kappa_R \rho} \frac{dT}{dz}. \tag{3.35}$$

---

[5] For $h \ll r$ one may self-consistently assume that $F_z \gg F_r$, provided that the radial temperature profile remains smooth. The assumption of smoothness is violated if the disk is subject to thermal instabilities, but in this case one should be wary of using time-independent vertical structure models in any case.

Let us assume for simplicity that *all* of the energy dissipation occurs at $z = 0$. In that case $F_z(z) = \sigma T_{\text{disk}}^4$ is a constant with height. Integrating, assuming that the opacity is constant,

$$-\frac{16\sigma}{3\kappa_R} \int_{T_c}^{T_{\text{disk}}} T^3 dT = \sigma T_{\text{disk}}^4 \int_0^z \rho(z') dz', \qquad (3.36)$$

$$-\frac{16}{3\kappa_R} \left[\frac{T^4}{4} \right]_{T_c}^{T_{\text{disk}}} = T_{\text{disk}}^4 \frac{\Sigma}{2}, \qquad (3.37)$$

where for the final equality we have used the fact that for $\tau \gg 1$ almost all of the disk gas lies below the photosphere. For large τ we expect that $T_c^4 \gg T_{\text{disk}}^4$, and the equation simplifies to

$$\frac{T_c^4}{T_{\text{disk}}^4} \simeq \frac{3}{4} \tau. \qquad (3.38)$$

The implication of this result is that active disks with large optical depths are substantially hotter at the mid-plane than at the surface. For example, if at some radius the optical depth to the disk's thermal radiation is $\tau = 10^2$, then $T_c \approx 3 T_{\text{disk}}$. This is important since it is the *central* temperature that largely determines, for example, which ices or minerals can be present. Relatively modest levels of accretion can thus affect the thermal and chemical structure of the disk substantially.

It will often be the case that *both* stellar irradiation and accretional heating contribute significantly to the thermal balance of the disk. If we define $T_{\text{disk, visc}}$ to be the effective temperature that would result from accretional heating in the absence of irradiation (i.e. the quantity called T_{disk}, with no subscript, above) and T_{irr} similarly to be the irradiation-only effective temperature, then application of Eq. (3.37) yields

$$T_c^4 \simeq \frac{3}{4} \tau T_{\text{disk, visc}}^4 + T_{\text{irr}}^4, \qquad (3.39)$$

as an approximation for the central temperature, again valid for $\tau \gg 1$. Note that in disk models, such as those described in Section 2.5.4, in which the surface dust temperature exceeds the local blackbody temperature, the fraction of the irradiating flux that is thermalized within the disk is all that matters for the central temperature. The temperature T_{irr} that enters into the above formula is then the "interior" temperature T_i computed for passive radiative equilibrium disk models.

3.3.2 Shakura–Sunyaev α Prescription

Although the effective temperature of an actively accreting disk (Eq. 3.30) is independent of the magnitude of the viscosity, the time scale on which evolution occurs (Eq. 3.11) and the profile of the surface density depend directly on the viscosity and its variation with radius. If we want to make progress in understanding the evolution of disks we cannot evade discussion of the physical origin of angular momentum transport forever! The first candidate to consider is molecular collisions, which generate a viscosity in a shear flow because of the finite mean-free path λ in the gas. Molecular viscosity, which is the normal mechanism considered in terrestrial fluids modeled with the Navier–Stokes equations, is given, approximately, by

$$\nu_m \sim \lambda c_s, \tag{3.40}$$

where the mean-free path in a gas with number density n is

$$\lambda = \frac{1}{n\sigma_{mol}}. \tag{3.41}$$

Here σ_{mol} is the cross-section for molecular collisions, which is very roughly equal to the physical size of molecules. Computing the molecular viscosity in detail is not at all a trifling task, but for our purposes high accuracy is not required, as an order of magnitude estimate establishes that ν_m is far too small to matter in protoplanetary disks. Adopting

$$\sigma_{mol} \approx 2 \times 10^{-15}\,cm^2, \tag{3.42}$$

as the collision cross-section of molecular hydrogen, and taking the sound speed at 10 AU to be $0.5\,km\,s^{-1}$ and the number density to be $n = 10^{12}\,cm^{-3}$ we estimate that $\nu_m \sim 2.5 \times 10^7\,cm^2\,s^{-1}$. The implied viscous time scale (Eq. 3.11) is

$$t_\nu \simeq \frac{r^2}{\nu_m} = 3 \times 10^{13}\,yr. \tag{3.43}$$

This is approximately ten million times longer than the observed time scale for disk evolution. Molecular viscosity is *not* the source of angular momentum transport within disks.

The consequence of a small molecular viscosity is, of course, a large Reynolds number, which can be defined as

$$Re \equiv \frac{UL}{\nu_m}, \tag{3.44}$$

where U and L are characteristic velocity and length scales in the system. Taking $U = c_s$ and $L = h = 0.05r$, the fluid Reynolds number at 10 AU is

$$Re \sim 10^{10}, \tag{3.45}$$

a staggeringly large number! One concludes that in the presence of a physical instability the protoplanetary disk will be highly turbulent, with dissipation of fluid motions occurring on a scale that is small compared to the disk scale-height.

Let us assume for the time being that the protoplanetary disk *is* turbulent. The turbulent fluid motions will result in the macroscopic mixing of fluid elements at neighboring radii, which can act as an "effective" or "turbulent" viscosity. The magnitude of this turbulent viscosity can be estimated from dimensional arguments. If the turbulence is approximately isotropic, the outer scale of the turbulent flow is limited to be no larger than the smallest scale in the disk, which is generally the disk scale-height h. The velocity of the turbulent motions can be similarly limited to be no larger than the sound speed c_s, since supersonic motions result in shocks and rapid dissipation (this is true even if the fluid is magnetized). We therefore write the turbulent viscosity in the form

$$\nu = \alpha c_s h, \tag{3.46}$$

where α is a dimensionless quantity, known as the Shakura–Sunyaev α parameter, that measures the efficiency of angular momentum transport due to turbulence (Shakura & Sunyaev, 1973).

With the help of Eq. (3.46) it is possible to specify the viscosity in terms of local disk quantities, the sound speed c_s and the disk scale-height h, and thereby compute the local disk structure. We will follow this route shortly in Section 3.3.3. Before doing so, however, it is useful to note some of the limitations of any disk model constructed using the α prescription. First, the physical argument that we gave for Eq. (3.46) reasonably limits α to be less than unity, but does *not* give any basis for assuming that α is a constant. We will take α to be constant later only because it is impossible to proceed otherwise, not for any deeper reason. In fact, α may vary with the temperature, density, and composition of the disk gas, and there may even be regions that fail to satisfy the basic assumption by not developing turbulence at all. Second, although the turbulent viscosity has the same dimensions as a molecular viscosity, one should not forget that it arises from an entirely distinct physical process. In particular, studying the evolution of turbulent disks in two or three dimensions using the Navier–Stokes equations, valid for molecular viscosity, is not recommended.

3.3.3 Vertically Averaged Solutions

Using the α prescription we can compute the viscosity ν as a function of r, Σ, and α. This in turn determines the steady-state surface density profile $\Sigma(r, \alpha, \dot{M})$ through Eq. (3.20), and can be employed together with Eq. (3.6) to calculate the time-dependent evolution of an arbitrary initial surface density profile. Here we will compute the viscosity in the vertically averaged or "one zone" approximation. This approximation amounts to replacing the equation of radiative diffusion (Eq. 3.33) with the approximate result given as Eq. (3.38), and replacing the vertical dependence of all other quantities by their central values.

Consider an annulus of the disk with surface density Σ at a radius where the Keplerian angular velocity is Ω (in thin disk solutions r and M_* enter only via the combination $\Omega = \sqrt{GM_*/r^3}$). The disk is characterized by eight variables: the mid-plane temperature T_c, effective temperature T_{disk}, sound speed c_s, density ρ, vertical scale-height h, opacity κ_R, viscosity ν, and optical depth τ. Apart from the effective temperature, which is defined at the photosphere, and the optical depth, which is evaluated *between* the surface and the mid-plane, all of these quantities are to be considered as the values at $z = 0$. Collecting together results that we have either already derived or which are trivial, we have the following set of equations:[6]

$$\nu = \alpha c_s h, \tag{3.47}$$

$$c_s^2 = \frac{k_B T_c}{\mu m_p}, \tag{3.48}$$

$$\rho = \frac{1}{\sqrt{2\pi}} \frac{\Sigma}{h}, \tag{3.49}$$

$$h = \frac{c_s}{\Omega}, \tag{3.50}$$

$$T_c^4 = \frac{3}{4}\tau T_{disk}^4, \tag{3.51}$$

$$\tau = \frac{1}{2}\Sigma \kappa_R, \tag{3.52}$$

$$\nu\Sigma = \frac{\dot{M}}{3\pi}, \tag{3.53}$$

$$\sigma T_{disk}^4 = \frac{9}{8}\nu\Sigma\Omega^2. \tag{3.54}$$

Once the opacity $\kappa_R(\rho, T_c)$ is specified, there are eight equations in eight unknowns to solve in order to determine the disk structure. If the opacity can be approximated as a power-law in density and temperature the solution is analytic, and one obtains an expression for ν that is a power-law in r, α, and Σ.

As an explicit example, we may consider a disk in which the mid-plane opacity is due to icy particles. In this limit an approximate form for the opacity is

$$\kappa = \kappa_0 T_c^2, \tag{3.55}$$

with the constant having a numerical value in cgs units $\kappa_0 = 2.4 \times 10^{-4}$. For this opacity there is no dependence on density, and hence the equation for ρ is redundant. Eliminating variables in turn the remaining equations yield

$$\Sigma^3 = \frac{64}{81\pi} \frac{\sigma}{\kappa_0} \left(\frac{\mu m_p}{k_B}\right)^2 \alpha^{-2}\dot{M}. \tag{3.56}$$

---

[6] For simplicity we assume that $r \gg R_*$ and neglect terms that depend upon the inner boundary conditions. Note also that different numerical factors can enter into, for example, the definition of the central density to be used in these equations. Given the overriding uncertainty that originates from our poor knowledge of angular momentum transport in disks, worrying about the "correct" value of such factors is rarely profitable.

For an accretion rate $\dot{M} = 10^{-7} \, M_\odot \, \mathrm{yr}^{-1}$ and an $\alpha = 0.01$, we find that $\Sigma \approx 140 \, \mathrm{g \, cm}^{-2}$. The corresponding viscosity is $\nu \approx 5 \times 10^{15} \, \mathrm{cm}^2 \, \mathrm{s}^{-1}$, and the viscous time at 30 AU is about $1.3 \times 10^6 \, \mathrm{yr}$. This time scale is broadly consistent with protoplanetary disks evolving significantly over a few million years, and indeed most observational attempts to constrain α via measures of disk evolution return estimates $\alpha \sim 10^{-2}$ (e.g. Hartmann *et al.*, 1998).

3.4 Hydrodynamic Angular Momentum Transport

The Shakura–Sunyaev α prescription furnishes a formula for the viscosity but leaves unanswered some of the most basic questions. What is the origin of the postulated turbulence? How large is α, and what is its dependence (if any) on the physical conditions within the disk? These issues are at the forefront of research for protoplanetary disks and for other accreting systems.

For disks, such as those around black holes, in which the gas is well coupled to the magnetic field, the physical origin of turbulence is well understood. The magnetorotational instability (MRI), which we discuss in Section 3.5.1, operates efficiently under such conditions and generates vigorous, sustained, turbulence which transports angular momentum outward. Other hydrodynamic processes (such as convection) may modify the properties of the turbulence, and magnetized disk winds may even be more important in driving accretion, but turbulent accretion would occur even in their absence.

Protoplanetary disks are more complex and interesting. There are substantial regions within protoplanetary disks where the ionization fraction is so low that the magnetic field is effectively decoupled from the fluid dynamics of the gas. Under low ionization conditions the MRI does not operate in the same way as it does in ideal MHD, and the strength of turbulence and angular momentum transport can be heavily suppressed. It is then possible, but not guaranteed, that hydrodynamic processes generate the bulk of whatever turbulence and transport is present. The simplest disk flows are linearly stable, but there are loopholes if the disk has non-trivial radial or vertical structure, or a large mass, which allow purely hydrodynamic sources of turbulence.

3.4.1 The Rayleigh Criterion

The stability of a rotating flow to infinitesimal hydrodynamic perturbations can be derived by linearizing the fluid equations, setting the time dependence of perturbations to be proportional to $e^{i\omega t}$, and searching for exponentially growing or decaying modes for which ω is imaginary. For a nonmagnetized, non-self-gravitating disk (roughly speaking, one for which $M_{\mathrm{disk}}/M_* < h/r$) the appropriate stability criteria are due to Rayleigh (for a derivation see e.g. Pringle & King, 2007). Such a disk flow is linearly stable to axisymmetric perturbations if and only if the specific angular momentum increases with radius. For instability we require

$$\frac{dl}{dr} = \frac{d}{dr}\left(r^2\Omega\right) < 0. \tag{3.57}$$

In a Keplerian disk the specific angular momentum is an increasing function of radius, $l \propto \sqrt{r}$ (Eq. 3.1), and the flow is predicted to be hydrodynamically stable. Despite the enormous value of the Reynolds number, there is then no ready justification for invoking hydrodynamic turbulence as the origin of the turbulent viscosity needed to drive accretion.

The Rayleigh criterion is based upon a linear analysis, and says nothing about whether a disk might be unstable to finite-amplitude perturbations. It is very difficult to prove that a given hydrodynamic flow is non-linearly stable. Nonetheless, the results of both numerical and laboratory experiments (Hawley *et al.*, 1999; Ji *et al.*, 2006) support the view that the simplest hydrodynamic disk flow would be unable to sustain astrophysically relevant levels of turbulence. The use of the word "simplest" here is a very significant caveat. It implies a toy disk model in which there is no vertical stratification, no vertical shear, a stable radial entropy profile, and no self-gravity. Linear and nonlinear instabilities are known to exist when any of these restrictions are relaxed.

3.4.2 Self-Gravity

Up to now we have considered disks whose own gravity is negligible compared to the force from the central star. Such disks are axisymmetric, which implies that no torque is exerted on gas at one radius due to the gravitational force from matter elsewhere in the disk. As the disk becomes more massive there is an increasing tendency for its own self-gravity to result in the formation of over-dense "clumps." Gravitational forces between these clumps can result in angular momentum transport.

Self-gravity is an important potential mechanism for planetesimal formation, which is discussed in Chapter 4. We defer a formal mathematical analysis of the conditions for clump formation until then, in favor of a simple argument based on time scales. Within a disk, the self-gravity of the disk gas always has a tendency to form denser clumps, but this is opposed by both pressure forces and by shear, which tend to oppose clump formation. We can estimate the conditions under which self-gravity is strong enough to win out over these stabilizing effects by requiring that the time scale for collapse be shorter than the time scales on which sound waves can cross a clump, or shear destroy it.

Consider a forming clump of scale Δr and mass $m \sim \pi(\Delta r)^2\Sigma$. In isolation, such a clump would collapse on the free-fall time scale

$$t_{\text{ff}} \sim \sqrt{\frac{\Delta r^3}{Gm}} \sim \sqrt{\frac{\Delta r}{\pi G \Sigma}}. \tag{3.58}$$

Stabilizing influences that may prevent collapse are pressure and shear. The time scale for a sound wave to cross the clump is

$$t_{\text{p}} \sim \frac{\Delta r}{c_s}, \tag{3.59}$$

where c_s is the disk sound speed, while the shear time scale (the time scale required for a clump to be sheared azimuthally by an amount Δr) is

$$t_{\text{shear}} = \frac{1}{r} \left(\frac{d\Omega}{dr} \right)^{-1} \sim \Omega^{-1}. \tag{3.60}$$

One observes that the shear time scale is independent of the clump size Δr, whereas the sound crossing time increases linearly with Δr. Pressure tends to stabilize small regions of the disk against gravitational collapse, while shear stabilizes the largest scales. The disk will be marginally unstable to clump formation if the free-fall time scale on the scale where shear and pressure support match is shorter than either t_{p} or t_{shear}. Setting all three time scales equal, we obtain a condition for instability in the form

$$\pi G \Sigma \gtrsim c_s \Omega. \tag{3.61}$$

At a given radius (i.e. at fixed Ω) the disk will be unstable if it is massive (large Σ) and/or cool and thin (small c_s). A more formal analysis (Toomre, 1964) gives the same result. The stability of a disk of either gas or stars is controlled by a parameter

$$Q \equiv \frac{c_s \Omega}{\pi G \Sigma}, \tag{3.62}$$

known as the "Toomre Q" parameter.[7] Instability of the disk against self-gravity sets in once $Q \lesssim 1$. Using the fact that $h = c_s / \Omega$, and estimating the disk mass as $M_{\text{disk}} \sim \pi r^2 \Sigma$, we can write the instability threshold in the more intuitive form

$$\frac{M_{\text{disk}}}{M_*} \gtrsim \frac{h}{r}. \tag{3.63}$$

This is (up to numerical factors) the same condition that we derived earlier (Eq. 2.24) for when disk gravity becomes important for the vertical structure. Since $h/r \approx 0.05$ is a representative number for protoplanetary disks, we require fairly massive disks before the effects of self-gravity can be expected to become important. Such disks are more likely to have existed at early epochs, possibly prior to the optically visible T Tauri phase of YSO evolution.

---

[7] If the disk is extremely massive, then the angular velocity profile itself may depart significantly from the Keplerian form. In that case, Ω in the stability criterion must be replaced by the *epicyclic frequency* κ, defined via $\kappa^2 = (1/r^3)d(r^4\Omega^2)/dr$ (note that $\kappa = \Omega$ for a Keplerian disk).

If self-gravity sets in within a disk, there are two possible outcomes:

• Collapse may continue unhindered, destroying the disk and forming one or more bound objects. Disk *fragmentation* via this process is a mechanism for planetesimal formation (in the case where the collapse occurs in the solid component of the disk material) star or brown dwarf formation, and, perhaps, giant planet formation.

• Adiabatic heating as the clump contracts to higher density may yield enough pressure to prevent complete collapse. Numerical simulations show that the outcome is the development of spiral arms induced by the self-gravity within the disk. Gravitational forces set up by the spiral arms work to transport angular momentum outward and mass inward.

The boundary between these possibilities is set by the ability of the disk to radiate away its thermal energy. If the disk can cool on a short time scale, pressure cannot build up within contracting clumps, and fragmentation results. Slow cooling, which is physically more likely in most protoplanetary disks, results instead in stable angular momentum transport.

A priori it is far from obvious that angular momentum transport via self-gravity can be described using the local language of α disks that we have developed previously. If, however, one allows this sleight of hand, we can give a plausible argument for an important result derived more formally by Gammie (2001): namely that the equivalent α in a self-gravitating disk depends upon the cooling time of the gas. Let us assume that the disk does not fragment, in which case heating at a local rate,

$$Q_+ = \frac{9}{4}\nu\Sigma\Omega^2, \tag{3.64}$$

must on average balance cooling. If the disk is optically thick, the local cooling rate is

$$Q_- = 2\sigma T_{\text{disk}}^4. \tag{3.65}$$

Writing the "gravitational viscosity" in the form

$$\nu = \alpha_{\text{grav}}\frac{c_s^2}{\Omega} = \frac{\alpha_{\text{grav}}}{\Omega}\frac{k_B T_c}{\mu m_p}, \tag{3.66}$$

and setting $Q_+ = Q_-$, we find that the effective α depends upon the ratio $\sigma T_{\text{disk}}^4/(\Sigma T_c)$. We recognize this ratio as being proportional to the cooling rate from the disk surface divided by the thermal energy. We therefore define

$$t_{\text{cool}} = \frac{U}{2\sigma T_{\text{disk}}^4}, \tag{3.67}$$

where U is the thermal energy content of the disk gas per unit area. Noting that, approximately, $U \simeq c_p \Sigma T_c$, where c_p is the heat capacity of the gas, the thermal balance condition yields

$$\alpha_{\mathrm{grav}} \approx \frac{4}{9\gamma(\gamma - 1)\Omega t_{\mathrm{cool}}}. \tag{3.68}$$

In this expression γ is the two-dimensional adiabatic index (defined in terms of the two-dimensional pressure P and internal energy U via $P = (\gamma - 1)U$) and we have reinserted the correct numerical factors from the analysis by Gammie (2001). One finds that the strength of angular momentum transport from self-gravity depends upon the cooling time of the disk, measured in units of the local dynamical time. This is intuitively reasonable. If the disk cools rapidly, strong gravitational turbulence, characterized by large density contrasts, is required in order to generate enough heat to balance the cooling. The strong density contrasts result in larger torques due to self-gravity. Eventually, if the cooling is too rapid, the disk cannot be stabilized against collapse, and fragmentation ensues. Simulations show that stable angular momentum transport is possible only for $\alpha_{\mathrm{grav}} \lesssim 0.1$.

3.4.3 Vertical Shear Instability

The vertical shear instability (VSI) is a linear hydrodynamic instability that can be present in disks whenever the angular velocity of the equilibrium structure varies with height above the mid-plane. The instability, which is closely related to one studied in rotating stars (Goldreich & Schubert, 1967; Fricke 1968), was suggested as a possible source of disk turbulence by Urpin and Brandenburg (1998) and studied in detail by Nelson *et al.* (2013).

Figure 3.5 illustrates the type of fluid perturbations that are unstable to the VSI. Let us assume for simplicity that the vertical structure of the disk is neutrally buoyant, so that there is neither a driver of vertical motions (as would be the case if the vertical entropy profile were convectively unstable) nor a stabilizing effect (as would be true for a highly stable stratification). In that limit the fluid stability is entirely determined by a Rayleigh-like criterion, and in the absence of vertical shear the system is stable. In the presence of vertical shear, however, displacement of a fluid element along a near-vertical trajectory from point "A" to point "B" moves it *radially outward* to a place where the specific angular momentum is *lower*. This is unstable. The general criterion for instability for a mode with radial wavenumber k_r and vertical wavenumber k_z,

$$\frac{\partial l^2}{\partial r} - \frac{k_r}{k_z}\frac{\partial l^2}{\partial z} < 0, \tag{3.69}$$

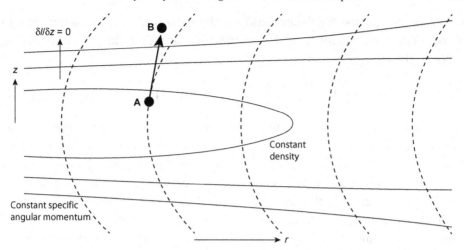

Figure 3.5 The vertical shear instability (VSI) is a hydrodynamic instability that can be present in disks where the specific angular momentum varies with height. This condition is generically satisfied in disks whose thermal structure is substantially set by stellar irradiation. Under neutrally buoyant conditions, fluid perturbations that are nearly vertical are unstable. In a stably stratified disk the operation of the instability depends upon the degree of stratification and the ability of the fluid to cool via radiative or thermal diffusion effects.

can always be satisfied by choosing "sufficiently vertical" modes with large k_r/k_z (Nelson *et al.* 2013).

The vertical shear that is the most basic prerequisite for the VSI is present wherever the disk has a radial temperature gradient, which is to say almost everywhere. Departures from neutral buoyancy pose a much more significant limitation. Disks typically have a stable vertical entropy profile, which means that buoyancy forces oppose the near-vertical motions that would be VSI unstable. Rapid radiative cooling or thermal diffusion is needed to allow the VSI to work. Considering these factors, Lin and Youdin (2015) find that the VSI could be a significant process in protoplanetary disks at intermediate radii (of the order of 5–50 AU), where moderate optical depths allow the efficient cooling that promotes the VSI.

3.4.4 Vortices

In Section 2.9.2 we noted that vortices are a special class of hydrodynamic structure that may be present within protoplanetary disks. Vortices can contribute to angular momentum transport, and are of particular interest for planet formation because they act to concentrate solid particles into dense regions at their core. It is therefore of interest to ask how vortices might form within disks. Defining the fluid *vorticity* as

$$\omega = \nabla \times \mathbf{v}, \tag{3.70}$$

we obtain an equation for the evolution of the *vortensity* ω/ρ by taking the curl of the momentum equation. In the absence of magnetic fields and microscopic viscosity, the result is

$$\frac{D}{Dt}\left(\frac{\omega}{\rho}\right) = \left(\frac{\omega}{\rho}\right) \cdot \nabla \mathbf{v} - \frac{1}{\rho}\nabla\left(\frac{1}{\rho}\right) \times \nabla P. \tag{3.71}$$

This equation exposes the most important physical consideration concerning the role of vortices within protoplanetary disks. In a *barotropic* fluid (one in which $P = P(\rho)$ only) surfaces of equal pressure are always parallel to surfaces of equal density. In this limit the second term on the right-hand-side vanishes, and any vortensity present within the disk is simply advected with the fluid motion. To the extent that the barotropic approximation is valid, vorticity is a conserved quantity that cannot be easily created, or, once present, destroyed. Indeed, in two-dimensional models of protoplanetary disks, vortices are found to be stable long-lived structures whose interactions can act to transport angular momentum[8] (Godon & Livio, 1999; Johnson & Gammie, 2005).

Vorticity can be spontaneously generated in more complete disk models, for example as a consequence of the subcritical baroclinic instability (Petersen *et al.*, 2007; Lesur & Papaloizou, 2010). This is a nonlinear (or "subcritical") instability that relies for its existence on the interplay of two distinct thermodynamic properties of the disk. First, it requires that the radial entropy gradient in the disk be unstable to the Schwarzschild criterion for the onset of convection. (This does not mean that convection sets in, because in a disk rotation exerts a strong stabilizing effect.) Second, radiative cooling (or thermal diffusion) must occur at a rate that promotes the generation of vorticity. These pre-conditions (which can be expressed quantitatively) mean that the subcritical baroclinic instability, rather like the vertical shear instability, is a process whose existence and operation in disks is strongly dependent on the detailed disk structure.

The Rossby wave instability (RWI; Lovelace *et al.*, 1999) is a second physical mechanism for producing vortices within at least some protoplanetary disks. The RWI is a linear instability that is present whenever there is "sufficiently sharp" radial structure in the disk. Specifically, for a two-dimensional disk model with angular velocity $\Omega(r)$, vertically integrated pressure $P(r)$, and adiabatic index γ, we can define the entropy S and epicyclic frequency κ via

$$S = \frac{P}{\Sigma^{\gamma}}, \tag{3.72}$$

$$\kappa^2 = \frac{1}{r^3}\frac{\partial}{\partial r}(r^4\Omega^2). \tag{3.73}$$

In terms of these quantities, the stability of the disk to the RWI is determined by the radial profile of a generalized potential vorticity,

---

[8] This is in accord with the observation that vortices can be long lived in strongly stratified, and thus physically quasi-two-dimensional, atmospheres. Jupiter's Great Red Spot is a particularly spectacular example.

$$\mathcal{L}(r) = \frac{\Omega\Sigma}{\kappa^2} \times S^{2/\gamma}. \tag{3.74}$$

A necessary condition for RWI is that $\mathcal{L}$ have an extremum. A precise sufficient condition is not easily stated, but variations in the potential vorticity of the order of 10% over radial scales $\approx h$ suffice to trigger instability. Once triggered, the RWI leads to the formation of anticyclonic vortices on time scales of the order of $10\Omega^{-1}$, which typically merge to form a single larger vortex on a longer time scale.

Localized radial structure can be produced in protoplanetary disks from the gravitational perturbation of a growing planet (see Section 7.1.4), and this is a specific scenario that is theoretically expected to initiate RWI. The instability could also occur in disks that do not contain planets, if some (obviously nondiffusive) process was able to create relatively pronounced ring-like structures.

3.5 Magnetohydrodynamic Angular Momentum Transport

Coupling a magnetic field to the gas grants the fluid additional degrees of freedom that violently destabilize the disk. The condition for a weakly magnetized disk flow to be linearly unstable is that the angular velocity (rather than the specific angular momentum as in the fluid case) decrease with radius,

$$\frac{\mathrm{d}}{\mathrm{d}r}\left(\Omega^2\right) < 0, \tag{3.75}$$

and this condition *is* satisfied by Keplerian disks. The linear instability of a disk coupled to a magnetic field is known as the magnetorotational (MRI) or Balbus–Hawley instability. The mathematical analysis of the instability has a long history (Velikhov, 1959; Chandrasekhar, 1961) but its importance in the context of accretion disks was only recognized at a much later date (Balbus & Hawley, 1991). This long hiatus is something of a puzzle, as Chandrasekhar's work was published in his classic textbook *Hydrodynamic and Hydromagnetic Stability*. Indeed, Safronov (1969) came agonizingly close to recognizing the significance of Chandrasekhar's result for the origin of turbulence within the protoplanetary disk. He noted correctly that the MHD stability criterion does not reduce to the Rayleigh criterion as the magnetic field tends toward zero, and that "for a weak magnetic field the cloud should be less stable than we found earlier in the absence of the field." Safronov then, however, dismisses the MRI on the (incorrect) grounds that the instability requires that the viscosity and diffusivity are identically zero.

3.5.1 Magnetorotational Instability

In the limit of ideal magnetohydrodynamics (MHD) the disk can be described by the equations of continuity,

$$\frac{\partial\rho}{\partial t} + \nabla \cdot (\rho\mathbf{v}) = 0, \tag{3.76}$$

momentum conservation,

$$\frac{\partial \mathbf{v}}{\partial t} + (\mathbf{v} \cdot \nabla)\,\mathbf{v} = -\frac{1}{\rho}\nabla\left(P + \frac{B^2}{8\pi}\right) - \nabla\Phi + \frac{1}{4\pi\rho}\,(\mathbf{B} \cdot \nabla)\,\mathbf{B}, \tag{3.77}$$

and magnetic induction,

$$\frac{\partial \mathbf{B}}{\partial t} = \nabla \times (\mathbf{v} \times \mathbf{B})\,. \tag{3.78}$$

The symbols have their conventional meanings, ρ is the density, $\mathbf{v}$ is the velocity, $\mathbf{B}$ is the magnetic field, P is the pressure, $B = |\mathbf{B}|$, and Φ is the gravitational potential. Magnetized disks have qualitatively different stability properties from fluid disks because of the presence of magnetic tension (described by the third term on the right-hand-side of Eq. 3.77), which exerts a force on the fluid that attempts to straighten curved field lines.

Demonstrating the existence of the MRI in the general case requires moderately lengthy algebra, which can be found in the review by Balbus and Hawley (1998). The essence of the instability, however, lies in the interplay of magnetic tension and flux freezing (Eq. 3.78) within a differentially rotating system, and this can be demonstrated in much simpler model systems. Consider an axisymmetric, incompressible disk that is threaded by a vertical magnetic field. In cylindrical polar coordinates (r, z, ϕ), the equations of motion of a small parcel of gas then read

$$\ddot{r} - r\dot{\phi}^2 = -\frac{d\Phi}{dr} + f_r, \tag{3.79}$$

$$r\ddot{\phi} + 2\dot{r}\dot{\phi} = f_\phi, \tag{3.80}$$

where the dots denote derivatives with respect to time, and f_r and f_ϕ are forces due to the coupling of the gas to the magnetic field, which we will specify shortly. We now concentrate attention on a small patch of the disk at radius r_0, co-rotating with the overall orbital motion at angular velocity Ω. We define a local Cartesian coordinate system (x, y) via

$$r = r_0 + x, \tag{3.81}$$

$$\phi = \Omega t + \frac{y}{r_0}, \tag{3.82}$$

and substitute these expressions into Eq. (3.80) above. Discarding quadratic terms, the result is

$$\ddot{x} - 2\Omega\dot{y} = -x\frac{d\Omega^2}{d\ln r} + f_x, \tag{3.83}$$

$$\ddot{y} + 2\Omega\dot{x} = f_y.$$

The second term on the left-hand-side of these equations represents the Coriolis force. In the absence of magnetic forces these equations describe the epicyclic motion of pressureless fluid perturbed from an initially circular orbit.

If the disk contains a vertical magnetic field B_z, perturbations to the fluid in the plane of the disk will be opposed by magnetic tension forces generated by the bending of the field lines. Consider in-plane perturbations varying with height z and time t as

$$\mathbf{s} \propto e^{i(\omega t - kz)}, \tag{3.84}$$

where $\mathbf{s}$ is the displacement vector. The corresponding perturbation to the magnetic field follows from the induction Eq. (3.78). It is

$$\delta \mathbf{B} = -ikB_z \mathbf{s}, \tag{3.85}$$

and this results in a magnetic tension force given by

$$\mathbf{f} = -(kv_A)^2 \mathbf{s}, \tag{3.86}$$

where $v_A = \sqrt{B_z^2/4\pi\rho}$ is the Alfvén speed. Using this expression for f_x and f_y, and recalling the $e^{i\omega t}$ time dependence, Eq. (3.83) becomes

$$-\omega^2 x - 2i\omega\Omega y = -x\frac{d\Omega^2}{d\ln r} - (kv_A)^2 x, \tag{3.87}$$

$$-\omega^2 y + 2i\omega\Omega x = -(kv_A)^2 y. \tag{3.88}$$

Combining these equations yields a dispersion relation (i.e. a relation between the wavenumber k and frequency ω of the perturbation) that is a quadratic in ω^2,

$$\omega^4 - \omega^2 \left[\frac{d\Omega^2}{d\ln r} + 4\Omega^2 + 2(kv_A)^2 \right] + (kv_A)^2 \left[(kv_A)^2 + \frac{d\Omega^2}{d\ln r} \right] = 0. \tag{3.89}$$

If $\omega^2 > 0$ then ω itself will be a real number, and the perturbation of the form $e^{i\omega t}$ will display oscillatory behavior. Instability occurs when $\omega^2 < 0$, since in this case ω is imaginary and the perturbation will have exponentially growing modes. The instability criterion is that

$$(kv_A)^2 + \frac{d\Omega^2}{d\ln r} < 0. \tag{3.90}$$

Taking the limit of a vanishingly weak field ($B_z \to 0$, $v_A \to 0$) we recover the aforementioned condition for the local linear instability of a differentially rotating disk in the presence of a weak magnetic field (Eq. 3.75). One may observe that the instability condition does not tend toward the Rayleigh criterion as the magnetic field goes to zero, but rather remains qualitatively distinct.

Two other important properties of the MRI – the growth rate of the instability and what it means for the magnetic field to be "weak" – can also be derived directly from Eq. (3.89). Specializing to a Keplerian rotation law $d\Omega^2/d\ln r = -3\Omega^2$ we find that the dispersion relation takes the form shown in Fig. 3.6. For a fixed value of

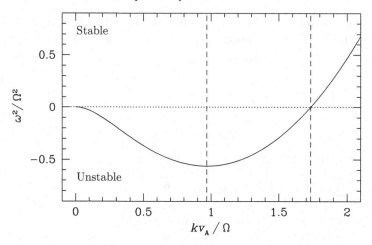

Figure 3.6 The unstable branch of the dispersion relation for the magnetorotational instability (Eq. 3.89) is plotted for a disk with a Keplerian angular velocity profile. Both ω and $k v_A$ have been plotted scaled to the orbital frequency Ω. The flow is unstable ($\omega^2 < 0$) for all spatial scales smaller than $k v_A < \sqrt{3}\Omega$ (rightmost dashed vertical line). The most unstable scale (shown as the dashed vertical line at the center of the plot) is $k v_A = (\sqrt{15}/4)\Omega$.

the magnetic field strength (and hence a fixed Alfvén speed v_A) the flow is unstable for wavenumbers $k < k_{\mathrm{crit}}$ (i.e. on sufficiently large spatial scales), where

$$k_{\mathrm{crit}} v_A = \sqrt{3}\Omega. \tag{3.91}$$

As the magnetic field becomes stronger, the *smallest* scale $\lambda = 2\pi/k_{\mathrm{crit}}$ that is unstable grows, until eventually it exceeds the disk's vertical extent, which we may approximate as $2h$, where h is the scale-height. For stronger vertical fields no unstable MRI modes fit within the thickness of the disk, and the instability will be suppressed. Noting that $h = c_s/\Omega$, we can express the condition that the vertical magnetic field be weak enough to admit the MRI (i.e. that $\lambda < 2h$) as

$$B^2 < \frac{12}{\pi}\rho c_s^2, \tag{3.92}$$

where c_s is the disk sound speed. If we define the plasma β parameter as the ratio of gas to magnetic pressure,

$$\beta \equiv \frac{8\pi P}{B^2}, \tag{3.93}$$

Eq. (3.92) can be expressed alternatively as

$$\beta > \frac{2\pi^2}{3}. \tag{3.94}$$

A magnetic field whose *vertical* component approaches equipartition with the thermal pressure ($\beta \sim 1$) will therefore be too strong to admit the existence of linear MRI modes, but a wide range of weaker fields is acceptable.

The maximum growth rate can be determined by setting $d\omega^2/d(kv_A) = 0$ for the unstable branch of the dispersion relation shown in Fig. 3.6. The most unstable scale for a Keplerian disk is

$$(kv_A)_{\max} = \frac{\sqrt{15}}{4}\Omega, \tag{3.95}$$

with a corresponding growth rate

$$|\omega_{\max}| = \frac{3}{4}\Omega. \tag{3.96}$$

The main point to notice about this last result is that it implies an *extremely* vigorous growth of the instability, with an exponential growth time scale that is a fraction of an orbital period. Practically, this means that if a disk is unstable to the MRI it is hard to envisage other physical processes that will operate on a shorter time scale and prevent the MRI from dominating the evolution.

The physical origin of the MRI is fairly transparent for this simple configuration. Consider the effect of perturbing a weak vertical magnetic field threading an otherwise uniform disk, so that some fluid (tied to the magnetic field due to flux freezing) moves slightly inward and some slightly outward. Due to the differential rotation, the perturbed field line – which now connects adjacent annuli in the disk – will be sheared by the differential rotation of the disk into a trailing spiral pattern. Provided that the field is weak enough, magnetic tension is inadequate to snap the field lines back to the vertical. What tension there is, however, acts to reduce the angular momentum of the inner fluid element, and boost that of the outer fluid element. The transfer of angular momentum further increases the separation, leading to an instability.

The very rapid linear growth of the MRI implies, of course, that there is no physical way to set up a disk with initial conditions that resemble those used for analytic convenience above. Within a real disk it is the *nonlinear* properties of the instability that are of interest for angular momentum transport, and these can only be studied in detail via numerical MHD simulations. Disentangling physical from numerical effects in such calculations is a nontrivial exercise, but one consensus result is that the nonlinear evolution of the MRI leads to a state of sustained MHD turbulence which is able to maintain disk magnetic fields in the presence of dissipation. A flow with this property is described as a magnetic dynamo. Denoting the velocity fluctuations in the radial and azimuthal directions by δv_r and δv_ϕ respectively, we can define an equivalent Shakura–Sunyaev α parameter in terms of an average over the fluctuating velocity and magnetic fields:

$$\alpha = \left\langle \frac{\delta v_r \delta v_\phi}{c_s^2} - \frac{B_r B_\phi}{4\pi\rho c_s^2} \right\rangle, \tag{3.97}$$

where the angle brackets denote a density-weighted average over time. This formula provides a way to connect the small-scale flow dynamics seen in numerical simulations with the gross angular momentum transport efficiency that is relevant for the evolution of the disk as a whole. Numerical work shows that the magnetic transport term (the second term on the right-hand-side, described as the Maxwell stress) is generically much larger than the fluid (or Reynolds) contribution in MHD turbulent flows initiated by the MRI, although both terms are positive. The predicted magnitude in the ideal MHD limit is of the order of $\alpha \sim 10^{-2}$ if there is no net vertical magnetic field threading the disk, and higher if there is.

3.6 Effects of Partial Ionization and Dead Zones

In Section 2.8 we showed that the protoplanetary disk at the radii of greatest interest for planet formation (0.1–50 AU) is expected to be predominantly neutral. Magnetic fields, which couple to charged particles, affect the neutral component of the fluid indirectly, via collisions between charged and neutral species. Accounting fully for the influence of partial ionization on the MRI (and, by extension, on the likely properties of disk turbulence and angular momentum transport) requires us to consider the role of three new physical effects that, together, make up *non-ideal* MHD: Ohmic diffusion, ambipolar diffusion, and the Hall effect.

3.6.1 Ohmic Dead Zones

A full analysis of the effect of the non-ideal terms on protoplanetary disks is quite involved, so it is useful to start with a deliberately oversimplified model that includes Ohmic diffusion only. The induction equation then takes the form

$$\frac{\partial \mathbf{B}}{\partial t} = \nabla \times [\mathbf{v} \times \mathbf{B} - \eta \nabla \times \mathbf{B}], \tag{3.98}$$

where $\mathbf{v}$ is the velocity of the neutral fluid and the second term on the right-hand-side represents the effect of Ohmic diffusion. The magnetic diffusivity, η, is inversely proportional to the electrical conductivity σ_c,

$$\eta = \frac{c^2}{4\pi \sigma_c}. \tag{3.99}$$

Under the assumption that the charge carriers in the disk are electrons and ions we can write down an explicit expression for η. The electrical conductivity (and, hence, the magnetic diffusivity) is determined by collisions between electrons and neutral species. If the cross-section for such collisions is σ, the collision frequency is

$$\nu_e = n_n \langle \sigma v \rangle_e, \tag{3.100}$$

where n_n is the number density of neutrals and the brackets denote a velocity-weighted average. The conductivity is

$$\sigma_c = \frac{e^2 n_e}{m_e \nu_e}. \tag{3.101}$$

Numerically, the collision rate is given by (Draine *et al.*, 1983)

$$\langle \sigma v \rangle_e = 8.28 \times 10^{-10} T^{1/2} \, \text{cm}^3 \, \text{s}^{-1}. \tag{3.102}$$

With this microphysical background we can estimate the minimum ionization fraction required for a protoplanetary disk to sustain the MRI by comparing the effect of the Ohmic dissipation term with the inductive term (the first term on the right-hand-side of Eq. 3.98). In the absence of dissipation, the MRI in the ideal limit results in a turbulent disk in which the magnetic field has a complex tangled structure. The development of a tangled field is opposed by the action of Ohmic dissipation, which acts diffusively to eliminate field gradients. We can estimate the critical ionization level needed for the MRI to operate by first calculating the magnitude of the magnetic diffusivity η for which the smoothing influence of Ohmic dissipation overcomes the competing tendency of the MRI to generate complex field structures.

To proceed, we first note that diffusion will erase small-scale structure in the magnetic field more readily than larger scale features. The largest scale MRI modes will therefore be the last to survive in the presence of Ohmic dissipation. Starting from the MRI dispersion relation (Eq. 3.89), we specialize to the case of a Keplerian disk and consider the weak-field/long-wavelength limit ($k v_A / \Omega \ll 1$). The growth rate of the MRI in this limit is

$$|\omega| \simeq \sqrt{3}(k v_A). \tag{3.103}$$

Writing this in terms of a spatial scale $\lambda = 2\pi / k$, we have that

$$|\omega| \simeq 2\pi \sqrt{3} \frac{v_A}{\lambda}. \tag{3.104}$$

Up to numerical factors, the MRI therefore grows on the Alfvén crossing time of the spatial scale under consideration. Equating this growth rate to the damping rate due to Ohmic diffusion,

$$|\omega_\eta| \sim \frac{\eta}{\lambda^2}, \tag{3.105}$$

and the condition for Ohmic dissipation to suppress the MRI on scale λ becomes

$$\eta \gtrsim 2\pi \sqrt{3} v_A \lambda. \tag{3.106}$$

To completely suppress the MRI, we demand that damping dominates growth on *all* scales, including the largest scale $\lambda \sim h$ available in a disk geometry. The limit on the diffusivity is then $\eta \gtrsim 2\pi \sqrt{3} v_A h$. It is instructive to express this result in a slightly different form. Recalling the definition of the fluid Reynolds number (Eq. 3.44), we define the *magnetic* Reynolds number Re_M via

$$Re_M \equiv \frac{UL}{\eta}, \tag{3.107}$$

where, as before, U is a characteristic velocity and L a characteristic scale of the system. Taking $U = v_A$ and $L = h$ for a disk, we can rewrite the condition for Ohmic dissipation to suppress the MRI in the form

$$Re_M \lesssim 1, \tag{3.108}$$

where numerical factors of the order of unity have been omitted.

The final step is to write the condition for the suppression of the MRI in terms of the ionization fraction $x \equiv n_e/n_n$. Making use of the expressions for the magnetic diffusivity and collision rate between electrons and neutrals (Eqs. 3.99 through 3.102) the magnetic diffusivity is

$$\eta \simeq 2.3 \times 10^2 x^{-1} T^{1/2} \, \text{cm}^2 \, \text{s}^{-1}. \tag{3.109}$$

Under the assumption that Maxwell stresses dominate the transport of angular momentum, Eq. (3.97) implies that

$$\alpha \sim \frac{v_A^2}{c_s^2}, \tag{3.110}$$

which allows us to write the Alfvén speed in the definition of the magnetic Reynolds number in terms of α and the sound speed. Since $h = c_s/\Omega$, we obtain

$$Re_M = \frac{v_A h}{\eta} = \frac{\alpha^{1/2} c_s^2}{\eta \Omega}. \tag{3.111}$$

Substituting for η and c_s^2, we obtain an expression for the predicted variation of the magnetic Reynolds number in a protoplanetary disk:

$$Re_M \approx 1.4 \times 10^{12} x \left(\frac{\alpha}{10^{-2}}\right)^{1/2} \left(\frac{r}{1 \, \text{AU}}\right)^{3/2} \left(\frac{T_c}{300 \, \text{K}}\right)^{1/2} \left(\frac{M_*}{M_\odot}\right)^{-1/2}. \tag{3.112}$$

This result is essentially identical to that derived by Gammie (1996), whose analysis we have followed in this discussion.[9] For the fiducial parameters, the critical ionization fraction below which Ohmic losses will suppress the MRI (i.e. for which $Re_M \lesssim 1$) is

$$x_{\text{crit}} \sim 10^{-12}. \tag{3.113}$$

The first thing to observe is that this is a *very* small number. A tiny ionization fraction is enough to couple the magnetic field to the gas well enough to allow the MRI to operate within the disk.

---

[9] The different temperature dependence arises from our use of a temperature-dependent cross-section for electron–neutral collisions.

Although x_{crit} is very small, it is far from obvious that even this level of ionization is attained throughout the disk. In Section 2.8 we found that thermal ionization of the alkali metals would yield $x > x_{\text{crit}}$ for $T_c \gtrsim 10^3$ K, but temperatures this high are only attained in the very inner disk. At larger distances from the star, the upper and lower surfaces of the disk will be well-enough ionized due to nonthermal processes (irradiation by the stellar X-ray flux, and/or cosmic rays) but the interior of the disk – shielded by the high column density from these external ionizing agents – may well have $x < x_{\text{crit}}$. We conclude that close to the mid-plane the MRI at the radii of greatest interest for planet formation may be suppressed by the low ionization fraction. A region in the disk where MHD turbulence is suppressed due to the damping effects of non-ideal terms is described as a *dead zone*. If Ohmic diffusion were really the only effect that mattered, the dead zone would stretch from an inner radius defined by thermal ionization of the alkali metals out to a distance of the order of 10 AU.

3.6.2 Non-ideal MHD Terms

Early studies of the effect that the low ionization levels in protoplanetary disks have on magnetic field evolution focused on the role of Ohmic diffusion. Ohmic diffusion is physically simple because it acts unconditionally to damp the MRI, but it is only one manifestation – and normally not even the most important one – of how weakly ionized gases interact with magnetic fields. The full set of "non-ideal terms" comprises Ohmic diffusion, the Hall effect, and ambipolar diffusion. The relative importance of these terms varies depending on how the dynamics of charge carriers is affected by magnetic fields and by collisions with neutral species. Imagine that we have a mostly neutral fluid, with a small sprinkling of charge carriers in the form of electrons and ions (dust particles can also be an important charge carrier, but we will ignore that possibility throughout). If we apply an external electric field **E** in the neutral frame, the charge carriers will start accelerating. Their drift velocity – and hence the amount of electric current they generate – can be limited either by collisions with neutrals or by magnetic forces. Three regimes can be distinguished:

- At high densities both electron and ion drift speeds are limited by collisions with neutral particles, irrespective of the magnetic field. This is the Ohmic regime.
- At intermediate densities the electron dynamics is controlled by magnetic forces, but the ions are unaffected by the field. This is the Hall regime.
- At low densities both electrons and ions are tied to the magnetic field, and both species drift together relative to the neutral fluid. This is the ambipolar diffusion regime.

Despite the similar-sounding physics underlying Ohmic, Hall, and ambipolar MHD, the mathematical description and consequences of the three effects are quite distinct. We will begin by deriving the non-ideal equivalent of the induction equation, and then use that to estimate the physical conditions under which each term is dominant within disks.

3.6.3 Non-ideal Induction Equation

The physics of how magnetic fields affect weakly ionized fluids is easy to visualize. Magnetic fields exert Lorentz forces on the charged species, here supposed to be electrons and ions, but not on the neutrals. Collisions between the neutrals and either the ions or the electrons lead to momentum exchange whenever the neutral fluid has a velocity differential with respect to the charged fluids. The strategy for describing this situation is to first write down coupled momentum equations for the three species, which together with Maxwell's equations for the electric and magnetic fields fully describe the evolution. Then, in the spirit of ideal MHD, we identify approximations that allow us to reduce the system to a *single* momentum equation for the bulk fluid plus an evolution equation for the time dependence of the magnetic field.

We begin by considering the momentum equation. For the neutrals we have

$$\rho \frac{\partial \mathbf{v}}{\partial t} + \rho (\mathbf{v} \cdot \nabla) \mathbf{v} = -\nabla P - \rho \nabla \Phi - \mathbf{p}_{nI} - \mathbf{p}_{ne}. \tag{3.114}$$

Here $\rho, \mathbf{v}$, and P (without subscripts) refer to the neutral fluid, and $\mathbf{p}_{nI}$ and $\mathbf{p}_{ne}$ are the rate of momentum exchange due to collisions between the neutrals and the ions/electrons respectively. Identical equations apply to the charged species, but for the addition of Lorentz forces,

$$\rho_e \frac{\partial \mathbf{v}_e}{\partial t} + \rho_e (\mathbf{v}_e \cdot \nabla) \mathbf{v}_e = -\nabla P_e - \rho_e \nabla \Phi$$
$$- en_e \left(\mathbf{E} + \frac{\mathbf{v}_e \times \mathbf{B}}{c} \right) - \mathbf{p}_{en},$$
$$\rho_I \frac{\partial \mathbf{v}_I}{\partial t} + \rho_I (\mathbf{v}_I \cdot \nabla) \mathbf{v}_I = -\nabla P_I - \rho_I \nabla \Phi$$
$$+ Zen_I \left(\mathbf{E} + \frac{\mathbf{v}_I \times \mathbf{B}}{c} \right) - \mathbf{p}_{In}. \tag{3.115}$$

In these equations $\mathbf{E}$ and $\mathbf{B}$ are the electric and magnetic fields, the ions have charge Ze, where $-e$ is the charge on an electron, and of course $\mathbf{p}_{ne} = -\mathbf{p}_{en}$ and $\mathbf{p}_{nI} = -\mathbf{p}_{In}$. Having three momentum equations looks complicated, but we can make a large simplification to the system by noting that the time scale for macroscopic evolution of the fluid is generally much longer than the time scale for collisional or magnetic forces to alter a charged particle's momentum. We can then ignore *everything* in the charged species' momentum equations, except for the Lorentz and collisional terms. For the ions we have,

$$Zen_I \left(\mathbf{E} + \frac{\mathbf{v}_I \times \mathbf{B}}{c} \right) - \mathbf{p}_{In} = 0, \tag{3.116}$$

with a similar equation for the electrons. Imposing charge neutrality, $n_e = Zn_I$, we eliminate the electric field between the ion and electron equations to find an expression for the sum of the momentum transfer terms,

$$\mathbf{p}_{In} + \mathbf{p}_{en} = \frac{en_e}{c}(\mathbf{v}_I - \mathbf{v}_e) \times \mathbf{B}. \qquad (3.117)$$

The current density

$$\mathbf{J} = en_e(\mathbf{v}_I - \mathbf{v}_e), \qquad (3.118)$$

so we can write this as,

$$\mathbf{p}_{In} + \mathbf{p}_{en} = \frac{\mathbf{J} \times \mathbf{B}}{c}. \qquad (3.119)$$

Finally we go to Maxwell's equations, two of which read

$$\nabla \times \mathbf{E} = -\frac{1}{c}\frac{\partial \mathbf{B}}{\partial t}, \qquad (3.120)$$

$$\frac{4\pi}{c}\mathbf{J} = \nabla \times \mathbf{B} + \frac{1}{c}\frac{\partial \mathbf{E}}{\partial t}. \qquad (3.121)$$

The final term on the right-hand-side of the second equation is the displacement current. Ignoring this term is the key approximation that leads to both ideal and non-ideal MHD. To justify dropping the displacement current, consider some fluid motion that has a characteristic length scale L and characteristic time scale T. Dimensional analysis gives the ratio of the two terms on the right-hand-side of the equation above as

$$\frac{|(1/c)\partial \mathbf{E}/\partial t|}{|\nabla \times \mathbf{B}|} \sim \frac{|\mathbf{B}L/(c^2 T^2)|}{|\mathbf{B}/L|} \sim \frac{L^2}{c^2 T^2}. \qquad (3.122)$$

Since the characteristic velocity $v \sim L/T$ the displacement current is of the order of $\mathcal{O}(v^2/c^2)$ and consistently ignorable in nonrelativistic MHD flows.[10] Doing so, we substitute Eq. (3.119) in the neutral equation of motion to obtain

$$\rho\frac{\partial \mathbf{v}}{\partial t} + \rho(\mathbf{v} \cdot \nabla)\mathbf{v} = -\nabla P - \rho\nabla\Phi + \frac{1}{4\pi}(\nabla \times \mathbf{B}) \times \mathbf{B}. \qquad (3.123)$$

This is identical to the ideal MHD momentum equation that holds for a fully ionized plasma. The triplet of momentum equations have reduced consistently to a single equation that looks like the momentum equation for a neutral fluid, but for the addition of a magnetic force term whose form does not depend upon the makeup of the gas.

Deducing the induction equation for non-ideal MHD amounts to finding a generalized version of Ohm's law that relates the current density $\mathbf{J}$ to the electric field $\mathbf{E}$.

---

[10] MHD actually works perfectly well also in relativistic situations, but we have no cause to discuss the extra subtleties involved in demonstrating this given the extremely low velocities involved in protoplanetary disk applications.

Once this is done the time evolution of the magnetic field follows immediately from Faraday's law as expressed in the Maxwell equation $\nabla \times \mathbf{E} = -(1/c)\partial \mathbf{B}/\partial t$. To find Ohm's law, we first need to define the momentum coupling terms entering into Eq. (3.115), and then make controlled approximations that allow us to drop several of the resulting terms. The argument is somewhat delicate, but rests ultimately on the physical fact that neutral–ion and neutral–election interactions are not identical. The derivation below is based closely on that given by Balbus (2011).

The first step is to define the momentum exchange terms. The neutral–ion and neutral–electron momentum exchange rates are written as

$$\mathbf{p}_{nI} = n\mu_{nI}\nu_{nI}(\mathbf{v} - \mathbf{v}_I),$$
$$\mathbf{p}_{ne} = n\mu_{ne}\nu_{ne}(\mathbf{v} - \mathbf{v}_e), \tag{3.124}$$

with n being the number density of neutral particles. The reduced masses of the interacting species are given in terms of the particle masses,

$$\mu_{nI} = \frac{m_n m_I}{m_n + m_I}, \tag{3.125}$$

$$\mu_{ne} = \frac{m_n m_e}{m_n + m_e} \simeq m_e, \tag{3.126}$$

where the final simplification follows because the electron mass is much smaller than the mass of a neutral particle. The collision frequencies (denoted by ν) depend upon the cross-section σ for collisions and on the relative velocity w for collisions,

$$\nu_{nI} = n_I \langle \sigma_{nI} w_{nI} \rangle,$$
$$\nu_{ne} = n_e \langle \sigma_{ne} w_{ne} \rangle,$$
$$\nu_{In} = n \langle \sigma_{nI} w_{nI} \rangle,$$
$$\nu_{en} = n \langle \sigma_{ne} w_{ne} \rangle. \tag{3.127}$$

The angle brackets denote averages over the velocity distributions of the interacting species, which can be assumed to be Maxwellian. We have written all four permutations of the collision frequency to emphasize that, for example, $\nu_{ne} \neq \nu_{en}$. The former quantity is proportional to the *electron* number density while the latter scales with the *neutral* number density. The order of subscripts for most of the other quantities is irrelevant (i.e. $\sigma_{nI} = \sigma_{In}, w_{nI} = w_{In}, \mu_{nI} = \mu_{In}$ etc.), while as noted already $\mathbf{p}_{nI} = -\mathbf{p}_{In}$.

There is not yet any physics in the above definitions. For what comes later, however, the key point is that electron–neutral collisions have significant differences from ion–neutral collisions. There are two separate issues. First, in thermal equilibrium at temperature T the relative velocity of collisions scales as $w \propto T^{1/2}\mu^{-1/2}$. Electron–neutral collisions thus occur at higher w than ion–neutral collisions. Second, because electrons are point particles, the cross-section that enters into their collision term with neutrals is a geometric cross-section for the neutral species involved. Ion–neutral collisions, on the other hand, have a long-range contribution and a cross-section that exceeds the geometric one. A calculation by Draine *et al.* (1983) gives

$$\langle \sigma_{In} w_{In} \rangle = 1.9 \times 10^{-9} \text{ cm}^3 \text{ s}^{-1},$$

$$\langle \sigma_{en} w_{en} \rangle = 10^{-15} \left(\frac{128 k_B T}{9 \pi m_e} \right)^{1/2} \text{ cm}^2. \tag{3.128}$$

The distinction between the two types of collisions is evidenced by the fact that the electron–neutral expression has the expected temperature dependence, while the ion–neutral expression is independent of temperature. The quantity that will interest us is

$$\frac{\mu_{In}}{m_e} \frac{\langle \sigma_{In} w_{In} \rangle}{\langle \sigma_{en} w_{en} \rangle} \approx 4.2 \times 10^3 \left(\frac{\mu_{In}}{m_p} \right) T^{-1/2}, \tag{3.129}$$

where we have written the ion–neutral reduced mass in units of the proton mass. This quantity is securely much larger than unity for all temperatures and ion masses of interest. When it crops up, we can safely neglect the denominator in favor of the numerator.

With these preliminaries in hand we can proceed to derive a generalized Ohm's law. The starting point is the momentum equation for the electrons, which is exactly analogous to Eq. (3.116). It reads,

$$-e n_e \left(\mathbf{E} + \frac{\mathbf{v}_e \times \mathbf{B}}{c} \right) - \mathbf{p}_{en} = 0. \tag{3.130}$$

Recall that the key assumption that allowed us to write the momentum equation in this form was that the time scales for microscopic processes (magnetic and collisional forces) to change the electron momentum are short compared to bulk fluid time scales. Substituting for $\mathbf{p}_{en} = n_e m_e \nu_{en} (\mathbf{v}_e - \mathbf{v})$ yields

$$\mathbf{E} + \frac{\mathbf{v}_e \times \mathbf{B}}{c} + \frac{m_e \nu_{en}}{e} (\mathbf{v}_e - \mathbf{v}) = 0. \tag{3.131}$$

We seek to rewrite the second and third terms in the form of an "inductive" piece (proportional to $\mathbf{v} \times \mathbf{B}$, with $\mathbf{v}$ the bulk fluid velocity) plus other terms that represent current densities. To that end, we first express Eq. (3.131) in the equivalent but lengthier form,

$$\mathbf{E} + \frac{1}{c} \left[\mathbf{v} + \underbrace{(\mathbf{v}_e - \mathbf{v}_I)}_{A} + \underbrace{(\mathbf{v}_I - \mathbf{v})}_{B} \right]$$

$$\times \mathbf{B} + \frac{m_e \nu_{en}}{e} \left[\underbrace{(\mathbf{v}_e - \mathbf{v}_I)}_{A} + \underbrace{(\mathbf{v}_I - \mathbf{v})}_{C} \right] = 0. \tag{3.132}$$

Making use of Eq. (3.118), the relative velocity between the electron and ion fluids is

$$(\mathbf{v}_e - \mathbf{v}_I) = -\frac{\mathbf{J}}{en_e}. \tag{3.133}$$

The two terms labeled "A" in the above equation can therefore be immediately written in terms of the current density $\mathbf{J}$.

Term "B" and term "C" both involve $(\mathbf{v}_I - \mathbf{v})$ which is proportional to $\mathbf{p}_{In}$. To write this velocity difference as a current, we go back to Eq. (3.119),

$$\mathbf{p}_{In} + \mathbf{p}_{en} = \frac{\mathbf{J} \times \mathbf{B}}{c}, \tag{3.134}$$

and write out the terms on the left-hand-side explicitly,

$$n_I \mu_{In} n \langle \sigma_{In} w_{In} \rangle (\mathbf{v}_I - \mathbf{v}) + n_e m_e n \langle \sigma_{en} w_{en} \rangle \left[(\mathbf{v}_e - \mathbf{v}_I) + (\mathbf{v}_I - \mathbf{v}) \right]. \tag{3.135}$$

Using Eq. (3.129), together with the fact that the ratio $n_e/n_I = Z$ is not very large, we find that the second term involving $(\mathbf{v}_I - \mathbf{v})$ can be dropped. We then have

$$n_I \mu_{In} n \langle \sigma_{In} w_{In} \rangle (\mathbf{v}_I - \mathbf{v}) = \frac{\mathbf{J} \times \mathbf{B}}{c} + n_e m_e n \langle \sigma_{en} w_{en} \rangle (\mathbf{v}_I - \mathbf{v}_e), \tag{3.136}$$

from which it follows that

$$(\mathbf{v}_I - \mathbf{v}) = \frac{1}{n_I \mu_{In} n \langle \sigma_{In} w_{In} \rangle} \frac{\mathbf{J} \times \mathbf{B}}{c} + \frac{n_e}{n_I} \frac{m_e}{\mu_{In}} \frac{\langle \sigma_{en} w_{en} \rangle}{\langle \sigma_{In} w_{In} \rangle} (\mathbf{v}_I - \mathbf{v}_e). \tag{3.137}$$

We are now ready to substitute this expression into Eq. (3.132) but before doing so, we observe that the term involving $(\mathbf{v}_I - \mathbf{v}_e)$ will always be smaller than the two terms labeled "A" (by the identical argument given above). Keeping only the first term on the right-hand-side gives a version of Ohm's law that reads

$$\mathbf{E} + \frac{\mathbf{v} \times \mathbf{B}}{c} - \frac{\mathbf{J} \times \mathbf{B}}{en_e c} \left[1 - \frac{n_e m_e}{n_I \mu_{In}} \frac{\langle \sigma_{ne} w_{ne} \rangle}{\langle \sigma_{In} w_{In} \rangle} \right]$$

$$+ \frac{1}{n_I \mu_{In} n \langle \sigma_{In} w_{In} \rangle} \frac{(\mathbf{J} \times \mathbf{B}) \times \mathbf{B}}{c^2} - \frac{m_e v_{en}}{e^2 n_e} \mathbf{J} = 0. \tag{3.138}$$

For one last time we note that the second term in the square brackets is ignorable. Dropping that, we obtain the final form of the non-ideal Ohm's law. To write it in its standard form, we introduce some additional definitions. The conductivity σ (not to be confused with the collision cross-section, for which we have used the same symbol) is

$$\sigma = \frac{e^2 n_e}{m_e v_{en}}, \tag{3.139}$$

while the magnetic resistivity η is

$$\eta = \frac{c^2}{4\pi \sigma}. \tag{3.140}$$

Finally, we have to contend with the fact that the ion–neutral collision term from Eq. (3.124) is conventionally written in terms of mass rather than number densities in the astrophysical literature,

$$\mathbf{p}_{In} = \rho \rho_I \gamma (\mathbf{v}_I - \mathbf{v}). \tag{3.141}$$

The quantity γ is called the drag coefficient. Comparison of the above expression with Eq. (3.124) shows that it can be written as

$$\gamma = \frac{\langle \sigma_{nI} w_{nI} \rangle}{m_n + m_I}. \tag{3.142}$$

An advantage of writing the collision term in this form is that γ is approximately constant (Eq. 3.128).

Putting the various definitions together, the generalized form of Ohm's law is

$$\mathbf{E} + \frac{\mathbf{v} \times \mathbf{B}}{c} - \frac{\mathbf{J}}{\sigma} - \frac{\mathbf{J} \times \mathbf{B}}{e n_e c} + \frac{(\mathbf{J} \times \mathbf{B}) \times \mathbf{B}}{\gamma c^2 \rho \rho_I} = 0. \tag{3.143}$$

Using $\partial \mathbf{B}/\partial t = -c \nabla \times \mathbf{E}$ and $\mathbf{J} = (c/4\pi) \nabla \times \mathbf{B}$ we obtain the non-ideal MHD induction equation,

$$\frac{\partial \mathbf{B}}{\partial t} = \nabla \times \left[\underbrace{\mathbf{v} \times \mathbf{B}}_{\text{induction}} - \underbrace{\eta \nabla \times \mathbf{B}}_{\text{Ohmic}} - \underbrace{\frac{\mathbf{J} \times \mathbf{B}}{e n_e}}_{\text{Hall}} + \underbrace{\frac{(\mathbf{J} \times \mathbf{B}) \times \mathbf{B}}{\gamma c \rho \rho_I}}_{\text{ambipolar}} \right]. \tag{3.144}$$

The three different non-ideal MHD effects – Ohmic diffusion, ambipolar diffusion, and the Hall effect – are labeled, though with the derivation we have given the correspondence of the mathematical terms to the different regimes of fluid–magnetic field coupling is not very transparent.

3.6.4 Density and Temperature Dependence of Non-ideal Terms

The induction equation for non-ideal MHD that we have derived,

$$\frac{\partial \mathbf{B}}{\partial t} = \nabla \times \left[\mathbf{v} \times \mathbf{B} - \eta \nabla \times \mathbf{B} - \frac{\mathbf{J} \times \mathbf{B}}{e n_e} + \frac{(\mathbf{J} \times \mathbf{B}) \times \mathbf{B}}{\gamma c \rho \rho_I} \right], \tag{3.145}$$

describes the behavior of weakly-ionized magnetized fluids under conditions appropriate to protoplanetary disks. As may be surmised from the complexity of the equation, not all of that behavior is simple or intuitive. Nonetheless a few simple properties stand out. The strength of the Ohmic term scales with $\eta \propto n_e^{-1}$, as does the Hall term, while ambipolar diffusion is proportional to ρ_I^{-1}. The importance of non-ideal effects relative to the inductive term therefore scales with the ionization degree in the same way for each term. The magnetic field dependence, on the other hand, is quite different. Noting that $\mathbf{J} \propto \nabla \times \mathbf{B}$ we see that the Ohmic, Hall, and ambipolar terms involve successively higher powers of $B = |\mathbf{B}|$. Moreover, all of the terms in the equation *apart* from the Hall term are indifferent to the *sign* of the magnetic field, in the sense that making the substitution $\mathbf{B} \to -\mathbf{B}$ leaves the equation as a whole unchanged. This is not true in circumstances where the Hall term matters, and indeed qualitatively different behavior occurs in Hall-dominated MHD depending on the sign of large-scale magnetic fields.

The common scaling of the three non-ideal effects with ionization degree means that it is possible to estimate their *relative* importance without needing to know just how weakly ionized the fluid is. To do so we replace the vectors in Eq. (3.145) with scalars, and replace $\nabla \times \mathbf{B}$ with the dimensionally equivalent B/L, where L is some characteristic length scale. For example, the ratio of the estimated magnitudes of the Ohmic and Hall terms, which we will denote by O/H, is,

$$\frac{O}{H} \sim \frac{\eta B/L}{B^2/(Len_e)} \sim \eta e n_e B^{-1}. \tag{3.146}$$

On substituting for η from Eqs. (3.127), (3.128), and (3.140) we find that O/H $\propto$ $nT^{1/2}B^{-1}$, and hence in general we need to specify the density, temperature, and magnetic field strength in order to estimate the relative importance of these two terms. A convenient parameterization is to express the magnetic field strength in dimensionless form via the ratio of the Alfvén speed $v_A = \sqrt{B^2/4\pi\rho}$ to the sound speed c_s. Doing so we find

$$\frac{O}{H} = \left(\frac{n}{8 \times 10^{17}\ \text{cm}^{-3}}\right)^{1/2} \left(\frac{v_A}{c_s}\right)^{-1}, \tag{3.147}$$

$$\frac{A}{H} = \left(\frac{n}{9 \times 10^{11}\ \text{cm}^{-3}}\right)^{-1/2} \left(\frac{T}{100\ \text{K}}\right)^{1/2} \left(\frac{v_A}{c_s}\right), \tag{3.148}$$

$$\frac{A}{O} = \left(\frac{n}{3 \times 10^{15}\ \text{cm}^{-3}}\right)^{-1} \left(\frac{T}{10^3\ \text{K}}\right)^{1/2} \left(\frac{v_A}{c_s}\right)^2. \tag{3.149}$$

These expressions allow us to calculate planes in the three-dimensional space of density, temperature, and field strength, defined for example by O/H = 1, on which any two of the non-ideal terms have nominally equal strength. Figure 3.7 shows a two-dimensional slice through this space, obtained by fixing the ratio $(v_A/c_s) = 0.1$. As expected based on both the physical origin and mathematical representation of the non-ideal terms, Ohmic diffusion dominates at high densities and ambipolar diffusion at low densities. The Hall effect is expected to be most important across several orders of magnitudes of intermediate density conditions. Increased temperatures favor a greater role for ambipolar diffusion over the Hall term.

3.6.5 Application to Protoplanetary Disks

Given a disk model and some assumed or calculated magnetic field structure we can use the preceding analysis to predict where in protoplanetary disks each of

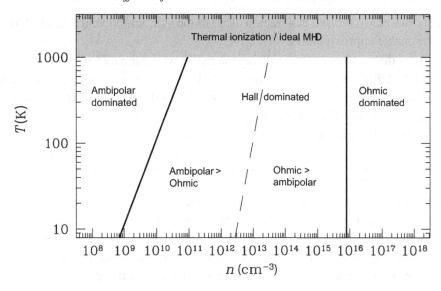

Figure 3.7 Approximate delineation of where the different non-ideal terms dominate in the plane of number density n and temperature T, assuming that the ratio of the Alfvén to sound speed $(v_A/c_s) = 0.1$. Above about 10^3 K the ionization fraction in protoplanetary disks is usually high enough that ideal MHD is a reasonable approximation.

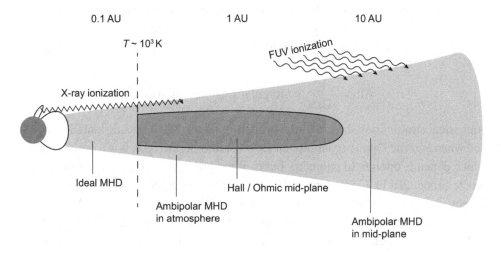

Figure 3.8 Illustration of where different non-ideal MHD effects are expected to be important in protoplanetary disks.

the different non-ideal terms ought to matter for the dynamics. This is an inexact science. Besides any uncertainties in the disk model, the calculation of the O/H and A/H ratios is crude and ignores the fact that the non-ideal terms have differing mathematical structures. Generically, however, density is the most important variable and we will end up with a picture something like Fig. 3.8. Three main regions can be identified:

- A hot thermally ionized disk region close to the star, within which ideal MHD is a reasonable approximation.
- An inner disk region, extending from the radius where the mid-plane temperature is around 10^3 K out to at least several AU, where the mid-plane dynamics is controlled by the Hall (and possibly Ohmic) term. In the lower density atmosphere ambipolar diffusion is important.
- An outer disk region where the mid-plane density is low enough that ambipolar diffusion dominates even in the mid-plane. Far ultraviolet ionization of a thin surface layer may modify the atmosphere dynamics.

Determining the importance of these different regions for the structure and evolution of protoplanetary disks, and for planet formation more generally, requires study of the linear properties of Eq. (3.145) together with numerical simulations of the nonlinear evolution of the modified MRI. In summarizing the results of such investigations it is useful to rewrite the induction equation in a different form:

$$\frac{\partial \mathbf{B}}{\partial t} = \nabla \times \left[\mathbf{v} \times \mathbf{B} - \eta_O \nabla \times \mathbf{B} - \eta_H \frac{\mathbf{J} \times \mathbf{B}}{B} + \eta_A \frac{(\mathbf{J} \times \mathbf{B}) \times \mathbf{B}}{B^2} \right]. \qquad (3.150)$$

Written in this form η_O, η_H and η_A all have the dimensions of a diffusivity, although only the Ohmic term truly behaves as a simple diffusion process. From the diffusivities we can construct two dimensionless quantities known as *Elsasser numbers*:

$$\Lambda_O = \frac{v_A^2}{\Omega \eta_O} \text{ (Ohmic Elsasser number)}, \qquad (3.151)$$

$$\text{Am} = \frac{v_A^2}{\Omega \eta_A} \text{ (ambipolar Elsasser number)}. \qquad (3.152)$$

Numerical simulations show that the properties of the MRI are substantially modified whenever the Ohmic or ambipolar Elsasser numbers drop below a critical value, which depends upon field geometry but is of the order of 1–10 (Turner *et al.*, 2007; Bai & Stone, 2011). Consistent with the original dead zone concept, we therefore expect changes to MHD turbulence both in the mid-plane around 1 AU (where Ohmic diffusion is the dominant dissipative process) and in the disk atmosphere and at large radii of the order of 100 AU (where ambipolar diffusion dominates). In particular, turbulence in the outer regions of the disk due to MHD processes is likely to be damped by the effects of ambipolar diffusion, particularly if there is little or no net magnetic field threading the disk (Simon *et al.*, 2013).

A Hall Elsasser number can be defined in an analogous fashion, but this parameterization is less useful because the Hall effect – unlike either the Ohmic or ambipolar terms – is nondissipative and poorly characterized as a diffusive process. The Hall term more closely resembles the inductive term $\nabla \times (\mathbf{v} \times \mathbf{B})$, and its strength can alternatively be characterized by the ratio of the Hall to the inductive term. In the linear regime the Hall term gives rise to a *Hall shear instability* (Kunz, 2008) which differs qualitatively from the usual MRI. In the nonlinear regime, simulations show that the combination of the Hall effect and a weak net vertical magnetic field generates a magnetic stress whose strength depends upon whether the magnetic field is aligned or anti-aligned with the axis defined by the angular momentum of the disk material (Lesur *et al.*, 2014).

3.7 Disk Winds

Thus far we have assumed that the evolution of the disk is driven by redistribution of angular momentum within the protoplanetary disk. A qualitatively different possibility is that evolution is driven instead by angular momentum *loss* mediated by open magnetic field lines which thread the disk, as shown schematically in Fig. 3.9. If the magnetic field at the disk surface has vertical and azimuthal components B_z^s and B_ϕ^s, respectively, then the torque per unit area exerted on the disk (counting both the upper and lower surfaces) is

$$T_m = \frac{B_z^s B_\phi^s}{2\pi} r. \tag{3.153}$$

One might worry about how to define the "surface" of the disk, but in practice the sharp drop in density with increasing height (Eq. 2.18) means that there is a reasonably well-defined transition between the (normally) gas pressure dominated disk interior and the magnetically dominated outer layers. If the magnetic field is weak enough that the angular velocity remains Keplerian (i.e. magnetic pressure gradients are negligible) then the angular momentum per unit area is $\Sigma\sqrt{GM_*r}$,

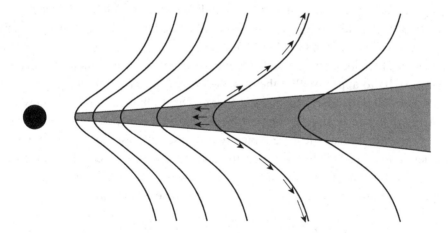

Figure 3.9 Showing the geometry assumed in disk wind models for protoplanetary disk evolution. Open magnetic field lines connect the disk to an almost force-free disk corona. A small fraction of the accreting matter is accelerated outward along the field lines to form a magnetically launched disk wind, while the magnetic torque exerted by the field at the disk surface robs the disk of angular momentum and drives accretion.

and the loss of that angular momentum via the magnetic torque results in a radial velocity

$$|v_{r,\mathrm{m}}| = \frac{B_z^{\mathrm{s}} B_\phi^{\mathrm{s}}}{\pi \Sigma \Omega}.$$
(3.154)

This inflow velocity can be compared to that which results from internal redistribution of angular momentum due to viscosity,

$$v_{r,\mathrm{visc}} = -\frac{3}{2}\frac{\nu}{r}.$$
(3.155)

Making use of Eqs. (3.46) and (3.97), and noting that for MRI-generated transport the Maxwell stress is dominant over the Reynolds stress,

$$\frac{|v_{r,\mathrm{m}}|}{|v_{r,\mathrm{visc}}|} \sim \frac{B_z^{\mathrm{s}} B_\phi^{\mathrm{s}}}{B_r B_\phi} \left(\frac{h}{r}\right)^{-1},$$
(3.156)

where the magnetic fields in the denominator are the turbulent fields evaluated near the mid-plane and we have ignored numerical factors of the order of unity. One obtains the result that if organized large-scale fields thread the disk and have comparable strengths to the turbulent fields, a wind can carry away angular momentum more efficiently than internal processes can redistribute it, especially if the disk is thin.

3.7.1 Condition for Magnetic Wind Launching

A simple model of a magnetohydrodynamic disk wind that can be analyzed quantitatively is the Blandford–Payne wind (Blandford & Payne, 1982). The geometry of the wind is illustrated in Fig. 3.10. We envisage a Keplerian disk that is threaded by a large-scale poloidal magnetic field, and assume that there is enough ionization for ideal MHD to apply. Within the disk the energy density in the magnetic field, $B^2/8\pi$, is smaller than ρc_{s}^2, the thermal energy. Due to flux conservation, however, the energy in the vertical field component, $B_z^2/8\pi$, is roughly constant with height for $z < r$, while the gas pressure typically decreases at least exponentially with a scale height $h \ll r$. This leads to a region above the disk surface where magnetic forces dominate. The magnetic force per unit volume can be written as the sum of a magnetic pressure gradient and a force due to magnetic tension,

$$\frac{\mathbf{J} \times \mathbf{B}}{c} = -\nabla \left(\frac{B^2}{8\pi}\right) + \frac{\mathbf{B} \cdot \nabla \mathbf{B}}{4\pi},$$
(3.157)

where the current

$$\mathbf{J} = \frac{c}{4\pi} \nabla \times \mathbf{B}.$$
(3.158)

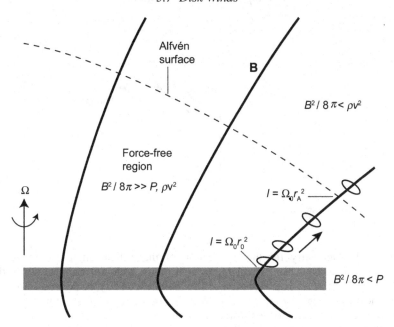

Figure 3.10 Illustration of the different physical regions of a Blandford–Payne disk wind (Spruit, 1996). We determine the condition for magnetic launching of a wind by noting that the magnetic field in the force-free region behaves as a rigid wire. Gas tied to the field lines behaves mechanically as if it is a bead strung to the wire, and can escape the disk if the centrifugal force exceeds that of gravity.

In the disk wind region where magnetic forces dominate, the requirement that they exert a finite acceleration on the low density gas can only be satisfied if the force approximately vanishes, i.e. that

$$\mathbf{J} \times \mathbf{B} \approx 0. \tag{3.159}$$

The structure of the magnetic field in the magnetically dominated region is then described as being "force-free," and in the disk wind case (where B changes slowly with z) the field lines must be approximately straight to ensure that the magnetic tension term is small. If the field lines support a wind, the force-free structure persists up to where the kinetic energy density in the wind, ρv^2, first exceeds the magnetic energy density. This criterion defines the *Alfvén surface*. Beyond the Alfvén surface, the inertia of the gas in the wind is sufficient to bend the field lines, which wrap up into a spiral structure as the disk below them rotates.

Magneto-centrifugal driving can launch a wind from the surface of a cold gas disk if the magnetic field lines are sufficiently inclined to the disk normal. The critical inclination angle in ideal MHD can be derived via a mechanical analogy (which is in fact exact). To proceed, we note that in the force-free region the magnetic field lines are (i) basically straight lines, and (ii) enforce rigid rotation out to the Alfvén surface at an angular velocity equal to that of the disk at the field

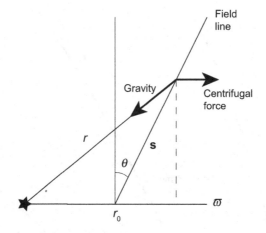

Figure 3.11 Geometry for the calculation of the critical angle for magneto-centrifugal wind launching. A magnetic field line **s**, inclined at angle θ from the disk normal, enforces rigid rotation at the angular velocity of the foot point, at cylindrical radius $\varpi = r_0$ in the disk. Working in the rotating frame we consider the balance between centrifugal force and gravity.

line's footpoint. The geometry is shown in Fig. 3.11. We consider a field line that intersects the disk at radius r_0, where the angular velocity is $\Omega_0 = \sqrt{GM_*/r_0^3}$, and that makes an angle θ to the disk normal. We define the spherical polar radius r, the cylindrical polar radius ϖ, and measure the distance along the field line from its intersection with the disk at $z = 0$ as s. In the frame co-rotating with Ω_0 there are no magnetic forces along the field line to affect the acceleration of a wind; the sole role of the magnetic field is to constrain the gas to move along a straight line at constant angular velocity. Following this line of argument, the acceleration of a wind can be fully described in terms of an effective potential,

$$\Phi_{\text{eff}}(s) = -\frac{GM_*}{r(s)} - \frac{1}{2}\Omega_0^2\varpi^2(s), \qquad (3.160)$$

that is the sum of the gravitational potential and the centrifugal potential in the rotating frame.

Written out explicitly the effective potential is

$$\Phi_{\text{eff}}(s) = -\frac{GM_*}{(s^2 + 2sr_0\sin\theta + r_0^2)^{1/2}} - \frac{1}{2}\Omega_0^2(r_0 + s\sin\theta)^2. \qquad (3.161)$$

This function is plotted in Fig. 3.12 for various values of the angle θ. If we consider first a vertical field line ($\theta = 0$) the effective potential is a monotonically increasing function of distance s. For modest values of θ there is a potential barrier defined by a maximum at some $s = s_{\text{max}}$, while for large enough θ the potential decreases monotonically from $s = 0$. In this last case purely magneto-centrifugal forces suffice to accelerate a wind off the disk surface, even in the absence of any thermal

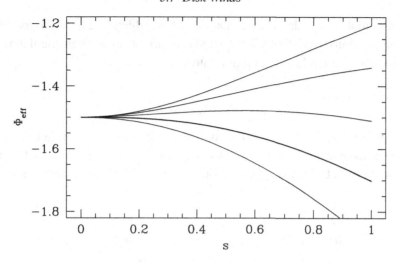

Figure 3.12 The variation of the disk wind effective potential Φ_{eff} (in arbitrary units) with distance s along a field line. From top downwards, the curves show field lines inclined at $0°, 10°, 20°, 30°$ (in bold) and $40°$ from the normal to the disk surface. For angles of $30°$ and more from the vertical, there is no potential barrier to launching a cold MHD wind directly from the disk surface.

effects. We compute the critical inclination angle θ_{crit}, defined as the minimum angle that allows magneto-centrifugal wind driving, via the condition

$$\left.\frac{\partial^2 \Phi_{eff}}{\partial s^2}\right|_{s=0} = 0. \tag{3.162}$$

Evaluating this condition, we find,

$$1 - 4\sin^2 \theta_{crit} = 0$$
$$\Rightarrow \theta_{crit} = 30°, \tag{3.163}$$

as the minimum inclination angle from the vertical needed for unimpeded wind launching in ideal MHD.

The rigid rotation of the field lines interior to the Alfvén surface means that gas being accelerated along them increases its specific angular momentum. The magnetic field, in turn, applies a torque to the disk that removes a corresponding amount of angular momentum. If a field line, anchored to the disk at radius r_0, crosses the Alfvén surface at (cylindrical) radius r_A, it follows that the angular momentum flux is

$$\dot{L}_w = \dot{M}_w \Omega_0 r_A^2, \tag{3.164}$$

where $\dot{M}_w$ is the mass loss rate in the wind. Removing angular momentum at this rate from the disk results in a local accretion rate $\dot{M} = 2\dot{L}_w/\Omega_0 r_0^2$. The ratio of the disk accretion rate to the wind loss rate is

$$\frac{\dot{M}}{\dot{M}_w} = 2\left(\frac{r_A}{r_0}\right)^2. \tag{3.165}$$

If r_A substantially exceeds r_0 (by a factor of a few, which is reasonable for detailed disk wind solutions) a relatively weak wind can carry away enough angular momentum to support a much larger accretion rate.

3.7.2 Net Flux Evolution

It is straightforward to generalize the surface density evolution Eq. (3.6) to include the effect of angular momentum loss in a disk wind. If the angular velocity remains Keplerian the effect of the wind on the disk can be modeled as an additional advective term,

$$\frac{\partial \Sigma}{\partial t} = \frac{3}{r} \frac{\partial}{\partial r} \left[r^{1/2} \frac{\partial}{\partial r} \left(\nu \Sigma r^{1/2} \right) \right] - \frac{1}{r} \frac{\partial}{\partial r} \left(r \Sigma v_{r,m} \right). \tag{3.166}$$

The potential presence of MHD disk winds has two important implications. First, winds can make a qualitative change to the predicted evolution of the outer disk radius with time. A disk in which accretion is driven solely by angular momentum redistribution expands with time, while one driven entirely by angular momentum loss in a wind shrinks. Second, including the possibility of winds breaks any linkage between disk evolution and disk turbulence. If the disk accretes due to angular momentum redistribution, then knowing the time scale for disk evolution gives an idea – admittedly a rough one – on the level of turbulence, which is important for many aspects of planet formation. In contrast, a wind-driven disk could still be turbulent, but it could in principle be almost entirely laminar.

A complete model for the evolution of the protoplanetary disk under the action of MHD disk winds requires understanding how much mass is lost from the disk at each radius, and how the poloidal magnetic fields that support the wind themselves evolve. Neither of these are easy problems. High energy radiation can heat the surface layers of the disk to a point where the gas is unbound, and can escape. In isolation this process, described as photoevaporation (Section 3.8.1), is efficient enough to explain the dispersal of low-mass disks at the end of the Class II phase. If poloidal magnetic fields are present, the rate of mass loss will then be determined by the interplay of thermal and magneto-centrifugal effects. The field evolution is also hard to calculate robustly. Since the flux threading the disk,

$$\Phi = \oint B_z dS, \tag{3.167}$$

is a conserved quantity, development of net vertical field cannot occur as a result of local dynamo processes, but rather requires that magnetic flux of one sign either be dragged into or preferentially expelled from the disk. In modern day star formation, disks form from molecular cloud cores that contain magnetic fields, and it is reasonable to think that young disks, at least, are threaded by some net flux. Over time this flux can be advected inward by the accretion flow, on a time scale given by the viscous time scale $\tau_\nu = r^2/\nu$ (Eq. 3.12). Flux can also escape outward if there is a large enough effective magnetic diffusivity. The simplest quantitative description of the competition between inward advection and outward diffusion was developed by Lubow *et al.* (1994). For $B_z^s \sim B_r^s$ they found that the outward diffusion time scale is given by

$$t_\eta \simeq \frac{r^2}{\eta}\left(\frac{h}{r}\right), \qquad (3.168)$$

where η is the magnetic diffusivity within the disk. If the diffusivity and the viscosity are both generated from the same turbulent processes within the disk, we expect that to order of magnitude $\eta \sim \nu$, and hence $t_\eta \sim (h/r)t_\nu$. More detailed calculations by Guilet and Ogilvie (2014) suggest that the difference between t_η and t_ν is smaller, and that the magnetic field in the inner disk can find an equilibrium value of the plasma β that measures the ratio of the mid-plane thermal pressure to the magnetic pressure. The actual value of β that results is, however, hard enough to compute that it remains unclear (on first principles grounds) whether net magnetic fields have a dominant influence on disk evolution.

3.8 Disk Dispersal

Depletion of the gaseous protoplanetary disk due to stellar accretion is predicted to be a gradual process. If we consider the self-similar solution (Eq. 3.25), for example, we find that for a time-independent viscosity scaling with radius as $\nu \propto r^\gamma$ the late-time behavior of the surface density is

$$\Sigma \propto t^{-(5/2-\gamma)/(2-\gamma)}, \qquad (3.169)$$

which is a power-law in time ($\Sigma \propto t^{-3/2}$ for the case of $\nu \propto r$). A disk that evolved due only to accretion would steadily become optically thin over an increasing range of radii, in the process transitioning rather slowly from a Class II to a Class III source. Observationally this is *not* what is observed. Although a number of candidate transition disks are known – typically YSOs that lack a near-IR excess despite the presence of a robust mid-IR excess – the scarcity of such systems suggests that the dispersal phase of protoplanetary disks lasts of the order of 10^5 yr (Simon & Prato, 1995; Wolk & Walter, 1996). The relative brevity of the dispersal time scale suggests that additional physical processes beyond viscous evolution contribute to the loss of gas from the disk. Plausibly the additional evolutionary agent is *photoevaporation*, a process in which UV or X-ray radiation heats the disk surface to the point at which it becomes hot enough to escape the gravitational potential as a thermally driven wind. Photoevaporative flows from young stars with disks are observed in the Orion nebula cluster (O'Dell *et al.*, 1993; Johnstone *et al.*, 1998), where an intense ionizing radiation field is provided by massive stars. Weaker UV irradiation originating from low mass stars themselves would suffice to disperse disks more generally on time scales consistent with those inferred observationally.

3.8.1 Photoevaporation

The basic physics of photoevaporation is simple: UV or X-ray radiation heats the disk surface to a temperature high enough that the thermal energy of the gas exceeds its gravitational binding energy. A pressure gradient drives this unbound gas away from the star, dispersing the disk. The details of photoevaporation depend upon the source of the irradiating flux (which can be the disk-bearing star itself or external stars in a cluster) and the energy of the photons. Extreme ultraviolet (EUV) radiation ($E > 13.6\,\text{eV}, \lambda < 912\,\text{Å}$) ionizes hydrogen atoms, producing a sharply defined layer of hot gas whose temperature, around 10^4 K, is almost independent of the density of the disk at the radius under consideration. Far ultraviolet radiation ($6\,\text{eV} < E < 13.6\,\text{eV}$) dissociates H_2 molecules, creating a neutral atomic layer whose temperature depends upon the precise balance of heating and cooling processes. Typical values are 100–5000 K. X-rays likewise produce a surface region of heated gas with a complex vertical structure.

Complete models of photoevaporative flows need at a minimum to include the source (either the central star or external stars) of high-energy radiation, and its complete spectrum. We can gain some insight by considering a simple model of photoevaporation due to EUV irradiation by the central star, which is easier to study (both analytically and numerically) than the general case.

Let us consider a disk exposed to EUV flux from the central star, as shown in Fig. 3.13. The photoionized gas has a temperature $T \simeq 10^4$ K, which corresponds to a sound speed $c_s \simeq 10\,\text{km s}^{-1}$. We define the gravitational radius r_g to be the location where the sound speed in the ionized gas equals the orbital velocity,

$$r_g = \frac{GM_*}{c_s^2} = 8.9 \left(\frac{M_*}{M_\odot}\right) \left(\frac{c_s}{10\,\text{km s}^{-1}}\right)^{-2} \text{AU}. \qquad (3.170)$$

In the simplest analysis, gas at $r < r_g$ is bound, and forms an extended atmosphere above the neutral disk surface with a scale-height $h \propto r^{3/2}$. Beyond r_g the ionized gas is unbound, and escapes the disk at a rate

$$\dot{\Sigma}_\text{wind} \sim 2\rho_0(r)c_s, \qquad (3.171)$$

where ρ_0, the density at the base of the ionized layer, is determined by the intensity of the irradiating EUV flux. Radiative transfer calculations by Hollenbach *et al.* (1994) suggest that, while the inner disk is present, the flux at $r > r_g$ is dominated by photons generated from recombinations in the atmosphere at smaller radii, and that the base density scales with radius roughly as $\rho_0(r) \propto r^{-5/2}$. Most of the mass loss is then concentrated at radii near r_g.

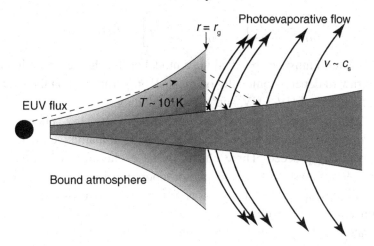

Figure 3.13 The geometry assumed in simple models of disk photoevaporation by a central stellar EUV source (Hollenbach *et al.*, 1994). The EUV flux incident on the disk ionizes the surface layers, heating them to a temperature of around 10^4 K. Inside a critical radius r_g (set very approximately by the condition that the sound speed in the ionized gas be smaller than the escape velocity) the ionized layer forms a bound atmosphere that absorbs the stellar flux and reradiates recombination radiation. At larger radii the gas escapes as a thermally driven disk wind.

More sophisticated analyses and numerical simulations result in a refinement of the simple picture of photoevaporation given above (Begelman *et al.*, 1983; Font *et al.*, 2004). These calculations suggest that mass loss starts at a radius of about 0.2 r_g (about 2 AU for a solar mass star), and that the gas escapes the disk at a modest fraction of the sound speed. The total mass loss rate from a disk exposed to an ionizing EUV flux Φ is estimated as

$$\dot{M}_{\text{wind}} \approx 1.6 \times 10^{-10} \left(\frac{\Phi}{10^{41} \text{ s}^{-1}} \right)^{1/2} \left(\frac{M_*}{1 \, M_\odot} \right)^{1/2} M_\odot \text{ yr}^{-1}. \tag{3.172}$$

This is an interesting value. For an ionizing photon flux $\Phi \sim 10^{41}$ s^{-1} the predicted mass loss rate from the disk is essentially negligible at early times (when the stellar accretion rate is three orders of magnitude greater), but comes to dominate after a few viscous times of disk evolution.

3.8.2 Viscous Evolution with Photoevaporation

Photoevaporation can be incorporated into simple time-dependent models of non-magnetized protoplanetary disk evolution. The flow away from the disk carries the same specific angular momentum as the disk at the launch point, and hence the effect of photoevaporation can be captured as a simple mass sink in the disk evolution equation:

$$\frac{\partial \Sigma}{\partial t} = \frac{3}{r}\frac{\partial}{\partial r}\left[r^{1/2}\frac{\partial}{\partial r}\left(\nu \Sigma r^{1/2}\right)\right] + \dot{\Sigma}_{\text{wind}}(r,t). \qquad (3.173)$$

The mass loss term must be specified via a model of the photoevaporation process, which in turn will depend nonlocally on the disk properties and on the strength and time evolution of the FUV, EUV, and X-ray irradiation.

Models that combine viscous disk evolution with EUV photoevaporation (Clarke *et al.*, 2001; Alexander *et al.*, 2006) provide a qualitative picture for how disk dispersal may proceed. Three phases in the evolution can be distinguished. In the first phase, mass loss due to photoevaporation is negligible compared to the mass flux flowing through the disk as a result of viscous transport of angular momentum, and the disk evolves as if there were no mass loss. Eventually, the mass accretion rate drops to become comparable to the wind mass loss rate, and mass flowing in toward the star from large disk radii is diverted into the wind rather than reaching the inner disk. An annular gap develops in the disk near r_g, and the inner disk, now cut off from resupply, drains rapidly on to the star on its own viscous time scale of the order of 10^5 yr. Figures 3.14 and 3.15 show the evolution of the mass accretion rate and disk surface density up to this point. Finally, the disk interior to r_g becomes optically thin to the stellar EUV flux, which then directly illuminates the inner edge of the outer disk. This results in an increased rate of mass loss due

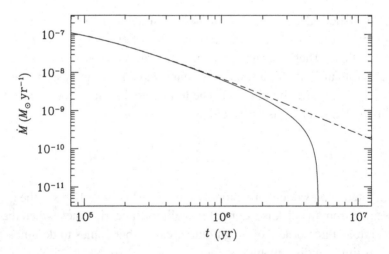

Figure 3.14 The numerically computed stellar accretion rate for protoplanetary disk models with (solid line) and without (dashed line) photoevaporative mass loss. The model, which is based on that in Clarke *et al.* (2001), assumes a viscosity $\nu \propto r$, and photoevaporative mass loss that scales as $\dot{\Sigma}_{\text{wind}} \propto r^{-5/2}$ outside $r_g = 5$ AU. No mass loss is assumed to occur within r_g. The mass loss rate integrated from r_g to 25 AU is 10^{-9} $M_\odot$ yr^{-1}. The initial surface density profile matches the self-similar solution with an initial accretion rate of 3×10^{-7} $M_\odot$ yr^{-1} and a cut-off at 10 AU. The viscosity is normalized such that the viscous time scale at 10 AU is 3×10^5 yr. The presence of a wind from the outer disk results in a sharply defined epoch of disk dispersal.

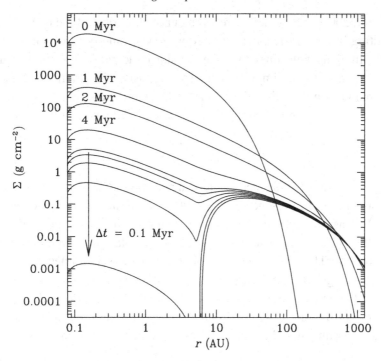

Figure 3.15 The evolution of the surface density within a protoplanetary disk model that includes photoevaporative mass loss. As in Fig. 3.14, the model assumes a viscosity $\nu \propto r$, and photoevaporative mass loss that scales as $\dot{\Sigma}_{\text{wind}} \propto r^{-5/2}$ outside $r_g = 5\,\text{AU}$. No mass loss is assumed to occur within r_g. The disk evolves slowly on Myr time scales until the stellar accretion rate drops to $\dot{M} \sim \dot{M}_{\text{wind}}$, at which point the inner disk becomes cut off from the outer disk and drains rapidly (on a 10^5 yr time scale) on to the star.

to photoevaporation, which disperses the entire outer disk on a time scale of a few 10^5 yr.

In principle, protoplanetary disks could be dispersed by purely thermal winds driven by an appropriate mix of UV and X-ray radiation (Alexander *et al.*, 2014; Ercolano & Pascucci, 2017). However, it is also possible that dispersal occurs through the combined action of photoevaporative (thermal) and magnetic winds. This will be the case if disks retain a significant net magnetic flux, which can support a magnetic wind, at late times.

3.9 Magnetospheric Accretion

If the star is unmagnetized, the protoplanetary disk will extend inward until it meets the stellar surface. This is the boundary layer geometry of accretion discussed in Section 3.2.2 and used to justify the zero-torque boundary condition typically assumed for protoplanetary disks. Observations, however, show that T Tauri stars are not unmagnetized, with kG magnetic fields being quite common

(Johns-Krull, 2007). These fields are strong enough to disrupt the disk close to the star, resulting in a *magnetospheric* accretion flow geometry in which gas at small radii flows in along field lines to strike the star in accretion hot spots away from the stellar equator.[11] The simplest geometry to consider is one in which the stellar magnetic field is dipolar, aligned with the stellar rotation axis, and perpendicular to the disk plane. The unperturbed field has a vertical component at the disk surface,

$$B_z = B_* \left(\frac{r}{R_*} \right)^{-3}, \qquad (3.174)$$

which falls off rapidly with increasing distance (note that $B_* R_*^3 = \mu$, the stellar dipole moment). In the presence of a disk, the vertical field will thread the disk gas and – as long as the coupling between the magnetic field and the matter is good enough – the field will be distorted by differential rotation between the Keplerian disk and the star. The differential rotation will twist the field lines that couple the disk to the star, generating an azimuthal field component at the disk surface B_ϕ. Computing this field accurately is not an easy matter (it may depend, for example, on the nature of the turbulence within the disk), but to an order of magnitude it is reasonable to assume that the rapid differential rotation results in a field

$$B_\phi \sim B_z, \qquad (3.175)$$

which is limited by the onset of instabilities that afflict very strongly twisted fields (Lynden-Bell & Boily, 1994). If we define the *co-rotation radius* r_{co} as the radius where the field lines – rotating with stellar rotation period P – have the same angular velocity as that of Keplerian gas in the disk,

$$r_{co} = \left(\frac{GM_* P^2}{4\pi^2} \right)^{1/3}, \qquad (3.176)$$

we can identify two regions of star–disk magnetic interaction:

- Interior to co-rotation ($r < r_{co}$) the disk gas has a greater angular velocity than the field lines. Field lines that link the disk and the star in this region are dragged forward by the disk, and exert a braking torque that tends to remove angular momentum from the disk gas.
- Outside co-rotation ($r > r_{co}$) the disk gas has a smaller angular velocity than the field lines. The field lines are dragged backward by the disk, and the coupling results in a positive torque on the disk gas.

Young stars are typically rapid rotators, with periods of a week or less, so the co-rotation radius that separates these two regimes lies in the inner disk. For $P = 7$ days, for example, the co-rotation radius around a solar mass star is at $r_{co} \simeq 15\,R_\odot$ or 0.07 AU.

---

[11] This accretion geometry was initially studied for accretion on to magnetized compact objects (Pringle & Rees, 1972; Ghosh & Lamb, 1979) and subsequently applied to accretion on to T Tauri stars (König, 1991).

The radius within which the external torque due to the star–disk magnetic interaction dominates over the internal viscous torque is estimated by comparing the time scales on which each mechanism removes the angular momentum of the disk gas. The stellar magnetosphere threading the disk will remove angular momentum (interior to co-rotation) on a time scale that is the angular momentum content per unit area divided by the magnetic torque (given by Eq. 3.153),

$$t_m \simeq 2\pi \frac{\Sigma \sqrt{GM_* r}}{B_z^s B_\phi^s r},$$ (3.177)

with the magnetic fields being evaluated at the disk surface. Equating this to the viscous time scale ($t_\nu \simeq r^2/\nu$) defines the magnetospheric radius

$$r_m \simeq \left(\frac{3 B_*^2 R_*^6}{2\dot{M} \sqrt{GM_*}} \right)^{2/7}.$$ (3.178)

In deriving this expression we have made use of the steady-state disk relation $\nu\Sigma = \dot{M}/(3\pi)$ and assumed the undistorted dipolar magnetic field geometry discussed above. The scaling of the magnetospheric radius with the magnetic field strength and mass accretion rate is robust – in fact the same scalings apply to the Alfvén radius derived for *spherical* accretion – but equally valid physical arguments result in different numerical factors. Adopting the expression above, the magnetospheric radius for fiducial T Tauri parameters ($B_* = 1\,\mathrm{kG}, R_* = 2\,R_\odot$, $\dot{M} = 10^{-8}\,M_\odot\,\mathrm{yr}^{-1}, M_* = M_\odot$) is

$$r_m \simeq 16\,R_\odot.$$ (3.179)

One concludes that kG-strength stellar magnetic fields are strong enough to dominate the dynamics of protoplanetary disks in the innermost regions, and that a stellar magnetosphere that extends out to roughly the co-rotation radius is plausible.

The presence of magnetospheric rather than boundary layer accretion means that protoplanetary disks around magnetic stars are predicted to have inner edges that lie at $r \simeq r_m$ rather than at the stellar equator. At smaller radii there is a low density magnetospheric cavity within which gas is channeled on to the star along field lines at roughly the free-fall velocity. This geometry has a number of observational implications:

- The innermost region of the disk that would otherwise produce strong near-IR emission is missing. This alters the IR colors of classical T Tauri stars.
- The final accretion on to the stellar surface occurs at roughly the free-fall velocity. This explains the P Cygni profiles (indicative of infall) that are seen in some spectral lines (Hartmann *et al.*, 1994).
- The zero-torque inner boundary condition that is justified for boundary layer accretion need no longer apply. Instead, the star's spin angular momentum can be coupled to the disk angular momentum via the torque exerted by the magnetic field. This torque can modify the stellar rotation rate.

Although important in their own right most of these effects apply at radii that are small enough ($r \lesssim 0.1$ AU) to be relatively unimportant for planet formation. The change to the inner boundary condition formally alters the steady-state disk surface density profile everywhere, but the fractional change becomes small well away from the inner boundary at $r \simeq r_m$. The presence of a low density magnetospheric cavity is of greater interest for models of planet migration (discussed in Chapter 7), which commonly invoke angular momentum loss to the gaseous protoplanetary disk as the driver of orbital decay. Such migration is likely to slow and eventually halt once the planet enters the magnetospheric cavity.

3.10 Further Reading

Accretion Power in Astrophysics, J. Frank, A. King, & D. Raine (2002), Cambridge: Cambridge University Press.

"Accretion Discs in Astrophysics," J. E. Pringle (1981), *Annual Review of Astronomy & Astrophysics*, **19**, 137.

"Instability, Turbulence, and Enhanced Transport in Accretion Disks," S. A. Balbus & J. F. Hawley (1998), *Reviews of Modern Physics*, **70**, 1.

"Dynamics of Protoplanetary Disks," P. J. Armitage (2011), *Annual Review of Astronomy & Astrophysics*, **49**, 195.

"Transport and Accretion in Planet-Forming Disks," N. J. Turner, *et al.* (2014), in *Protostars and Planets VI*, H. Beuther, R. S. Klessen, C. P. Dullemond, and T. Henning (eds.), Tucson: University of Arizona Press, p. 411.

4

Planetesimal Formation

The formation of terrestrial planets from micrometer-sized dust particles requires growth through at least 12 orders of magnitude in size scale. It is conceptually useful to divide the process into three main stages that involve different dominant physical processes:

- **Planetesimal formation**. Planetesimals are primordial bodies with sizes that, depending upon how they formed, may have ranged between 100 m and a few hundred km in size. Reaching this size is doubly consequential. For an individual body it marks a transition between bodies whose strength is determined entirely by their material properties, and larger ones that resist disruption because of self-gravity. For a population of bodies, the planetesimal scale matters because at a similar scale orbital evolution is dominated by mutual gravitational interactions rather than aerodynamic coupling to the gas disk. Material properties (loosely how "sticky" small particles are upon collision), aerodynamic and gravitational physics all play a key role in solid evolution up to the formation of planetesimals.
- **Terrestrial planet formation**. Once a population of planetesimals has formed within the disk their dynamical evolution is dominated by gravitational interactions. Growth can occur from a combination of planetesimal–planetesimal (and later protoplanet–planetesimal) collisions, and from the aerodynamically assisted accretion of smaller solid particles that were not previously incorporated into planetesimals. The latter process is called "pebble accretion."
- **Giant planet formation and core migration**. Once planets have grown to about an Earth mass, coupling to the gas disk becomes significant once again, though now it is *gravitational* rather than aerodynamic forces that matter. For $M_p \sim M_\oplus$ this coupling can result in eccentricity damping and exchange of orbital angular momentum with the gas (a process described as *migration*), while for $M_p \sim 10\ M_\oplus$ the interaction becomes strong enough that the planet can start to capture an envelope from the protoplanetary disk.

The boundaries between these regimes are somewhat arbitrary and inconsistently defined, but it is useful to keep this ordering of the most important physics in mind as we discuss planet formation.

4.1 Aerodynamic Drag on Solid Particles

Consider a spherical particle of solid material of radius s and material density ρ_{m}. The first step to understanding how such a particle evolves within the protoplanetary disk is to calculate the aerodynamic force experienced by the particle when it moves at a velocity v *relative to the local velocity* of the gas disk. In calculating the force there are two physical regimes to consider. If $s \lesssim \lambda$, the mean-free path of gas molecules within the disk, then the fluid on the scale of the particle is effectively a collisionless ensemble of molecules with a Maxwellian velocity distribution. The drag force in this regime – which is normally the most relevant for small particles within protoplanetary disks – is called Epstein drag. In the alternate Stokes drag regime, which applies for $s \gtrsim \lambda$, the disk gas flows as a fluid around the obstruction presented by the particle. In either regime the force scales with the frontal area πs^2 that the particle presents to the gas. This means that the acceleration caused by gas drag – which is proportional to the drag force divided by the particle mass – *decreases* with particle size (as s^{-1} for spherical particles) and eventually becomes negligible once bodies of planetesimal size have formed.

4.1.1 Epstein Drag

Epstein drag is felt by solid particles that are smaller than the mean-free path of gas molecules within the disk. The form of the drag law in this regime can be derived by considering the frequency of collisions between the particle and gas molecules, given by elementary arguments as the product of the collision cross-section, relative velocity, and molecule number density. We model the solid particle as a sphere of radius s moving with velocity v relative to the disk gas. Within the gas the mean thermal speed of the molecules is

$$v_{\mathrm{th}} = \sqrt{\frac{8 k_{\mathrm{B}} T}{\pi \mu m_{\mathrm{H}}}}. \tag{4.1}$$

The gas temperature and density are T and ρ, and the mean molecular weight is μ. Up to factors of the order of unity, the frequency with which gas molecules collide with the "front" side of the particle is

$$f_+ \approx \pi s^2 (v_{\mathrm{th}} + v) \frac{\rho}{\mu m_{\mathrm{H}}}, \tag{4.2}$$

while the collision frequency on the back side is

$$f_- \approx \pi s^2 (v_{\mathrm{th}} - v) \frac{\rho}{\mu m_{\mathrm{H}}}. \tag{4.3}$$

Noting that the momentum transfer per collision is approximately given by $2\mu m_H v_{th}$, we find that the net drag force in the Epstein regime scales as

$$F_D \propto -s^2 \rho v_{th} v. \tag{4.4}$$

The force is linear in the relative velocity and proportional to the surface area of the particle and to the thermal speed of molecules in the disk gas. A more accurate derivation, valid for $s < \lambda$, $v \ll v_{th}$, and a Maxwellian distribution of molecular speeds, yields

$$\mathbf{F}_D = -\frac{4\pi}{3}\rho s^2 v_{th}\mathbf{v}. \tag{4.5}$$

The drag force, of course, acts in the opposite direction to the vector describing the relative velocity between the particle and the gas. Extensions to this formula to describe the case where the particle moves supersonically with respect to the gas ($v \gtrsim v_{th}$) can be found in Kwok (1975), though it is normally the subsonic regime that is relevant for protoplanetary disks.

4.1.2 Stokes Drag

Once particles grow to a size much larger than the molecular mean-free path the interaction with the gas can be treated in classical fluid terms, without reference to the molecular nature of the gas. Drag in this regime is called Stokes drag. Naively, we might guess that the drag force would scale with the ram pressure experienced by the particle, and hence we write the force as

$$\mathbf{F}_D = -\frac{C_D}{2}\pi s^2 \rho v \mathbf{v}, \tag{4.6}$$

where C_D, the *drag coefficient*, describes how aerodynamic the particle is. In general, C_D will depend upon the shape of the particle, but for spherical particles it depends only upon the fluid Reynolds number (Eq. 3.44), which we define here on the scale of the particle as

$$\text{Re} = \frac{2sv}{v_m}. \tag{4.7}$$

Note that the viscosity here is the (small) molecular viscosity of the gas, rather than any turbulent viscosity within the disk. In terms of the Reynolds number, Weidenschilling (1997b) quotes a piecewise expression for scaling of the drag coefficient:

$$C_D \simeq 24\text{Re}^{-1}, \qquad \text{Re} < 1, \tag{4.8}$$

$$C_D \simeq 24\text{Re}^{-0.6}, \qquad 1 < \text{Re} < 800, \tag{4.9}$$

$$C_D \simeq 0.44, \qquad \text{Re} > 800. \tag{4.10}$$

Comparison of the expressions for Epstein and Stokes drag shows that they are equal for a particle of size $s = 9\lambda/4$, and this can be taken as the transition size when constructing a smooth drag law that encompasses both regimes.

4.2 Dust Settling

Aerodynamic drag on particles is important for understanding both the vertical distribution and radial motion of dust and larger bodies within the protoplanetary disk. More subtle issues concern the interaction between aerodynamic forces and turbulence, which we have already noted is likely to be a ubiquitous feature (albeit with poorly determined properties) of the disk. To begin with, we ignore turbulence and consider the vertical settling and growth of dust particles suspended in a laminar disk. We quantify the coupling between the solid and gas components of the disk by defining the *friction time scale* for a particle of mass m as

$$t_{\text{fric}} = \frac{mv}{|F_{\text{D}}|}, \tag{4.11}$$

where, as before, v is the relative velocity between the particle and the gas. For many purposes, a dimensionless version of the friction time that is obtained by multiplying by the local Keplerian angular velocity is more physically meaningful:

$$\tau_{\text{fric}} \equiv t_{\text{fric}}\Omega_{\text{K}}. \tag{4.12}$$

The dimensionless friction time is also called the dimensionless stopping time and the Stokes number.

The friction time scale measures the time in which drag modifies the relative velocity significantly. Writing the particle mass $m = (4/3)\pi s^3 \rho_{\text{m}}$ in terms of the material density ρ_{m}, t_{fric} takes on a simple form in the Epstein drag regime:

$$t_{\text{fric}} = \frac{\rho_{\text{m}}}{\rho} \frac{s}{v_{\text{th}}}. \tag{4.13}$$

Adopting conditions appropriate for a particle at the mid-plane of the protoplanetary disk at $r = 1$ AU ($\rho = 10^{-9}\,\text{g cm}^{-3}$, $\rho_{\text{m}} = 3\,\text{g cm}^{-3}$, $v_{\text{th}} = 10^5\,\text{cm s}^{-1}$) we obtain an estimate for the friction time scale for a particle of size $s = 1\,\mu\text{m}$:

$$t_{\text{fric}} \approx 3\,\text{s}. \tag{4.14}$$

Small dust particles are thus very tightly coupled to the gas.

We now consider the forces acting on a small dust particle at height z above the mid-plane of a laminar disk. Concentrating for now just on the vertical forces, the z component of the stellar gravity yields a downward force

$$|F_{\text{grav}}| = m\Omega^2 z, \tag{4.15}$$

where $\Omega = \sqrt{GM_*/r^3}$ is the local Keplerian angular velocity (cf. Section 2.3). The *gas* in the disk is supported against this force by an upwardly directed pressure gradient, but no such force acts on a solid particle. If started at rest a particle will therefore accelerate downward until the gravitational force is balanced by aerodynamic drag. In the Epstein regime we have

$$|F_D| = \frac{4\pi}{3} \rho s^2 v_{th} v. \tag{4.16}$$

In practice – given the very short friction time – force balance is attained almost instantaneously and the particle drifts toward the disk mid-plane with a terminal velocity given by equating $|F_D|$ and $|F_{grav}|$:

$$v_{settle} = \frac{\rho_m}{\rho} \frac{s}{v_{th}} \Omega^2 z. \tag{4.17}$$

Inserting numerical values roughly appropriate for a 1 μm particle at $z \sim h$ at 1 AU from a solar mass star ($\rho = 6 \times 10^{-10}$ g cm^{-3}, $z = 3 \times 10^{11}$ cm, $v_{th} = 10^5$ cm s^{-1}) one finds that $v_{settle} \approx 0.06$ cm s^{-1} and that the settling time, defined as

$$t_{settle} = \frac{z}{|v_{settle}|}, \tag{4.18}$$

is about 1.5×10^5 yr. In the absence of turbulence we would therefore expect that micrometer-sized particles ought to sediment out of the upper layers of the disk on a time scale that is short compared to the disk lifetime.

Returning to Eq. (4.17), we note that the terminal velocity of a dust particle is inversely proportional to the gas density. Settling will therefore be faster at high z where the gas is tenuous. Using the Gaussian density profile appropriate for a vertically isothermal disk (Eq. 2.18), and noting that the mean thermal speed differs from the sound speed that determines the vertical scale-height h only by a numerical factor, we obtain a general expression for the settling time as a function of z:

$$t_{settle} = \frac{2}{\pi} \frac{\Sigma}{\rho_m s \Omega} \exp\left[-\frac{z^2}{2h^2}\right]. \tag{4.19}$$

The strong z dependence implied by this formula means that in the absence of turbulence dust particles would be expected to settle out of the uppermost disk layers rather quickly.

4.2.1 Single Particle Settling with Coagulation

Even if the neglect of turbulence were justified, the estimate of the dust settling time given above would be incomplete because it ignores the likelihood that dust particles will collide with one another and grow during the settling process. The settling velocity increases with particle size, so any such coagulation hastens the collapse of the dust toward the disk mid-plane.

To estimate how fast particles could grow during sedimentation we appeal to a simple single particle growth model due to Safronov (1969) (see also Dullemond & Dominik, 2005). Imagine that a single "large" particle of radius s and mass $m = (4/3)\pi s^3 \rho_m$ is settling toward the disk mid-plane at velocity v_{settle} through a background of much smaller solid particles. By virtue of their small size the settling of the small particles can be neglected. If every collision leads to coagulation the large particle grows in mass at a rate that reflects the amount of solid material in the volume swept out by its geometric cross-section,

$$\frac{dm}{dt} = \pi s^2 |v_{settle}| f\rho(z),$$ (4.20)

where f is the dust to gas ratio in the disk. Substituting for the settling velocity one finds

$$\frac{dm}{dt} = \frac{3}{4} \frac{\Omega^2 f}{v_{th}} zm.$$ (4.21)

Since $z = z(t)$ this equation cannot generally be integrated immediately,[1] but rather must be solved in concert with the equation for the height of the particle above the mid-plane,

$$\frac{dz}{dt} = -\frac{\rho_m}{\rho} \frac{s}{v_{th}} \Omega^2 z.$$ (4.22)

Solutions to these coupled equations provide a very simple model for particle growth and sedimentation in a nonturbulent disk.

Figure 4.1 shows numerical solutions to Eqs. (4.21) and (4.22) for initial particle sizes of $0.01\,\mu m$, $0.1\,\mu m$, and $1\,\mu m$. The particles settle from an initial height $z_0 = 5h$ through a disk whose parameters are chosen to be roughly appropriate to a (laminar) Solar Nebula model at 1 AU from the Sun. Both particle growth and vertical settling are found to be extremely rapid. With the inclusion of coagulation, particles settle to the disk mid-plane on a time scale of the order of 10^3 yr – more than two orders of magnitude faster than the equivalent time scale in the absence of particle growth. By the time that the particles reach the mid-plane they have grown to a final size of a few mm, irrespective of their initial radius.

The single particle model described above is very simple, both in its neglect of turbulence and because it assumes that the only reason that particle–particle collisions occur is because the particles have different vertical settling velocities. Other drivers of collisions include Brownian motion, turbulence, and differential *radial* velocities. The basic result, however, is confirmed by more sophisticated models (e.g. Dullemond & Dominik, 2005), which show that, if collisions lead to particle adhesion, growth from sub-μm scales up to small macroscopic scales (of the order of a mm) occurs rapidly. This means that there is no time scale

---

[1] Note, however, that if the particle grows rapidly (i.e. more rapidly than it sediments) then the form of the equation implies exponential growth of m with time.

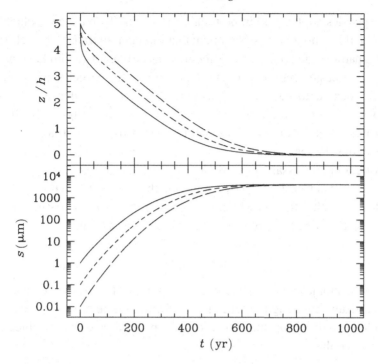

Figure 4.1 The settling and growth of a single particle in a laminar (nonturbulent) protoplanetary disk. The model assumes that a single particle (with initial size $s = 1\,\mu m$ (solid line), $0.1\,\mu m$ (dashed line), or $0.01\,\mu m$ (long-dashed line)), accretes all smaller particles it encounters as it settles toward the disk mid-plane. The smaller particles are assumed to be at rest. The upper panel shows the height above the mid-plane as a function of time, the lower panel the particle radius s. For this example the disk parameters adopted are: orbital radius $r = 1\,\mathrm{AU}$, scale-height $h = 3 \times 10^{11}\,\mathrm{cm}$, surface density $\Sigma = 10^3\,\mathrm{g\,cm^{-2}}$, dust to gas ratio $f = 10^{-2}$, and mean thermal speed $v_{th} = 10^5\,\mathrm{cm\,s^{-1}}$. The dust particle is taken to have a material density $\rho_m = 3\,\mathrm{g\,cm^{-3}}$ and to start settling from a height $z_0 = 5h$.

problem associated with the very earliest phases of particle growth. Indeed, what is more problematic is to understand how the population of small grains – which are unquestionably present given the IR excesses characteristic of classical T Tauri stars – survive to late times. The solution to this quandary involves the inclusion of particle *fragmentation* in sufficiently energetic collisions, which allows a broad distribution of particle sizes to survive out to late times. Fragmentation is not likely given collisions at relative velocities of the order of a cm s^{-1} – values typical of settling for μm-sized particles – but becomes more probable for collisions at velocities of a m s^{-1} or higher.

4.2.2 Settling in the Presence of Turbulence

It is a matter of everyday experience that dust, which settles out of the air readily in an unused room, can remain suspended in the presence of vigorous air currents.

The same physics applies within protoplanetary disks. Turbulence – which may be present even if it is not responsible for angular momentum transport – acts to stir up small solid particles and this prevents them from settling into a thin layer at the disk mid-plane. The easiest regime to treat is that in which the particles are small enough to be well coupled to the gas (mathematically we require that a dimensionless version of the friction time $\Omega t_{\text{fric}} \ll 1$) and represent a negligible fraction of the total disk mass. These conditions are met for dust particles in protoplanetary disks with typical dust to gas ratios.

The conditions necessary for turbulence to stir up the dust enough to oppose vertical settling can be estimated by comparing the settling time (Eq. 4.19) with the time scale on which diffusion will erase spatial gradients in the particle concentration. To diffuse vertically across a scale z requires a time scale

$$t_{\text{diffuse}} = \frac{z^2}{D}, \tag{4.23}$$

where D is an anomalous (i.e. turbulent) diffusion coefficient whose magnitude we will discuss later. Equating the settling and diffusion time scales at $z = h$ we find that turbulence will inhibit the formation of a particle layer with a thickness less than h provided that

$$D \gtrsim \frac{\pi e^{1/2}}{2} \frac{\rho_{\text{m}} s h^2 \Omega}{\Sigma}. \tag{4.24}$$

This result is not terribly transparent. We can cast it into a more interesting form if we assume that the turbulence stirring up the particles is the same turbulence responsible for angular momentum transport within the disk. In that case it is plausible that the anomalous diffusion coefficient has the same magnitude and scaling as the anomalous viscosity,[2] which motivates us to write

$$D \sim \nu = \frac{\alpha c_{\text{s}}^2}{\Omega}. \tag{4.25}$$

With this form for D, the minimum value of α required for turbulence to oppose settling becomes

$$\alpha \gtrsim \frac{\pi e^{1/2}}{2} \frac{\rho_{\text{m}} s}{\Sigma}, \tag{4.26}$$

which is roughly the ratio between the column density through a single solid particle and that of the whole gas disk. For small particles this critical value of α is extremely small. If we take $\Sigma = 10^2 \, \text{g cm}^{-2}$, $\rho_{\text{m}} = 3 \, \text{g cm}^{-3}$, and $s = 1 \, \mu\text{m}$, for example, we obtain $\alpha \gtrsim 10^{-5}$. This implies that small particles of dust will remain

---

[2] A great deal of interesting complexity is being swept under the carpet here. Although D and ν have the same dimensions ($\text{cm}^2 \, \text{s}^{-1}$), and in the broadest sense arise "from the same turbulent processes," there is no detailed reason why the *vertical* diffusion of a trace scalar contaminant should be equivalent to the *radial* diffusion of angular momentum. For example, in magnetized disks angular momentum transport is dominated by Maxwell rather than fluid stresses. Nonetheless, numerical simulations suggest that taking $D = \nu$ is typically reasonable to within a numerical factor of a few.

suspended throughout much of the vertical extent of the disk in the presence of turbulence with any plausible strength. For larger particles the result is different. If we consider particles of radius 1 mm – a size that we argued above might form very rapidly – we find that the critical value of $\alpha \sim 10^{-2}$. This value is comparable to some large-scale estimates of α for protoplanetary disks. Particles of this size and above will therefore not have the same vertical distribution as the gas in the disk.

To proceed more formally, we can consider the solid particles as a separate fluid that is subject to the competing influence of settling and turbulent diffusion. If the "dust" fluid with density[3] ρ_d can be treated as a trace species within the disk (i.e. that $\rho_d/\rho \ll 1$) then it evolves according to an advection–diffusion equation of the form (Dubrulle *et al.*, 1995; Fromang & Papaloizou, 2006)

$$\frac{\partial \rho_d}{\partial t} = D \frac{\partial}{\partial z} \left[\rho \frac{\partial}{\partial z} \left(\frac{\rho_d}{\rho} \right) \right] + \frac{\partial}{\partial z} \left(\Omega^2 t_{\text{fric}} \rho_d z \right). \tag{4.27}$$

Simple steady-state solutions to this equation can be found in the case where the dust layer is thin enough that the *gas* density varies little across the dust scale-height. In that limit, the dimensionless friction time Ωt_{fric} is independent of z and we obtain

$$\frac{\rho_d}{\rho} = \left(\frac{\rho_d}{\rho} \right)_{z=0} \exp \left[-\frac{z^2}{2h_d^2} \right], \tag{4.28}$$

where h_d, the scale-height describing the vertical distribution of the particle concentration ρ_d/ρ, is

$$h_d = \sqrt{\frac{D}{\Omega^2 t_{\text{fric}}}}. \tag{4.29}$$

If, as previously, we assume that $D \sim \nu$, we can write a compact expression for the ratio of the concentration scale-height to the usual gas scale-height:

$$\frac{h_d}{h} \simeq \sqrt{\frac{\alpha}{\Omega t_{\text{fric}}}}. \tag{4.30}$$

The condition for solid particles to become strongly concentrated toward the disk mid-plane is then that the dimensionless friction time is substantially greater than α. For any reasonable value of α this implies that substantial particle growth is required before settling takes place.

4.3 Radial Drift of Solid Particles

The fact that solid particles do not experience the same pressure forces as the gas has even more important consequences for the *radial* dynamics of solids

---

[3] The dust density ρ_d is the mass of solid particles per unit volume within the disk. It should not be confused with either the gas density ρ or the material density ρ_m which expresses the density of the matter that makes up the particles.

(Weidenschilling, 1997b). As we showed previously in Section 2.4, the gas in the disk is normally partially supported against gravity by an outward pressure gradient, and as a result orbits the star at a slightly sub-Keplerian velocity. If locally the mid-plane pressure can be written as a power-law in radius, $P \propto r^{-n}$, the actual gas orbital velocity can be written in terms of the Keplerian velocity $v_K = \sqrt{GM_*/r}$ as

$$v_{\phi,\text{gas}} = v_K (1 - \eta)^{1/2}, \tag{4.31}$$

where $\eta = nc_s^2/v_K^2$ (Eq. 2.30). Let us now consider the implications of this sub-Keplerian rotation for solid particles of different sizes embedded within the gas. For a small dust particle, aerodynamic coupling to the gas is very strong ($\Omega t_{\text{fric}} \ll 1$). To a good approximation the dust will be swept along with the gas, and its azimuthal velocity will equal that of the disk gas. Since this is sub-Keplerian, the centrifugal force will be insufficient to balance gravity, and the particle will spiral inward at its radial terminal velocity. Inward radial drift also occurs for large rocks that are poorly coupled to the gas ($\Omega t_{\text{fric}} \gg 1$). In this case the aerodynamic forces can be regarded as perturbations to the orbital motion of the body, which orbits the star with an azimuthal velocity that is close to the Keplerian speed. This is faster than the motion of the disk gas, and as a result the rock experiences a headwind that tends to remove angular momentum from the orbit. The loss of angular momentum again results in inward drift.

With this physical understanding in mind we proceed to calculate the rate of radial drift as a function of the friction time of particles located at the disk mid-plane (the analogous calculation for $z \neq 0$ can be found in Takeuchi & Lin, 2002). If the radial and azimuthal velocities of the particle are v_r and v_ϕ respectively, the equations of motion including the aerodynamic drag forces can be written as

$$\frac{dv_r}{dt} = \frac{v_\phi^2}{r} - \Omega_K^2 r - \frac{1}{t_{\text{fric}}} \left(v_r - v_{r,\text{gas}} \right), \tag{4.32}$$

$$\frac{d}{dt} \left(r v_\phi \right) = -\frac{r}{t_{\text{fric}}} \left(v_\phi - v_{\phi,\text{gas}} \right). \tag{4.33}$$

We simplify the azimuthal equation by noting that the specific angular momentum always remains close to Keplerian (i.e. the particle spirals in through a succession of almost circular, almost Keplerian orbits):

$$\frac{d}{dt} \left(r v_\phi \right) \simeq v_r \frac{d}{dr} \left(r v_K \right) = \frac{1}{2} v_r v_K. \tag{4.34}$$

This yields

$$v_\phi - v_{\phi,\text{gas}} \simeq -\frac{1}{2} \frac{t_{\text{fric}} v_r v_K}{r}. \tag{4.35}$$

Turning now to the radial equation, we substitute for Ω_K using Eq. (4.31). Retaining only the lowest order terms,

$$\frac{dv_r}{dt} = -\eta \frac{v_K^2}{r} + \frac{2v_K}{r}\left(v_\phi - v_{\phi,\text{gas}}\right) - \frac{1}{t_{\text{fric}}}\left(v_r - v_{r,\text{gas}}\right). \qquad (4.36)$$

The dv_r/dt term is negligible. Dropping that term, and eliminating $(v_\phi - v_{\phi,\text{gas}})$ between Eqs. (4.35) and (4.36), we obtain

$$v_r = \frac{(r/v_K)t_{\text{fric}}^{-1}v_{r,\text{gas}} - \eta v_K}{(v_K/r)t_{\text{fric}} + (r/v_K)t_{\text{fric}}^{-1}}. \qquad (4.37)$$

This result can be cast into a more intuitive form using the dimensionless stopping time

$$\tau_{\text{fric}} = t_{\text{fric}}\Omega_K, \qquad (4.38)$$

in terms of which the particle radial velocity is

$$v_r = \frac{\tau_{\text{fric}}^{-1}v_{r,\text{gas}} - \eta v_K}{\tau_{\text{fric}} + \tau_{\text{fric}}^{-1}}. \qquad (4.39)$$

Let us note some special cases of this general result. For small particles that are tightly coupled to the gas ($\tau_{\text{fric}} \ll 1$), radial drift occurs at a speed

$$v_r \simeq v_{r,\text{gas}} - \eta \tau_{\text{fric}} v_K. \qquad (4.40)$$

Such particles are dragged in with the gas, on top of which they experience a radial drift relative to the gas at a rate that is linear in the dimensionless stopping time. For very large particles, conversely, the radial drift velocity decreases linearly with the stopping time.

Figure 4.2 shows the radial drift velocity as a function of the dimensionless stopping time for parameters $(\eta, v_{r,\text{gas}}/v_K)$ that are approximately appropriate for the protoplanetary disk at a radius of 5 AU. The drift velocity peaks when $\tau_{\text{fric}} \simeq 1$ at a value

$$v_{r,\text{peak}} \simeq -\frac{1}{2}\eta v_K, \qquad (4.41)$$

which depends only upon the pressure gradient in the disk via the dependence of η on the sound speed and surface density gradients. The particle size that corresponds to $\tau_{\text{fric}} \simeq 1$ can be computed based on the formulae for the friction time in the appropriate drag regime (either Epstein or Stokes). In the Epstein regime, for example, a dimensionless stopping time of unity occurs for a particle size

$$s(\tau = 1) = \frac{\rho v_{\text{th}}}{\rho_m \Omega_K}. \qquad (4.42)$$

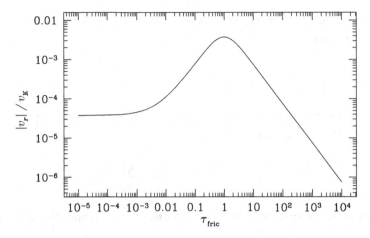

Figure 4.2 The radial drift velocity of particles at the mid-plane of the protoplanetary disk is plotted as a function of the dimensionless stopping time
$\tau_{\mathrm{fric}} = t_{\mathrm{fric}}\Omega_{\mathrm{K}}$. The model plotted assumes that $\eta = 7.5 \times 10^{-3}$ and that
$v_{r,\mathrm{gas}}/v_{\mathrm{K}} = -3.75 \times 10^{-5}$. These values are approximately appropriate for a
disk with $h/r = 0.05$ and $\alpha = 10^{-2}$ at 5 AU.

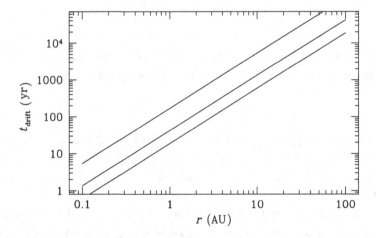

Figure 4.3 The *minimum* time scale for the radial drift of solid particles as a
function of radius, for disk models in which $\Sigma \propto r^{-1}$ and $h/r = 0.025$
(uppermost line), $h/r = 0.05$, or $h/r = 0.075$ (bottom line).

At 5 AU in a disk with $\Sigma = 10^2 \, \mathrm{g \, cm}^{-2}$ and $h/r = 0.05$ the fastest drift occurs for
a particle size $s \simeq 20 \, \mathrm{cm}$.[4] This is typical – in the inner disk (1–10 AU) the most
rapid radial drift coincides with particle sizes in the 10 cm to a few m range.

Figure 4.3 plots the minimum radial drift time scale (i.e. the time scale evaluated
at the peak of the curve in Fig. 4.2) $t_{\mathrm{drift}} = r/|v_{r,\mathrm{peak}}|$ as a function of radius in

---

[4] With these parameters the mean-free path is of the order of a meter, so it is consistent to use the Epstein
formula.

the disk. Throughout the main planet-forming regions of the disk this time scale is extremely short – of the order of 10^3 yr or less for reasonable disk parameters. This is an important result, from which flow two robust conclusions:

- **Planetesimal formation must be rapid**, or very inefficient. This conclusion follows from the fact that if growth through the cm–m-size scale were not very rapid the vast majority of the solid material in the disk would drift toward the star to be evaporated in the hot inner regions of the protoplanetary disk.
- **Radial redistribution of solids is very likely to occur**. Radial flow of solid particles on a time scale shorter than the disk lifetime occurs not just at the peak of the radial velocity drift curve, but also for substantially smaller and larger particles. Local enhancements or depletions of solids (relative to the gas surface density) will occur as a result.

These inferences follow from rather simple and well-understood physics, namely the action of aerodynamic drag on particles orbiting within a sub-Keplerian gas disk.

4.3.1 Radial Drift with Coagulation

The size dependence of the radial drift velocity introduces a relative velocity between particles of different sizes, which can promote collisions and (possibly) growth via coagulation. We can use the same arguments that we employed to study coagulation during vertical settling to assess whether this potentially beneficial aspect of radial drift outweighs the deleterious effects of radial drift in depleting the solid surface density. As previously, we note that, in the limit where all collisions are adhesive, the growth rate of a particle of radius s that collides primarily with smaller particles[5] as it drifts inward at the disk mid-plane is approximately

$$\frac{dm}{dt} = \pi s^2 |v_r| f \rho_0, \tag{4.43}$$

where f is the ratio of particle to gas density at $z = 0$, and ρ_0, the mid-plane gas density, is given by $\rho_0 = (1/\sqrt{2\pi})\Sigma/h$. Comparing the growth time scale $t_{\text{grow}} = m/(dm/dt)$ to the drift time scale $t_{\text{drift}} = r/|v_r|$ we find that $t_{\text{grow}} < t_{\text{drift}}$ for particles of size

$$s \lesssim \frac{3f}{4\sqrt{2\pi}} \left(\frac{h}{r}\right)^{-1} \frac{\Sigma}{\rho_{\text{m}}}. \tag{4.44}$$

If we assume that a modest amount of vertical settling has already taken place, appropriate values for the parameters in the inner disk might be $f = 0.1$, $\Sigma = 10^3 \, \text{g cm}^{-2}$, $\rho_{\text{m}} = 3 \, \text{g cm}^{-3}$, and $h/r = 0.05$. With these values we find

---

[5] We assume that the particles in question have settled to $z \approx 0$, and that their size is such that they lie on the left-hand-side of the peak in the radial drift velocity curve plotted in Fig. 4.2.

that particles with a size up to $s \simeq 2$ m would collide with at least their own mass of other particles during their inward drift. This implies that the high peak value of the radial drift speed is not, in and of itself, an insurmountable barrier to ongoing growth, provided that vertical settling has taken place. What *is* a serious problem is the fact that the resultant high relative collision velocities between large rocks will very likely invalidate the assumption that collisions will lead to growth. Indeed, at the peak of the radial velocity drift curve it is possible that collisions will actually break up particles.

4.3.2 *Particle Concentration at Pressure Maxima*

Since both the mid-plane density and the mid-plane sound speed typically decrease with radius, it is normally the case that the disk pressure has a maximum at or close to the star. This results in sub-Keplerian gas velocities across most of the disk, and inward radial drift. The more general rule, however, is that particles drift in the direction of the pressure gradient. Outward drift is therefore possible if the disk possesses a local pressure maximum. This possibility may be of interest if, for example, turbulence in the disk is strong enough to create strong zonal flows and local pressure maxima. In that situation, which is illustrated in Fig. 4.4, particles would be expected to flow toward the maximum from both smaller and larger radii. The time scale for this process is even faster than the global drift time scale. If the pressure maximum has some radial scale Δr, the *local* pressure gradient $\sim P/\Delta r$ exceeds the global gradient $\sim P/r$. The time scale for solids to pile up at the pressure maximum is then shorter than the global drift time scale by a factor of

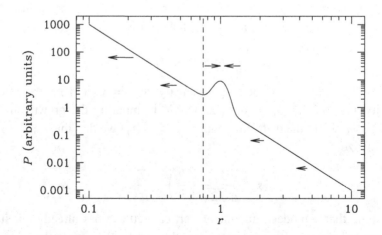

Figure 4.4 How a nonmonotonic pressure profile in the disk can result in particle pile-up. If the pressure at the disk mid-plane has a local maximum, particles drift toward the location of the maximum from both sides, resulting in a rapid concentration of solids at that point. The potential importance of this effect for early-stage planet formation was recognized in an early paper by Whipple (1972).

$(r/\Delta r)^2$. Another relevant circumstance in which inward drift can be halted occurs in the presence of a massive planet, which as we will show later tends to create an annular gap in the disk surface density near the location of its orbit. A disk with such a gap possesses a pressure maximum at the inner edge of the disk that is exterior to the planet. Solids with some range of sizes would be expected to accumulate at that location.

4.3.3 Particle Pile-up

Solids that experience radial drift can become concentrated ("pile up") in the inner disk (Youdin & Shu, 2002; Youdin & Chiang, 2004). The effect is present in the simplest case where diffusion and inward drag by the mean flow are small, and feedback of the particles on the gas can be neglected. The radial drift speed is then given by Eq. (4.40) as

$$v_r \simeq -\eta \tau_{\text{fric}} v_{\text{K}}, \tag{4.45}$$

with $\eta \propto (h/r)^2$. For particles that are in the Epstein regime of drag, the stopping time in the mid-plane is (from Eqs. 4.13 and 2.19)

$$t_{\text{fric}} = \frac{\rho_{\text{m}}}{\rho_0} \frac{s}{v_{\text{th}}} = \sqrt{2\pi} \frac{\rho_{\text{m}} h}{\Sigma} \frac{s}{v_{\text{th}}}. \tag{4.46}$$

Since $h = c_{\text{s}}/\Omega$, and c_{s} and v_{th} differ only by a numerical factor, we obtain

$$\tau_{\text{fric}} = \frac{\pi}{2} \frac{\rho_{\text{m}}}{\Sigma} s. \tag{4.47}$$

Suppose now that the surface density profile of solids has attained a steady state, such that the mass flux $\dot{M}_{\text{d}}$ is independent of radius. Then,

$$\dot{M}_{\text{d}} = -2\pi r \Sigma_{\text{d}} v_r = \text{constant}, \tag{4.48}$$

and substituting for v_r we find

$$\frac{\Sigma_{\text{d}}}{\Sigma} \propto \left(\frac{h}{r}\right)^{-2} r^{-1/2}. \tag{4.49}$$

For a disk with constant (h/r) the steady-state concentration of solids increases closer to the star as $r^{-1/2}$. In the case of a flaring disk with mid-plane temperature profile, say, $T_{\text{c}} \propto r^{-1/2}$, the effect is stronger. A constant α model of such a disk has a steady-state gas surface density profile $\Sigma \propto r^{-1}$, with the solids following $\Sigma_{\text{d}} \propto r^{-2}$.

4.3.4 Turbulent Radial Diffusion

The presence of turbulence within the disk modifies the radial transport of solid particles, though the differences between the laminar and turbulent cases are less dramatic for radial drift than for vertical settling. Whereas turbulence essentially precludes settling for all but the largest particles, it does not alter the *mean* sub-Keplerian flow that is responsible for radial drift. Substantial rocks will still drift inwards rapidly, unless, as discussed above, the turbulence is so strong as to create local pressure maxima. Where turbulence matters most is for small particles that are well coupled to the gas. Such particles can diffuse in a turbulent flow as well as being advected with the mean gas motion.

In general there are three processes to consider when modeling the radial transport of solids within a turbulent disk: advection with the mean flow, radial drift relative to the gas due to aerodynamic drag, and turbulent diffusion. Let us consider the limit where advection and diffusion are the dominant processes. This limit is valid for trace gas species and (approximately) for small particles that are sufficiently well coupled to the gas that advection dominates over aerodynamic drag in Eq. (4.39). Writing the surface density of the gas or dust (generically the "contaminant") as Σ_d, we define the concentration of the contaminant as

$$f = \frac{\Sigma_d}{\Sigma}. \tag{4.50}$$

This is the dust to gas ratio that we have discussed before, except that now we seek to determine how f evolves with radius and time in the disk. If the contaminant is neither created nor destroyed within the region of the disk under consideration, continuity demands that

$$\frac{\partial \Sigma_d}{\partial t} + \nabla \cdot \mathbf{F}_d = 0, \tag{4.51}$$

where $\mathbf{F}_d$, the flux, can be decomposed into two parts: an advective part describing transport of the dust or gas with the mean disk flow, and a diffusive part describing the tendency of turbulence to equalize the concentration of the contaminant across the disk. For $f \ll 1$ we can reasonably assume that the diffusive properties of the disk depend only on the *gas* surface density, in which case the flux can be written as

$$\mathbf{F}_d = \Sigma_d \mathbf{v} - D\Sigma \nabla \left(\frac{\Sigma_d}{\Sigma} \right). \tag{4.52}$$

Here $\mathbf{v}$ is the mean velocity of gas in the disk and D is the usual turbulent diffusion coefficient. We note that the diffusive term vanishes if f is constant. Combining this equation with the continuity equation for the gaseous component, we obtain an evolution equation for f in an axisymmetric disk. In cylindrical polar coordinates

$$\frac{\partial f}{\partial t} = \frac{1}{r\Sigma} \frac{\partial}{\partial r} \left(Dr\Sigma \frac{\partial f}{\partial r} \right) - v_r \frac{\partial f}{\partial r}. \tag{4.53}$$

In common with the equation describing the settling of solid particles in a turbulent disk (Eq. 4.27), this is an advection–diffusion equation, though here the advective component is due to the radial flow of the disk gas rather than settling. It is straightforward to generalize this equation to account for the radial drift of larger particles that are imperfectly coupled to the gas – all that is required is to add an additional flux to Eq. (4.52).

Equation (4.53) expresses a competition between diffusion, whose strength depends upon the turbulent diffusion coefficient D, and radial advection at a velocity v_r. For a steady disk away from the boundaries, the radial velocity can be written in terms of the viscosity as

$$v_r = -\frac{3\nu}{2r},$$
(4.54)

so, as was the case with vertical settling, it is the ratio of the two transport coefficients that is critical. This ratio is called the Schmidt number:

$$\mathrm{Sc} \equiv \frac{\nu}{D}.$$
(4.55)

Diffusion becomes increasingly more important for low values of the Schmidt number.

In most cases of interest the contaminant equation (Eq. 4.53), where necessary modified to allow for radial drift of particles, needs to be solved numerically along with the evolution equation for the gas surface density (Eq. 3.173). We can gain considerable insight into the general behavior, however, by examining the properties of analytic solutions available for some special cases (Clarke & Pringle, 1988). Let us imagine a steady disk in which the surface density profile $\Sigma \propto r^{-\gamma}$ (correspondingly, the viscosity scales with radius as $\nu \propto r^{\gamma}$). At some instant a ring of contaminant is injected into the disk at $r = r_0$. Writing $x = r/r_0$, we define $P(> x)$ to be the maximum fraction of contaminant that is ever at a disk radius of x or larger. For $\gamma = 2$ this quantity can be expressed analytically in terms of the complementary error function[6]

$$P(> x) = \frac{1}{2}\,\mathrm{erfc}\left[\left(\frac{3}{2}\,\mathrm{Sc}\,\ln x\right)^{1/2}\right].$$
(4.56)

The solution is plotted for a variety of Schmidt numbers in Fig. 4.5. The extent to which contaminant can diffuse "upstream" (radially outward for a steady disk in which $v_r < 0$) is a rather sensitive function of Sc. If the Schmidt number is significantly greater than unity (i.e. the anomalous viscosity is greater than the turbulent diffusivity) almost no upstream diffusion occurs. Values of $\mathrm{Sc} = 0.33$ or lower, on the other hand, result in significant amounts of dust or gas diffusing

---

[6] The complementary error function $\mathrm{erfc}(x) \equiv 1 - \mathrm{erf}(x)$, where, $\mathrm{erf}(x) \equiv \frac{2}{\sqrt{\pi}} \int_0^x e^{-t^2}\,dt$ is the error function.

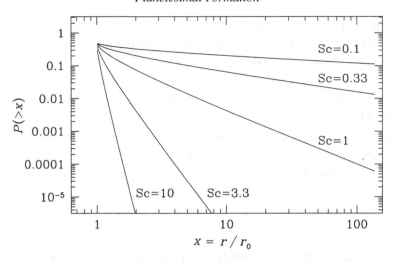

Figure 4.5 Demonstration of how the Schmidt number (Sc $\equiv \nu/D$, the ratio of the turbulent viscosity to the turbulent diffusivity) affects the ability of a trace contaminant (a minor gas species, or tightly coupled dust particles) to diffuse upstream against the inward flow of a steady accretion disk. The quantity $P(> x)$ is the maximum fraction of contaminant, released at $x = r/r_0 = 1$ in a steady disk with $\Sigma \propto r^{-2}$, that ever reaches radius x or larger. Upstream diffusion is significant only for Sc ≤ 1.

outward against the mean flow to large distances. Although the specific disk model for which the analytic solution applies is not very realistic ($\gamma = 1$ or $\gamma = 3/2$ are more typical values than $\gamma = 2$) the basic conclusion holds more generally – the extent of outward diffusion depends on the Schmidt number.

4.4 Diffusion of Large Particles

From the preceding discussion it ought to be clear that determining the value of the Schmidt number is critical if we want to quantify either the vertical distribution of particles or their radial transport through the disk. If we restrict ourselves to considering gaseous species, or very small dust particles for which the dimensionless friction time $\tau_{\text{fric}} \ll 1$, the Schmidt number is best thought of as an intrinsic property of the turbulence present within the disk. To an order of magnitude it is normally reasonable to assume that Sc ~ 1, but deviations from a Schmidt number of unity can be a factor of several in either direction and may differ between the vertical and radial directions. Numerical simulations of the relevant turbulent processes are needed in order to quantify such departures.

For larger particles, the diffusion coefficient D_p is predicted to vary with the particle size according to the value of the dimensionless friction time τ_{fric}.[7] This

---

[7] Confusingly, different authors use the term "Schmidt number" to refer to either ν/D or D/D_p. These are different physical quantities. Here we define Sc to be the ratio of the kinematic viscosity in the disk to the

is easiest to visualize if we consider a large rock for which $\tau_{\text{fric}} \gg 1$. A rock is not swept up by turbulence in the relatively diffuse gas of the disk. Instead one can envisage the rock as orbiting the star with a well-defined semi-major axis, eccentricity, and inclination. The rock is buffeted by random aerodynamic forces due to the presence of turbulence, and these cause a slow drift in the orbital elements. Particles cannot plausibly diffuse *faster* than gas molecules, so we anticipate that the ratio D/D_p is bounded from below at unity. As particles become larger and the coupling becomes weaker, the ratio will increase.

We can illustrate the dependence of D_p on τ_{fric} using simple arguments borrowed from Youdin and Lithwick (2007). We assume that the disk is turbulent, and that the turbulence can be described using the hydrodynamic picture of interacting eddies developed by Kolmogorov. The characteristic turnover time scale of the eddies is written in dimensionless form as

$$\tau_{\text{eddy}} \equiv \Omega t_{\text{eddy}}, \tag{4.57}$$

exactly analogous to the definition of the dimensionless friction time $\tau_{\text{fric}} \equiv \Omega t_{\text{fric}}$. The characteristic velocity of the turbulent fluctuations δv_g is defined such that the *gaseous* diffusion coefficient

$$D = \delta v_g^2 t_{\text{eddy}}. \tag{4.58}$$

For simplicity we focus on the limit in which $\tau_{\text{fric}} \gg 1$ and $\tau_{\text{eddy}} \ll 1$. In this regime of friction time a particle on a noncircular or noncoplanar orbit executes (to leading order) epicyclic oscillations with a frequency $\kappa = \Omega$. During each oscillation of duration Ω^{-1} the particle receives $N \sim \tau_{\text{eddy}}^{-1}$ independent impulses, each imparting a velocity kick

$$\delta v_p \sim \frac{\tau_{\text{eddy}}}{\tau_{\text{fric}}} \delta v_g. \tag{4.59}$$

The kicks are independent, so they accumulate as a random walk. After a time Ω^{-1}

$$\delta v_p \sim \frac{\tau_{\text{eddy}}}{\tau_{\text{fric}}} \delta v_g \sqrt{N} \sim \frac{\sqrt{\tau_{\text{eddy}}}}{\tau_{\text{fric}}} \delta v_g, \tag{4.60}$$

and the particle has drifted a distance $\delta l \sim \delta v_p \Omega^{-1}$. Noting that the diffusion coefficient (Eq. 4.58) can be generically rewritten in the form $D_p \sim \delta l^2 \Omega$ we obtain, in the limit of large τ_{fric}, that $D_p \sim D/\tau_{\text{fric}}^2$. Since we have argued that $D_p \to D$ for $\tau_{\text{fric}} \ll 1$ we can guess a general expression for the ratio of the gas to the particle diffusion coefficients:

$$\frac{D}{D_p} \sim 1 + \tau_{\text{fric}}^2. \tag{4.61}$$

---

gaseous diffusion coefficient. In this section we discuss the relation between the gas and particle diffusivities, but refrain from giving that ratio a name.

This agrees to within order unity factors with a more formal analysis of radial diffusivities (Youdin & Lithwick, 2007).

The suppression of particle diffusivity for $\tau_{\text{fric}} \gg 1$ means that radial diffusion is an important effect only for relatively small particles ($s \lesssim 1$ mm) – and even then only in the case where the Schmidt number is relatively low. For larger particles, radial drift due to the mean aerodynamic drag dominates and can result in large-scale redistribution of solid material.

4.5 Particle Growth via Coagulation

The population of dust grains within the disk is simultaneously affected by processes that modify their spatial distribution (vertical settling, radial drift, and turbulent diffusion), and by those that modify their size distribution (coagulation, and destructive collisions). These two facets of the problem are coupled. Settling and radial drift depend upon the sizes of the particles, while the rate and outcome of collisions depend on the global particle distribution and on the rate of drift and the strength of turbulence. Models that aspire to being complete are necessarily complex. Here, we break the problem down into three more manageable questions. What determines the collision velocities between particles? What are the determining factors for collision outcomes? And how do we treat the situations where particles of disparate sizes are colliding, coagulating, and fragmenting, in a consistent way?

4.5.1 Collision Rates and Velocities

The speed and extent of particle growth via coagulation depend upon two factors: how frequently do particles collide, and what is the outcome of those collisions? For a single population of particles the collision time scale is

$$t_{\text{collide}} = \frac{1}{n\sigma \Delta v}, \tag{4.62}$$

where n is the particle number density, σ the cross-section for collisions, and Δv the relative velocity. If the particles are spheres of radius s with mass $m = (4/3)\pi s^3 \rho_m$ the cross-section is given by

$$\sigma = \pi (2s)^2, \tag{4.63}$$

while the number density is

$$n = \frac{f\rho}{m}. \tag{4.64}$$

Here ρ is the gas density and f, the ratio of the density of solid particles to the gas density, incorporates any enhancement or depletion of solids due to vertical settling, radial drift, or small-scale concentration processes. Determining the appropriate value of Δv is trickier. For small dust particles the dominant effect will be due to Brownian motion, which, for a particle in thermal equilibrium with a gas at temperature T, introduces a random velocity of the order of $(1/2)mv^2 \sim k_B T$. For particles of masses m_1 and m_2 the collision velocity in the Brownian motion regime is

$$\Delta v = \sqrt{\frac{8k_B T (m_1 + m_2)}{\pi m_1 m_2}}. \tag{4.65}$$

The velocity is of the order of 0.1 cm s^{-1} for µm-sized particles in a gas at $T = 300$ K. In the simple case where all the particles have the same mass $m_1 = m_2 = m$, Eqs. (4.62) through (4.65) yield as an estimate of the collision time scale

$$t_{\text{collide}} = \frac{\pi \rho_m^{3/2} s^{5/2}}{6\sqrt{3k_B T} f\rho}. \tag{4.66}$$

Substituting typical numerical values for conditions in the inner protoplanetary disk (a gas density $\rho = 10^{-10}$ g cm^{-3}, a dust to gas ratio $f = 10^{-2}$, a material density $\rho_m = 3$ g cm^{-3}, and a temperature $T = 300$ K) we find

$$t_{\text{collide}} \simeq 24 \left(\frac{s}{1 \text{ µm}}\right)^{5/2} \text{ yr.} \tag{4.67}$$

This is a very short time scale. Small dust particles will inevitably collide in the environment of the protoplanetary disk even if Brownian motion is the only process that yields a relative velocity between particles. If the collisions are adhesive, particle growth will be rapid.

For larger particles the relative velocity is usually determined by a combination of differential radial drift and velocity induced by imperfect coupling to turbulence within the disk. We can readily estimate the maximum values of these velocities. For radial drift, the peak drift speed for a single particle is $(1/2)\eta v_K$ (Eq. 4.41), where η depends on the disk scale height through $\eta = n(h/r)^2$ and $n \sim 3$ is a pre-factor that is of the order of unity. A rough estimate for the maximum collision speed due to differential radial drift between two large, but non-identically sized particles, is

$$\Delta v_{\text{drift, max}} \sim \frac{1}{4}\eta v_K. \tag{4.68}$$

In a disk with a mid-plane temperature profile $T \propto r^{-1/2}$ the radial dependence of η cancels the radial dependence of v_K. We obtain

$$\Delta v_{\text{drift, max}} \sim 25 \text{ m s}^{-1}, \tag{4.69}$$

as an estimate of the peak collision velocity for two particles with $\tau_{\text{fric}} \sim 1$ drifting radially through the disk.

Similar values of the maximum collision velocity occur due to aerodynamic coupling between particles and turbulent disk gas. A calculation by Ormel and Cuzzi (2007) shows that the peak collision velocity in this limit is

$$\Delta v_{t,\,max} \sim \alpha^{1/2} c_s. \tag{4.70}$$

In a disk with a mid-plane temperature profile $T \propto r^{-1/2}$ and a sound speed at 1 AU of 1 kms^{-1} we have

$$\Delta v_{t,\,max} \sim 30 \left(\frac{\alpha}{10^{-3}} \right)^{1/2} \left(\frac{r}{AU} \right)^{-1/4} \text{m s}^{-1}. \tag{4.71}$$

Consideration of these limiting cases makes clear that the range of particle collision velocities realized within protoplanetary disks is substantial. For micron-sized particles, velocities of the order of 0.1–1 cm s^{-1} are reasonable, while particles with $\tau_{fric} \sim 1$ are expected to collide at speeds that are of the order of 10 m s^{-1} (with at most a weak radial dependence). Translating $\tau_{fric} \sim 1$ to a physical size requires specifying a disk model, but typically we would be thinking about $s \sim 1$ m objects at 1 AU (and throughout the disk interior to the snow line) and $s \sim 1$ cm particles at 100 AU.

4.5.2 Collision Outcomes

Determining the outcome of particle collisions as a function of particle size, composition, and relative velocity is one of the few aspects of planet formation where laboratory experiments – rather than theoretical calculations or astronomical observations – have primacy (a review of such experiments has been given by Blum & Wurm, 2008). The parameter space of interest is large. As we have noted, the relative velocities of bodies upon collision range from $\Delta v \sim 0.1$ cm s^{-1} (for Brownian motion of μm-sized dust particles) up to $\Delta v \sim 10 - 100$ m s^{-1} (for the relative velocity of meter-sized rocks subject to radial drift). Of equal importance is the composition of the colliding objects, which here means not just their chemical makeup (silicates, water ice, etc), but also their internal structure (solid particles, or aggregates with varying shapes and strengths).

At an elementary level the considerations that determine whether two particles stick upon collision are readily understood (e.g. Youdin, 2010). We consider two identical particles of mass m, which have a relative velocity (when they are well separated) of Δv. Upon collision we assume that short-range adhesive forces act to bind the particles together with an energy ΔE_s, and that a fraction f of the impact energy is dissipated. Requiring that the particles remain bound (total energy less than zero) upon rebounding from each other, then yields a simple expression for the maximum impact velocity that results in sticking:

$$\frac{1}{4} m \Delta v^2 \leq \frac{f}{(1-f)} |\Delta E_s|. \tag{4.72}$$

For a given impact velocity, particles can therefore adhere to each other for one of two reasons. First, they may stick as a consequence of strong surface forces (i.e. the particles are physically "sticky"). This is most important for small particles. The surface forces that determine ΔE_s act only over the small area that is physically in contact when the two particles collide, and hence increase with particle size more slowly than the kinetic energy on the left-hand-side. For larger particles, the declining ratio of surface area to mass means that surface forces become less important as a determining factor in collision outcomes, which are instead controlled increasingly by the ability of particles to dissipate energy upon collision. Solid bodies fail to dissipate sufficient energy and tend to rebound rather than stick at typical collision velocities. It is likely, however, that small macroscopic particles (with sizes in the range between µm and cm) resemble a loosely bound *aggregate* of smaller dusty or icy particles. Such aggregates can dissipate a greater fraction of the kinetic energy of impact internally (i.e. $f \sim 1$), and experiments confirm that this can result in sticking probabilities that are substantially more favorable for ongoing particle growth.

Güttler *et al.* (2010) mapped out the outcome of particle collisions under expected protoplanetary disk conditions, using constraints from laboratory experiments that were available at the time. Full or partial coagulation, bouncing, and fragmentation are all possible for particle mass ratios and collision velocities found in disks. In broad terms, however, there is a fragmentation threshold for silicate particles at

$$\Delta v_f \approx 1 \text{ m s}^{-1}, \tag{4.73}$$

and net growth is hard to achieve if collision velocities substantially exceed this value. Instead, a quasi-equilibrium is set up in which simultaneous coagulation and fragmentation processes balance. The resulting size distribution has most of the mass in the largest particles, which may be of mm- or cm-sized dimensions (Birnstiel *et al.*, 2011).

Laboratory experiments on aggregates made up of µm- or sub-µm-sized silicate particles are intended to represent conditions interior to the water snow line in protoplanetary disks. Further out, water and other ices (such as CO) are the dominant species. Knowledge of collision outcomes in this regime is scarcer than for silicates. Ices can have a fragmentation threshold that is substantially larger than silicates (perhaps of the order of 10 m s^{-1}, which approaches the peak collision speeds realized in disks; Gundlach & Blum, 2015). The fragmentation threshold for ice is, however, likely to be temperature dependent, and may be comparable to the silicate value in the outer regions of the disk where the temperature is far below the sublimation temperature (and where other ice species are present).

4.5.3 Coagulation Equation

It is an observed fact that particles of µm and mm size coexist within protoplanetary disks, and it is likely that at any time the size distribution includes physically interesting numbers of particles spanning an even wider range of scales. For representative particles of mass (m_1, m_2) collisions occur at a rate that depends upon the number density of the colliding particles and their relative velocity, with a velocity that has both a mean value and a distribution. Several outcomes, including full or partial coagulation, bouncing, and fragmentation, depend upon the parameters of the collision and the chemical and physical makeup of the particles. A mathematical way to handle this is based upon the *coagulation equation* developed by Smoluchowski (1916). The formalism that results is easy to state but difficult to solve except by numerical means.

Suppose that at some time the number of solid particles per unit volume with masses in the range between m and $(m + dm)$ is $n(m)dm$. As time proceeds, the number of particles of mass m increases whenever there is an adhesive collision between any two particles whose masses *sum* to m, and decreases whenever a particle of mass m coagulates with any other particle. Mathematically,

$$\frac{\partial n(m)}{\partial t} = \frac{1}{2} \int_0^m A(m', m - m')n(m')n(m - m')dm'$$
$$- n(m) \int_0^\infty A(m', m)n(m')dm', \qquad (4.74)$$

where the factor of one-half eliminates double counting of the collisions that increase the number of particles of mass m. As written this equation is extremely general, and it finds applications in physical chemistry, biology, and cosmology as well as in planet formation. The physics of any specific application is encoded in the *reaction kernel A*, which can be written in our case as

$$A(m_1, m_2) = P(m_1, m_2, \Delta v)\Delta v(m_1, m_2)\sigma(m_1, m_2). \qquad (4.75)$$

Here $P(m_1, m_2, \Delta v)$ is the probability that a collision between two particles of masses m_1 and m_2 leads to adhesion, Δv is the relative velocity at collision, and σ is the collision cross-section. It is straightforward to further generalize the coagulation equation to include fragmentation and/or populations of particles with different physical properties (e.g. some particles may be compact spheres, while others are porous or fractal aggregates).

The coagulation equation (Eq. 4.74) is also important in the theory of terrestrial planet formation, and we defer detailed discussion of analytic solutions to the equation until Section 5.5. Suffice to say that it has the form of an integro-differential equation for the time evolution of the particle mass distribution, and that it is difficult to solve. Two general points, however, are worth noting now. First, growth of particles of mass m is described by a weighted sum over all possible collisions that yield the correct total mass. This means that even if the kernel is near zero for some combinations of masses (perhaps for near-equal mass collisions of meter-scale bodies), growth is still possible if there is a high probability that other types of collision lead to adhesion. Second, the sticking probability P is evidently of central importance when assessing whether growth can occur and how rapid the process is. In general P will be a function of the masses of the particles involved, their collision velocities, and additional parameters describing the shape and strength of the objects.

The integro-differential nature of the coagulation equation is rather different from most equations encountered in astrophysics, but efficient numerical techniques are available to solve it. (This is not a trivial business, and a good literature search is recommended before trying to develop your own scheme!) The dimensionality of the problem becomes higher in situations where it is physically necessary to keep track of additional variables beyond mass, such as particle porosity or chemical composition. In these circumstances Monte Carlo methods, based on random sampling of representative particle collisions, are the simplest numerical schemes.

4.5.4 Fragmentation-Limited Growth

The experimental result that particle growth is limited by fragmentation above a threshold collision velocity can be converted, with the aid of a disk model, into an estimate of the maximum particle size. As an example, let us assume that we are dealing with small particles ($\tau_{\text{fric}} \lesssim 1$) whose collision velocities are set by aerodynamic coupling to turbulence in the disk gas. An approximation for the peak mutual collision velocity, valid if τ_{fric} is not too much less than unity, is (Ormel & Cuzzi, 2007)

$$\Delta v^2 \simeq \frac{3}{2}\alpha\tau_{\text{fric}}c_s^2. \tag{4.76}$$

Using the explicit relation between τ_{fric} and particle size (Eq. 4.13), and assuming that the maximum particle size s_{max} is set by the condition that $\Delta v = \Delta v_f$, the fragmentation threshold, we find

$$\Delta v_f^2 \simeq \frac{3}{2}\alpha\frac{\rho_m}{\rho}\frac{s_{\text{max}}}{v_{\text{th}}}\Omega_K c_s^2. \tag{4.77}$$

At the disk mid-plane $\rho = \rho_0 = (1/\sqrt{2\pi})(\Sigma/h)$. Using this expression, the fact that $h = c_s/\Omega_K$, and noting the numerical difference between the thermal speed v_{th} (Eq. 4.1) and the sound speed c_s, we obtain

$$s_{\text{max}} \simeq \frac{4}{3\pi}\frac{\Sigma}{\alpha\rho_m}\frac{\Delta v_f^2}{c_s^2}. \tag{4.78}$$

This is an estimate for how large particles can grow if (a) there is a strict velocity-dependent threshold to growth (due to fragmentation or bouncing), and (b) particles have enough time to reach the maximum before radial drift transports them to a place where the physical conditions are different (Birnstiel *et al.*, 2012; Pinilla *et al.*, 2012). The most important radial scaling comes from the surface density term, so the prediction of this model is that s_{max} ought to generally decrease with orbital radius. Superimposed on that decrease there may be one or more jumps at radial locations where different material properties lead to different threshold velocities.

In the absence of particle trapping at pressure maxima, radial drift can move particles inward on a time scale that is shorter than collisions grow them to the maximum size allowed by fragmentation. Because s_{max} increases closer to the star, the drifting (and growing) particles may *never* reach the sizes that are in principle permissible at any given radius. This situation – in which the particle size as a function of orbital distance is limited by *time* rather than by material properties – is described as *drift-limited growth*. Approximate expressions for the resulting particle sizes can be derived in an analogous manner to the fragmentation-limited case (Birnstiel *et al.*, 2012).

4.6 Gravitational Collapse of Planetesimals

Thus far we have assumed that the only important interactions between particles are physical collisions, and that those particles are dynamically unimportant for the evolution of the gas disk. These assumptions are valid for small dust particles distributed uniformly throughout the gas disk – since the total mass of solids is negligible compared to the mass of gas at the epoch when the disk forms – but they may be locally violated due to some combination of vertical settling, radial drift, and photoevaporation. If the solid particles start to play a dynamical role a number of new physical effects may occur:

- Gravitational instability within a dense layer of particles located close to the disk mid-plane. Safronov (1969) and, independently, Goldreich and Ward (1973) proposed that such an instability might result in the prompt formation of planetesimals. Although it is now known that the simplest version of their theory, known as the *Goldreich–Ward mechanism*, fails, closely related physical ideas remain important.
- Modification of the properties of turbulence within the gas due to feedback from the solid component.
- The existence of new two-fluid instabilities that arise because of the coupling between the solid and gaseous components. These might result in clumping of the solid particles (over and above over-densities that would occur if the solids were passive tracers within the turbulent gas flow), and promote planetesimal formation either via direct collisions or via gravitational collapse.

We begin by analyzing the stability of a thin particle layer to gravitational collapse. The physical considerations are identical to those discussed heuristically in Section 3.4 for the stability of a *gaseous* disk, with the particle velocity dispersion σ taking the place of the gas sound speed, and the control parameter is once again Toomre's Q (Eq. 3.62). Let us derive that result more formally.

4.6.1 Gravitational Stability of a Particle Layer

We consider a razor-thin disk of particles with uniform surface density Σ_0 and constant velocity dispersion σ orbiting the star in circular orbits in the $z = 0$ plane. In cylindrical polar coordinates (r, ϕ, z) the density of the disk is given by

$$\rho_0(r, \phi, z) = \Sigma_0 \delta(z), \tag{4.79}$$

where $\delta(z)$ is a Dirac delta function, while the velocity field is

$$\mathbf{v}_0(r, \phi, z) = (0, r\Omega, 0). \tag{4.80}$$

Here $\Omega = \Omega(r)$ is the angular velocity of the particles. The angular velocity need not be Keplerian, but for circular orbits we require that the centrifugal force balance gravity. If the gravitational potential is Φ_0,

$$\Omega^2 r = -\frac{d\Phi_0}{dr}. \tag{4.81}$$

Note that because of the assumptions of constant density and constant velocity dispersion, the equivalent of a pressure gradient force does not enter into the problem.

To proceed we now make a number of simplifications. First, we assume that coupling of the particles to any gas present can be ignored. Second, we assume that the particle layer can be treated as a *fluid* in which the two-dimensional sound speed, defined in terms of the pressure p and surface density Σ in the usual way via

$$c_s^2 \equiv \frac{dp}{d\Sigma}, \tag{4.82}$$

is equivalent to the particle velocity dispersion σ. This is not formally correct, and it is not obvious that it is correct at all. If the particles are collisionless, their "pressure" need not be isotropic and their dynamics should be described not by the fluid equations but by the collisionless Boltzmann equation. It turns out, however, that the fluid and collisionless stability criteria are practically indistinguishable (see, e.g. Binney & Tremaine, 1987), with the fluid version being much easier to derive. Adopting the fluid description, the basic equations describing the dynamics of the particle disk are the continuity and momentum equations, together with the Poisson equation for the gravitational field:

$$\frac{\partial \Sigma}{\partial t} + \nabla \cdot (\Sigma \mathbf{v}) = 0, \tag{4.83}$$

$$\frac{\partial \mathbf{v}}{\partial t} + (\mathbf{v} \cdot \nabla)\mathbf{v} = -\frac{\nabla p}{\Sigma} - \nabla \Phi, \tag{4.84}$$

$$\nabla^2 \Phi = 4\pi G \rho. \tag{4.85}$$

Our task is to determine, given these equations, the conditions under which the initial equilibrium state of the disk (defined by Eqs. 4.79, 4.80, and 4.81) is stable against the effect of disk self-gravity, which unopposed would tend to result in the particles gathering into dense clumps or rings. This can be accomplished with a standard linear stability analysis. We consider infinitesimal axisymmetric perturbations to the equilibrium state:

$$\Sigma = \Sigma_0 + \Sigma_1(r,t), \tag{4.86}$$

$$p = p_0 + p_1(r,t), \tag{4.87}$$

$$\Phi = \Phi_0 + \Phi_1(r,t), \tag{4.88}$$

$$\mathbf{v} = \mathbf{v}_0 + \left[v_r(r,t), \delta v_\phi(r,t), 0 \right], \tag{4.89}$$

that have a spatial and temporal dependence given by (using the surface density as an example)

$$\Sigma_1(r,t) \propto \exp[i(kr - \omega t)]. \tag{4.90}$$

Here k is the spatial wavenumber of the perturbation (related to the wavelength via $\lambda = 2\pi/k$) and ω is the temporal frequency. Ultimately, what we need to determine is whether there are values of k (i.e. spatial scales) for which ω is imaginary, since imaginary values of the frequency will result in exponentially growing perturbations. Making one further approximation, we assume that for the perturbations of interest

$$kr \gg 1. \tag{4.91}$$

This amounts to considering disturbances that are small compared to the radial extent of the disk.

We now substitute the expressions for the surface density, pressure, gravitational potential, and velocity into the fluid equations, discarding any terms we encounter that are quadratic in the perturbed quantities (this is a *linear* stability analysis). For the continuity equation this yields

$$-i\omega\Sigma_1 + v_r \Sigma_0 \left(\frac{1}{r} + ik \right) = 0, \tag{4.92}$$

which simplifies further in the local limit ($kr \gg 1$) to

$$-\omega\Sigma_1 + k v_r \Sigma_0 = 0. \tag{4.93}$$

Deriving the analogous algebraic equations from the momentum equation requires us to express the convective operator $(\mathbf{v} \cdot \nabla)\mathbf{v}$ in cylindrical coordinates. This takes the rather unwieldy form

$$(\mathbf{v} \cdot \nabla)\mathbf{v} = \left[v_r \frac{\partial v_r}{\partial r} + \frac{v_\phi}{r}\frac{\partial v_r}{\partial \phi} + v_z \frac{\partial v_r}{\partial z} - \frac{v_\phi^2}{r}, \right.$$
$$\left. v_r \frac{\partial v_\phi}{\partial r} + \frac{v_\phi}{r}\frac{\partial v_\phi}{\partial \phi} + v_z \frac{\partial v_\phi}{\partial z} + \frac{v_r v_\phi}{r}, v_r \frac{\partial v_z}{\partial r} + \frac{v_\phi}{r}\frac{\partial v_z}{\partial \phi} + v_z \frac{\partial v_z}{\partial z} \right]. \tag{4.94}$$

With this in hand, the momentum equation reduces immediately to

$$-i\omega v_r - 2\Omega \delta v_\phi = -\frac{1}{\Sigma_0}\frac{dp_1}{dr} - \frac{d\Phi_1}{dr}, \tag{4.95}$$

$$-i\omega \delta v_\phi + v_r \left[\Omega + \frac{d}{dr}(r\Omega)\right] = 0, \tag{4.96}$$

where the two equations come from the radial and azimuthal components respectively.

The next step is to relate the perturbations in pressure and gravitational potential expressed on the right-hand-side of Eq. (4.95) to perturbations in the surface density. For the pressure term this is straightforward. Equation (4.82) implies that

$$\frac{1}{\Sigma_0}\frac{dp_1}{dr} = \frac{1}{\Sigma_0}c_s^2 ik\Sigma_1. \tag{4.97}$$

Dealing with the potential perturbations requires more work. Starting from the linearized Poisson equation,

$$\nabla^2 \Phi_1 = 4\pi G\Sigma_1 \delta(z), \tag{4.98}$$

we write out the Laplacian explicitly and simplify making use of the fact that for short wavelength perturbations $kr \gg 1$. This yields a relation between the density and potential fluctuations:

$$\frac{d^2\Phi_1}{dz^2} = k^2\Phi_1 + 4\pi G\Sigma_1 \delta(z). \tag{4.99}$$

For $z \neq 0$ the only solution to this equation that remains finite for large $|z|$ has the form

$$\Phi_1 = C\exp[-|kz|], \tag{4.100}$$

where C remains to be determined. To do so we integrate the Poisson equation vertically between $z = -\epsilon$ and $z = +\epsilon$:

$$\int_{-\epsilon}^{+\epsilon} \nabla^2 \Phi_1 dz = \int_{-\epsilon}^{+\epsilon} 4\pi G\Sigma_1 \delta(z)dz. \tag{4.101}$$

Noting that both $\partial^2\Phi_1/\partial x^2$ and $\partial^2\Phi_1/\partial y^2$ are continuous at $z = 0$, whereas $\partial^2\Phi_1/\partial z^2$ is *not*, we obtain

$$\left.\frac{d\Phi_1}{dz}\right|_{-\epsilon}^{+\epsilon} = 4\pi G\Sigma_1. \tag{4.102}$$

Taking the limit $\epsilon \to 0$ we find that $C = -2\pi G\Sigma_1/|k|$, and hence that the general relation between potential and surface density fluctuations on the $z = 0$ plane is

$$\Phi_1 = -\frac{2\pi G\Sigma_1}{|k|}. \tag{4.103}$$

Taking the radial derivative

$$\frac{d\Phi_1}{dr} = -\frac{2\pi i k G \Sigma_1}{|k|},$$ (4.104)

which allows us to eliminate the potential from the right-hand-side of Eq. (4.95) in favor of the surface density. The result is

$$-i\omega v_r - 2\Omega \delta v_\phi = -\frac{1}{\Sigma_0} c_s^2 i k \Sigma_1 + \frac{2\pi i k G \Sigma_1}{|k|}.$$ (4.105)

Finally, we are ready to derive the functional relationship between ω and k, known as the dispersion relation. Eliminating v_r and δv_ϕ between Eqs. (4.93), (4.96), and (4.105) we find that

$$\omega^2 = \kappa^2 + c_s^2 k^2 - 2\pi G \Sigma_0 |k|,$$ (4.106)

where the *epicyclic frequency* κ is defined as

$$\kappa^2 \equiv 4\Omega^2 + 2r\Omega \frac{d\Omega}{dr}.$$ (4.107)

In a Keplerian potential $\kappa^2 = \Omega^2$.

The basic properties of the dispersion relation for a self-gravitating particle (or gas) disk (Eq. 4.106) are readily apparent. Recall that for instability to occur, we require that $\omega^2 < 0$ so that ω itself is imaginary and perturbations grow exponentially with time.[8] Both rotation and pressure – the first and second terms on the right-hand-side of the dispersion relation – are unconditionally positive and act as stabilizing influences, with rotation stabilizing all scales equally and pressure preferentially stabilizing short-wavelength (large k) perturbations. Self-gravity, the third term, is destabilizing with a spatial dependence (linear in k) that lies intermediate between that of rotation and pressure. These dependencies are shown in Fig. 4.6.

Setting $d\omega^2/dk = 0$, we find that the wavenumber of the minimum in $\omega^2(k)$ is given by

$$k_{min} = \frac{\pi G \Sigma_0}{c_s^2}.$$ (4.108)

If the minimum in the $\omega^2(k)$ curve falls into the unstable region, modes with $k \simeq k_{min}$ will display the fastest exponential growth. The condition for the disk to be marginally unstable to gravitational instability is then obtained by requiring that $\omega^2(k_{min}) = 0$. For a Keplerian disk instability requires that

$$Q \equiv \frac{c_s \Omega}{\pi G \Sigma_0} < 1,$$ (4.109)

---

[8] There will also be exponentially *decaying* solutions, but the growing modes will rapidly dominate.

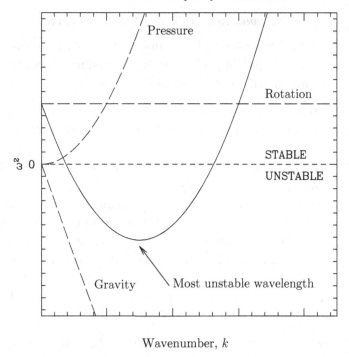

Wavenumber, k

Figure 4.6 The dispersion relation (solid curve) for axisymmetric perturbations to a self-gravitating razor-thin gaseous disk. The short-dashed line shows the boundary between instability ($\omega^2 < 0$) and stability ($\omega^2 > 0$). The long-dashed lines show the individual contributions from rotation (stabilizing at all wavenumbers), pressure (stabilizing at large wavenumber/short wavelength) and self-gravity (destabilizing).

where Q is customarily referred to as the "Toomre Q" parameter after Alar Toomre's 1964 paper. Up to numerical factors, the result matches that deduced using time scale arguments in Section 3.4. Additional numerical factors arise when the analysis is generalized to allow for the possibility of *nonaxisymmetric* instability within the disk. For nonaxisymmetric modes the control parameter is still Q, but instability sets in more easily (i.e. at higher Q) and results in the development of spiral arms within the disk rather than the rings that would be the endpoint of the purely axisymmetric instability. In most circumstances the subtleties introduced by the presence of nonaxisymmetric modes are of only moderate importance, and it suffices to note that the disk will become unstable to its own self-gravity when

$$Q \lesssim Q_{\text{crit}}, \qquad (4.110)$$

with Q_{crit} being of the order of, but slightly larger than, unity.

4.6.2 Application to Planetesimal Formation

With the mathematical formalities of disk instability in place we can proceed to consider how this process might play a role in planetesimal formation. The basic idea, illustrated schematically in Fig. 4.7 involves three stages:

(1) Initially the solid component of the disk is well mixed with gas. The Q of the solid component is very large and the effects of self-gravity play no role in the evolution.

(2) Over time the dust settles vertically to form a thin sub-disk of particles around the $z = 0$ plane. As we have already noted, even the slightest breath of turbulence suffices to stir up dust particles, so substantial settling requires at least some collisional growth to have occurred. *Radial* drift can also contribute to an increase in the mid-plane particle density in the inner disk.

(3) Due to some combination of high surface density and/or low velocity dispersion, the particle sub-disk becomes unstable according to the Q criterion. This may lead to the formation of bound clumps of particles, which rapidly agglomerate to form planetesimals.

Deferring for the moment the question of whether this sequence of events can plausibly occur, we can ask what would be the properties of planetesimals formed

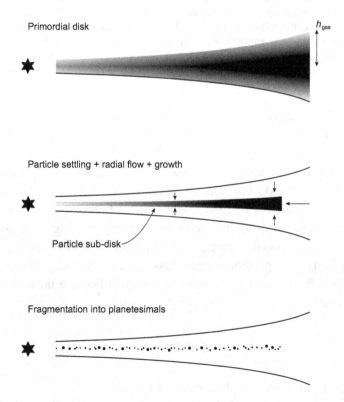

Figure 4.7 How planetesimals form within the classical Goldreich–Ward scenario.

via gravitational instability? If the particle layer has velocity dispersion σ and surface density Σ_s its vertical scale-height will be

$$h_d = \frac{\sigma}{\Omega},$$ (4.111)

while the most unstable wavelength

$$\lambda \sim \frac{2\pi}{k_{min}} = \frac{2\sigma^2}{G\Sigma_s}.$$ (4.112)

For $Q = Q_{crit}$ these scales are comparable, so instability, if it occurs, results in the collapse of relatively small patches of the particle layer into planetesimals.

The mass of clumps formed via gravitational instability is of the order of

$$m_p \sim \pi\lambda^2\Sigma_s \sim 4\pi^5 G^2 Q_{crit}^4 \frac{\Sigma_s^3}{\Omega^4}.$$ (4.113)

Adopting parameters that might be appropriate for the protoplanetary disk at 1 AU ($\Sigma_s = 10\,\mathrm{g\,cm^{-2}}$, $Q_{crit} = 1$) the estimated planetesimal mass is

$$m_p(1\ \mathrm{AU}) \sim 3 \times 10^{18}\ \mathrm{g},$$ (4.114)

which corresponds to a spherical body with a radius of 5–10 km. If we assume – rather unrealistically – that once gravitational instability of the particle layer sets in collapse occurs on the free-fall time scale, then the formation time is very short, being less than a year for the fiducial parameters above.

In principle a particle layer that fragments to form planetesimals could be composed of objects of any size, provided that the conditions for instability are met. The case that attracts greatest interest in the literature, however, is that where the particles have sizes of the order of a cm or smaller. Such particles are small enough not to suffer the potentially devastating rapid radial drift of larger bodies, and there does not appear to be any particular obstacle to forming them rapidly via pairwise collisions. Subsequent instability of a layer made of small particles then has the feature – understandably attractive to many authors – of rapidly forming planetesimals in a way that *bypasses* all of the potential hurdles involved in particle growth through the meter-scale regime.

4.6.3 Self-Excited Turbulence

Since the surface density in solid material is typically only of the order of 1% of the gas surface density, a very thin particle layer is required in order for gravitational instability to set in. For example, if the surface density in solids at 1 AU is $\Sigma_s = 10\,\mathrm{g\,cm^{-2}}$, we require a velocity dispersion of $\sigma \simeq 10\,\mathrm{cm\,s^{-1}}$ in order to attain $Q = 1$. If the *gas* disk has $h/r = 0.05$ at this radius, the relative thickness of the particle and gaseous disks is then

$$\frac{h_{\rm d}}{h} \sim 10^{-4}. \tag{4.115}$$

In the absence of gas, attaining such a razor-thin particle disk is not implausible – the vertical thickness of Saturn's rings, for example, is of the order of only 10 m. In the gas-rich environment of the protoplanetary disk, however, we need to consider carefully whether turbulence will preclude the particle layer from ever becoming thin enough to become unstable.

Intrinsic disk turbulence provides one barrier to the development of very thin particle layers. Equation (4.30) yields an estimate of the ratio of the thicknesses of the particle and gas disks as a function of α and the dimensionless friction time $\Omega t_{\rm fric}$. From that analysis it is obvious that the classical Goldreich–Ward instability cannot work for small particles (those with $\Omega t_{\rm fric} \ll 1$ that have not yet grown to the size where radial drift becomes rapid) unless α is very small. Fully turbulent regions of the disk are unpromising sites for planetesimal formation via gravitational instability.

One might argue, of course, that the mid-plane of the disk is *not* necessarily intrinsically turbulent (e.g. the discussion in Section 3.6). Unfortunately, even if the gas disk on its own were laminar the presence of the particle layer tend would to excite turbulence, via the mechanism depicted in Fig. 4.8. The crucial physical point is that for gravitational instability to occur we require that the local particle density *exceed* the local gas density (by as much as two orders of magnitude if we adopt the parameters given above). Within the particle layer the gas is therefore subdominant to the particles, and the orbital velocity of the flow will be comparable to the Keplerian velocity. Just above the particle layer, on the other hand, the gas dominated disk orbits at slightly less than the Keplerian speed due to the influence of radial pressure gradients. As a consequence there will exist a vertical shear which, if it is too large, will be unstable to the development of Kelvin–Helmholtz

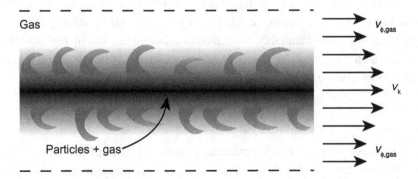

Figure 4.8 How the presence of a dense particle disk results in excitation of turbulence within the flow.

instabilities. This self-excited turbulence can prevent the particle layer from ever settling to the point where self-gravity could set in (Cuzzi *et al.*, 1993).

To estimate the seriousness of this impediment to planetesimal formation we need to consider the hydrodynamic stability of a stratified shear flow. The simplest case to analyze is that of a nonrotating flow (i.e. the Coriolis force is neglected) with a density profile $\rho(z)$ and a velocity profile $v_\phi(z)$.[9] The flow is assumed to be in hydrostatic equilibrium with a vertical pressure gradient being balanced by the vertical component of gravity g_z. Stability against shear instabilities in this situation is measured by the *Richardson number*

$$\mathrm{Ri} \equiv -\frac{g_z \mathrm{d}\ln\rho/\mathrm{d}z}{(\mathrm{d}v_\phi/\mathrm{d}z)^2}. \tag{4.116}$$

A necessary condition for instability is that $\mathrm{Ri} < 0.25$ somewhere within the flow.

Let us evaluate the Richardson number across a particle layer of vertical height h_d, at radius r within a gas disk of scale-height h and in which the mid-plane pressure varies as $P \propto r^{-n}$. We assume that g_z is dominated by the vertical component of the star's gravity, that the density in $\mathrm{d}\ln\rho/\mathrm{d}z$ is the total density (gas plus particles), and that the velocity shear reflects the difference between the Keplerian velocity and the gas disk velocity across the height of the layer. Collecting together previously derived results we then estimate these terms as

$$g_z = \Omega^2 h_\mathrm{d}, \tag{4.117}$$

$$\frac{\mathrm{d}\ln\rho}{\mathrm{d}z} = -\frac{1}{h_\mathrm{d}}, \tag{4.118}$$

$$\left(\frac{\mathrm{d}v_\phi}{\mathrm{d}z}\right)^2 = \frac{n^2}{4}\left(\frac{h}{r}\right)^4 \frac{\Omega^2 r^2}{h_\mathrm{d}^2}. \tag{4.119}$$

The resulting Richardson number is (adopting $n = 3$)

$$\mathrm{Ri} \simeq 0.25 \left(\frac{h/r}{0.05}\right)^{-2}\left(\frac{h_\mathrm{d}/h}{0.0375}\right)^2. \tag{4.120}$$

Clearly, attaining the very thin layers needed to induce gravitational instability is rather difficult. As the particle layer settles toward the mid-plane, self-excited turbulence is liable to set in long before $Q \sim 1$.

The roadblock of self-excited turbulence largely excludes the classical Goldreich–Ward mechanism as a viable planetesimal formation process. What survives in modern thinking is the key role of gravitational collapse, which can still occur if the particle density is enhanced by processes other than just vertical settling. Three potentially multiplicative processes can also occur:

---

[9] As several authors have recently emphasized, this analysis is considerably too simple to yield an accurate answer (Garaud & Lin, 2004; Gómez & Ostriker, 2005; Chiang, 2008). In more complete analyses the presence of the Coriolis force tends to further destabilize the disk flow, while the presence of *radial* shear is a stabilizing effect. The calculation presented here is intended only to illustrate the basic physical considerations at work.

- Large-scale enhancement (scales $\gg h$) due to the radial drift and pile-up of particles.
- Intermediate-scale ($\sim h$) concentration of solids, for example in zonal flows or vortices.
- Small-scale ($\ll h$) clustering from two-fluid instabilities present in disks.

We have already discussed the first two of these possibilities, which have in common the fact that they involve the particle response to features (either the overall surface density profile or localized structure) that are intrinsic properties of the gas disk. Two-fluid instabilities are qualitatively different, as they depend for their existence on the feedback momentum transport between solids and gas exerts on the gas itself.

4.7 Streaming Instability

The streaming instability (Youdin & Goodman, 2005) is an instability of disks that can be present whenever two phases or fluids (gas and solid particles) have a mutual velocity and interact via aerodynamic forces. The instability grows on a time scale that is often intermediate between the dynamical and radial drift time scales, and leads to small-scale particle concentration. Strictly construed, the streaming instability is a local linear instability of a highly simplified but mathematically well-defined system. The main interest in the instability, however, occurs because it offers a route to forming solid over-densities that are dense enough to collapse gravitationally to form planetesimals. As a result, the term "streaming instability" is also used more loosely to refer to two-phase instabilities in more complex systems (for example systems where the solids are settling, or where the gas is intrinsically turbulent), and even to self-gravitating systems (where conceptually distinct instabilities can be present).

4.7.1 Linear Streaming Instability

The radial drift calculation of Section 4.3 considered a single particle interacting with a background gas disk, and remains approximately valid for a population of drifting solids whose surface density is negligible compared to that of the gas. The drift solution needs to be modified if the surface density of the solids becomes non-negligible, because then the momentum transfer from the particles *to the gas* will matter for the background disk structure.

A model for the drift of both solids and gas due to their aerodynamic interactions can be derived by considering the gas to be incompressible, and the particles to form a fluid in which there are no pressure forces. The continuity and momentum equations for the aerodynamically coupled system are

$$\frac{\partial \rho_d}{\partial t} + \nabla \cdot (\rho_d \mathbf{v}_d) = 0, \tag{4.121}$$

$$\nabla \cdot \mathbf{v} = 0, \tag{4.122}$$

$$\frac{\partial \mathbf{v_d}}{\partial t} + \mathbf{v_d} \cdot \nabla \mathbf{v_d} = -\Omega_K^2 \mathbf{r} - \frac{1}{t_{\text{fric}}} (\mathbf{v_d} - \mathbf{v}), \tag{4.123}$$

$$\frac{\partial \mathbf{v}}{\partial t} + \mathbf{v} \cdot \nabla \mathbf{v} = -\Omega_K^2 \mathbf{r} + \frac{1}{t_{\text{fric}}} \frac{\rho_d}{\rho} (\mathbf{v_d} - \mathbf{v}) - \frac{\nabla P}{\rho}, \tag{4.124}$$

where the un-subscripted variables refer to the gas and the variables with a subscript "d" refer to the particles. There is no vertical component of stellar gravity in these equations, nor is there any self-gravity of the fluids themselves. Solving them for a steady state, one obtains an axisymmetric flow describing the drift of solids and gas that is called the Nakagawa–Sekiya–Hayashi (NSH) equilibrium (Nakagawa *et al.*, 1986).

Youdin and Goodman (2005) studied the linear stability of the NSH equilibrium to axisymmetric perturbations. The system exhibits instability for a wide range of parameters (τ_{fric}, and the dust-to-gas ratio or "metallicity" ρ_d/ρ), provided that there is a nonzero mutual velocity between the particles and the gas in the equilibrium solution. Growth is generally faster for larger τ_{fric}, and for values of ρ_d/ρ that are near but not equal to one. The instability does not have an intuitive physical explanation.

4.7.2 Streaming Model for Gravitational Collapse

The direct consequence of the streaming instability is that aerodynamically interacting solids in protoplanetary disks are likely to have a clumpy distribution on small spatial scales $\ll h$. The instability can play a role in planetesimal formation if it generates over-densities that are large enough to initiate gravitational collapse. A necessary condition for collapse is that the density exceed the Roche density, such that self-gravity wins out over the tidal gravitational field of the star. For a clump of solids of mass m and radius r, orbiting at distance a from a star of mass M_*, the tidal acceleration (i.e. the difference in force per unit mass between the edge and center of the clump) is

$$a_{\text{tidal}} = \frac{GM_*}{a^2} - \frac{GM_*}{(a+r)^2} \simeq \frac{2GM_*}{a^3} r. \tag{4.125}$$

Comparing this to the acceleration due to the clump's own self-gravity,

$$a_{\text{grav}} \sim \frac{Gm}{r^2}, \tag{4.126}$$

we find that collapse will occur if

$$r \lesssim \left(\frac{m}{M_*} \right)^{1/3} a. \tag{4.127}$$

Since the clump density $\rho \sim m/r^3$, an equivalent statement is that gravitational collapse requires that the solid density ρ must exceed the Roche density,

$$\rho_R \sim \frac{M_*}{a^3} = 6 \times 10^{-7} \left(\frac{M_*}{M_\odot}\right) \left(\frac{a}{\text{AU}}\right)^{-3} \text{ g cm}^{-3}. \qquad (4.128)$$

This is a high density. The *gas* density at the mid-plane of the disk at 1 AU might be of the order of 10^{-9} g cm^{-3} (assuming a gas surface density of 10^3 g cm^{-2} and a $h/r \sim 0.02$), and vertically well-mixed solids would have a density roughly two orders of magnitude less. Whether generated by the streaming instability or some other process, very high solid over-densities with $\Delta\rho/\rho \sim 10^4$ are needed before self-gravity becomes important enough to start gravitational collapse.

Yang *et al.* (2017) studied the properties of over-densities generated by the streaming instability using numerical simulations that included the vertical component of stellar gravity but did not incorporate the self-gravity of the particles themselves. They found that for a given size of particle (expressed via the dimensionless stopping time τ_{fric}) the vertically integrated solid-to-gas ratio, $Z \equiv \Sigma_d/\Sigma$, determined the strength of particle clumping. The strong particle clumping that is a prerequisite for gravitational collapse requires $Z > Z_{\text{crit}}(\tau_{\text{fric}})$. The functional form of Z_{crit} can be fit by an approximate formula,

$$\log Z_{\text{crit}} = 0.10(\log \tau_{\text{fric}})^2 + 0.20 \log \tau_{\text{fric}} - 1.76 \quad (\tau_{\text{fric}} < 0.1),$$
$$\log Z_{\text{crit}} = 0.30(\log \tau_{\text{fric}})^2 + 0.59 \log \tau_{\text{fric}} - 1.57 \quad (\tau_{\text{fric}} > 0.1), \qquad (4.129)$$

which is plotted in Fig. 4.9. The streaming instability is most effective at strongly clumping solids in a broad region around $\tau_{\text{fric}} \sim 0.1$, but even for favorable parameters requires solid-to-gas ratios somewhat above the fiducial value of 0.01 to trigger gravitational collapse (Johansen *et al.*, 2009b).

In more complete models, the extent of particle clumping and gravitational collapse is likely to depend on a number of parameters. In addition to the particle stopping time, solid-to-gas ratio, and the velocity differential between solids and gas (all control parameters for even the linear theory), the relative strength of self-gravity and tidal shear, the strength of intrinsic disk turbulence, and the particle size distribution may play a role in determining the outcome. This parameter space is not fully explored, but existing simulations, such as the one shown in Fig. 4.10, show that clumping can be strong enough to initiate gravitational collapse and the formation of planetesimals, with a broad size distribution that extends to large masses (Johansen *et al.*, 2007).

4.8 Pathways to Planetesimal Formation

Planetesimal formation is the key step that links what is in principle directly observable – the distribution of highly mobile small solid particles with $s \lesssim$ mm in protoplanetary disks – to the distribution of much less mobile km-scale bodies that

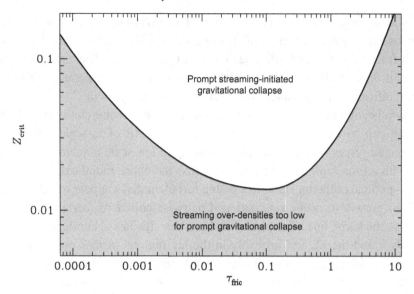

Figure 4.9 The critical solid to gas ratio Z_{crit} for prompt planetesimal formation is plotted as a function of the dimensionless stopping time τ_{fric}. For $Z > Z_{crit}$, the streaming instability generates solid over-densities that are large enough to rapidly trigger gravitational collapse. The critical curve is an analytic fit to numerical results by Yang *et al.* (2017).

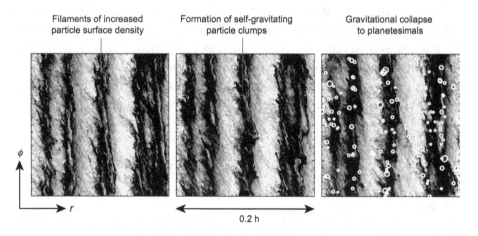

Figure 4.10 The formation of planetesimals from the combined action of the streaming instability and gravitational collapse. The streaming instability creates filaments and clumps of solid particles, which can become dense enough to collapse under self-gravity. From left to right, the panels show the time evolution of the solid surface density from a three-dimensional local simulation (Abod *et al*, 2019).

is the effective initial condition for subsequent planet formation. We do not fully understand how planetesimals form. To summarize the current state of knowledge, there appears to be no impediment to the rapid growth of dusty or icy particles up to small macroscopic dimensions ($s \sim 1$ mm). The growth mechanism at these scales is pairwise collisions that result in sticking, and the time scale in the inner disk is a small fraction of the disk lifetime. The fact that dust is still present in the inner regions of disks with ages of several Myr suggests that regeneration of dust, via erosive collisions, accompanies an overall trend toward growth.

Growth beyond the mm or cm size regime presents greater challenges. A strong argument can be made, based on the rapid radial drift of m-scale bodies due to aerodynamic forces, that growth all the way up to km-scale planetesimals must be rapid, with a time scale less than 10^5 yr across the entire radial extent of the disk. Particle–particle collision velocities within turbulent disks appear to be too large to allow this growth to occur via continued pairwise collisions, certainly in the disk interior to the snow line (if experiments based on silicates accurately represent the materials found there), and probably in the icy regions further out. Gravitational instability provides a pathway to planetesimal formation that can bypass some of the scales where collisional growth is uncertain. A variety of processes, including settling, radial drift, and the streaming instability, can combine to concentrate solids sufficiently to trigger collapse.

The most promising mechanisms for particle growth and planetesimal formation have short time scales, suggesting that the initial steps toward planet formation occur early in the disk lifetime. It is less clear whether they operate with comparable efficiency everywhere in the disk, at specific discrete locations (such as near ice lines), or at random places (such as in zonal flows or vortices). In the next chapter we will adopt as a working hypothesis the idea that planetesimals typically form rapidly with a smooth radial distribution, and study how those planetesimals grow into larger bodies. This is the classical approach, but distinctly different models could also be consistent with what we know about planetesimal formation.

4.9 Further Reading

"The Multifaceted Planetesimal Formation Process," A. Johansen *et al.* (2014), in *Proto-stars and Planets VI*, H. Beuther, R. S. Klessen, C. P. Dullemond, & T. Henning (eds.), Tucson: University of Arizona Press, p. 547.

"Forming Planetesimals in Solar and Extrasolar Nebulae," E. Chiang & A. N. Youdin (2010), *Annual Review of Earth and Planetary Sciences*, **38**, 493.

"The Growth Mechanisms of Macroscopic Bodies in Protoplanetary Disks," J. Blum & G. Wurm (2008), *Annual Review of Astronomy & Astrophysics*, **46**, 21.

5

Terrestrial Planet Formation

Once planetesimals have formed, further growth proceeds via a combination of two processes. The first, planetesimal-driven growth, is a purely gravitational process in which the only role of the gas disk is to provide a modest degree of aerodynamic damping of protoplanetary eccentricity and inclination. Planetesimal driven growth can be studied using N-body simulations, which are indispensable for the late phases of final assembly of the Solar System's terrestrial planets. Earlier phases can also be approached using statistical methods, similar to that used in the kinetic theory of gases. In either treatment, the key questions are to determine the evolution of self-consistent distributions of mass, eccentricity, and inclination, which are used to determine the rate and outcome of collisions.

Classical models for terrestrial planet formation start from an initial condition in which the surface density of planetesimals is specified, and include only planetesimal-driven growth. This is not, however, the only possible way in which growth may proceed. If a significant mass of small solid particles co-exist with planetesimals and protoplanets – as seems very likely – bodies can grow in mass as a consequence of the aerodynamically assisted accretion of the small solids. This process, called pebble accretion, is qualitatively distinct from planetesimal accretion.

5.1 Physics of Collisions

The formation of terrestrial planets from planetesimals involves a cascade of pair-wise collisions. For the most part the gravity of growing planets is strong enough that we can assume that most of the mass of the two colliding bodies ends up agglomerating into a single larger object. For masses and collision velocities for which this is true we can gloss over the detailed physics of the collisions, and the primary input to models of growth is the collision cross-section. The cross-section is enhanced by the gravity of the bodies ("gravitational focusing"), and modified as a result of the tidal gravitational field of the star. For small bodies and large impact velocities the assumption of perfect accretion can fail, and we need to consider the strength of the bodies explicitly to determine whether collisions lead to agglomeration or fragmentation.

5.1.1 Gravitational Focusing

For sufficiently small bodies, the effects of gravity can be ignored for the purposes of determining whether they will physically collide. A massive planet, on the other hand, will deflect the trajectories of other bodies toward it, and as a result has a collision cross-section that is much larger than its physical cross-section.

To evaluate the magnitude of this *gravitational focusing*, consider two bodies of mass m, moving on a trajectory with impact parameter b, as shown in Fig. 5.1. The relative velocity at infinity is σ. At closest approach, the bodies have separation R_c and velocity v_{max}. Equating energy in the initial (widely separated) and final (closest approach) states we have

$$\frac{1}{4}m\sigma^2 = mv_{max}^2 - \frac{Gm^2}{R_c}. \tag{5.1}$$

Noting that there is no radial component to the velocity at the point of closest approach, angular momentum conservation gives

$$v_{max} = \frac{1}{2}\frac{b}{R_c}\sigma. \tag{5.2}$$

If the sum of the physical radii of the bodies is R_s, then for $R_c < R_s$ there will be a physical collision, while larger R_c will result in a harmless flyby. The *largest* value of the impact parameter that will lead to a physical collision is thus

$$b^2 = R_s^2 + \frac{4GmR_s}{\sigma^2}, \tag{5.3}$$

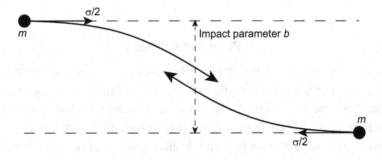

Figure 5.1 Gravitational focusing. If the random velocity σ of bodies is smaller than the escape speed from their surface, the cross-section for physical collisions is boosted beyond the geometric cross-section by gravitational focusing.

which can be expressed in terms of the escape velocity from the point of contact, $v_{esc}^2 = 4Gm/R_s$, as

$$b^2 = R_s^2 \left(1 + \frac{v_{esc}^2}{\sigma^2}\right). \tag{5.4}$$

The cross-section for collisions is then

$$\Gamma = \pi R_s^2 \left(1 + \frac{v_{esc}^2}{\sigma^2}\right), \tag{5.5}$$

where the term in brackets represents the enhancement to the physical cross-section due to gravitational focusing. Clearly a planet growing in a "cold" planetesimal disk for which $\sigma \ll v_{esc}$ will grow much more rapidly as a consequence of gravitational focusing.

5.1.2 Shear versus Dispersion Dominated Encounters

Our derivation of the gravitational focusing term assumed that the only significant forces acting upon the colliding bodies were those due to their mutual gravity. This is a good approximation for studies, for example, of stellar collisions within a star cluster, but its legitimacy must always be evaluated carefully for planetary accretion where the gravity of the star is often important. What matters is not so much the total gravitational force due to the star, but rather the *difference* in the force that is experienced by two bodies on similar orbits (i.e. the tidal gravitational field). Three-body dynamics is more complex than the two-body case, and in general is not amenable to fully analytic treatments.

The condition for three-body dynamics to be important can be estimated from a time scale argument. We first estimate the radius within which the gravity of the protoplanet, with mass M_p and orbital radius a, dominates over the stellar tidal field. To do this we equate the orbital frequency of a protoplanet around the star ($\sqrt{GM_*/a^3}$) to that of a test particle orbiting the planet at radius r ($\sqrt{GM_p/r^3}$). This yields an estimate of the radius of the *Hill sphere*:

$$r_H \sim \left(\frac{M_p}{M_*}\right)^{1/3} a. \tag{5.6}$$

Within r_H two-body effects provide an adequate description of the dynamics. Likewise, we can define a characteristic velocity (the Hill velocity) as the orbital velocity around the protoplanet at the distance of the Hill radius:

$$v_H \sim \sqrt{\frac{GM_p}{r_H}}. \tag{5.7}$$

If the random velocity σ of small bodies is large compared to the Hill velocity ($\sigma > v_H$), then the rate with which they enter the Hill sphere and collide is determined by two-body dynamics. This regime is described as *dispersion dominated*.

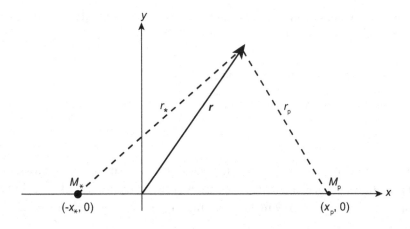

Figure 5.2 Geometry for the derivation of Hill's equations.

If, conversely, $\sigma < v_H$ then three-body effects must be considered and the system is said to be *shear dominated*.

The concept of the Hill sphere is important enough to be worth deriving more rigorously. To do so, we consider the motion of a test particle (a body whose mass is negligible) in a binary system consisting of a star of mass M_* and a protoplanet of mass M_p. We assume that the protoplanet orbits the star in a circular orbit with angular frequency Ω, and work in a co-rotating coordinate system such that the line joining the star to the planet coincides with the x-axis. The origin of the coordinates is taken to be the center of mass. This geometry is shown in Fig. 5.2. The motion of the test particle then obeys

$$\ddot{\mathbf{r}} = -\nabla\Phi - 2\left(\mathbf{\Omega} \times \dot{\mathbf{r}}\right) - \mathbf{\Omega} \times \left(\mathbf{\Omega} \times \mathbf{r}\right), \tag{5.8}$$

$$\Phi = -\frac{GM_*}{r_*} - \frac{GM_p}{r_p}, \tag{5.9}$$

where the dots denote time derivatives. Expressed in terms of components we have

$$\ddot{x} - 2\Omega\dot{y} - \Omega^2 x = -G\left[\frac{M_*(x + x_*)}{r_*^3} + \frac{M_p(x - x_p)}{r_p^3}\right], \tag{5.10}$$

$$\ddot{y} + 2\Omega\dot{x} - \Omega^2 y = -G\left[\frac{M_*}{r_*^3} + \frac{M_p}{r_p^3}\right] y, \tag{5.11}$$

$$\ddot{z} = -G\left[\frac{M_*}{r_*^3} + \frac{M_p}{r_p^3}\right] z. \tag{5.12}$$

These equations are valid for arbitrary masses M_* and M_p. We can simplify them considerably, however, by considering the limit $M_p \ll M_*$. In this limit $|x_*| \ll |x_p|$

and $\Omega^2 = GM_*/x_p^3$. We also shift coordinates so that $x = 0$ coincides with the position of the protoplanet, and consider motion at a distance $\Delta \equiv (x^2 + y^2)^{1/2}$ from the protoplanet that is small compared to the orbital radius. Simple algebraic manipulation of the equations then yields an approximate set of equations of motion that take the form

$$\ddot{x} - 2\Omega\dot{y} = \left(3\Omega^2 - \frac{GM_p}{\Delta^3}\right)x, \tag{5.13}$$

$$\ddot{y} + 2\Omega\dot{x} = -\frac{GM_p}{\Delta^3}y. \tag{5.14}$$

These are known as Hill's equations. They describe the motion of a small body (in our case a planetesimal) in the vicinity of a larger body (a protoplanet) that has a circular orbit around the star.

The left-hand-side of Hill's equations merely describes epicyclic motion of a test particle around the star. Looking at the right-hand-side, we note that the radial (in the x-direction) force vanishes for $\Delta = r_H$, where r_H, the Hill sphere radius, is given in terms of the orbital radius a as

$$r_H = \left(\frac{M_p}{3M_*}\right)^{1/3} a. \tag{5.15}$$

This is a more formal derivation of the distance from the protoplanet within which the gravitational attraction of the protoplanet dominates over the tidal gravitational field of the star.[1]

Hill's equations can be used to compute the trajectories of test particles that move in the vicinity of a larger body. The resulting trajectories are shown schematically in Fig. 5.3 for the case of test particles with almost circular orbits. As one might expect, particles on near circular orbits that pass more than a few r_H from the protoplanet are essentially unperturbed by the presence of the protoplanet, and will not collide. More interestingly, particles that are on orbits that are *too close* in radius to the protoplanet follow what are referred to as horseshoe or tadpole orbits, which also fail to enter the Hill sphere and do not contribute to the collision cross-section. The dynamics of these orbits is the same as that of the Trojan asteroids in the Solar System, of which the largest population is that in 1:1 resonance with Jupiter. Only for a range of intermediate separations is the perturbation from the protoplanet able to overcome the tidal gravitational force and bring the test particle into the region where a collision can occur. We will discuss the resulting collision rate later, but for now the main point to note is simply that the dynamics of collisions is qualitatively different in the shear and dispersion dominated cases.

---

[1] In the astrophysical literature you will sometimes find reference to the *Roche lobe* of the protoplanet, which is essentially the same concept as the Hill sphere. Less frequently you may find discussion of the *Tisserand sphere of influence*, which is the region within which motion of test particles is better described by the two-body dynamics of the planet plus the test particle than by the two-body dynamics of the star plus the test particle. The radius of the Tisserand sphere of influence scales slightly differently with planet mass than the Hill sphere radius, but the distinction is immaterial for our purposes.

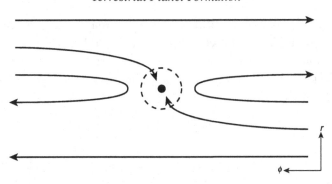

Figure 5.3 Trajectories of test particles on almost circular orbits that encounter a protoplanet, as viewed in the frame co-rotating with the protoplanet. In the shear dominated regime particles whose orbits are *too close* to that of the protoplanet never enter the latter's Hill sphere and collide with the planet.

5.1.3 Accretion versus Disruption

When two initially solid bodies physically collide the outcomes can be divided broadly into three categories:

- **Accretion**. All or most of the mass of the impactor becomes part of the mass of the final body, which remains solid. Small fragments may be ejected, but overall there is net growth.
- **Shattering**. The impact breaks up the target body into a number of pieces, but these pieces remain part of a single body (perhaps after reaccumulating gravitationally). The structure of the shattered object resembles that of a *rubble pile*.
- **Dispersal**. The impact fragments the target into two or more pieces that do not remain bound.

These possibilities are illustrated in Fig. 5.4. If the target is itself a rubble pile then the first possibility – ending up with a solid body – is unlikely, but the collision could still either disperse the pieces or merely rearrange them into a larger but still shattered object.

To delineate the boundaries between these regimes quantitatively, we consider an impactor of mass m colliding with a larger body of mass M at velocity v. We define the specific energy Q of the impact via

$$Q \equiv \frac{mv^2}{2M},$$
(5.16)

and express the thresholds for the various collision outcomes in terms of Q. Conventionally, we define the threshold for catastrophic disruption Q_D^* as the minimum specific energy needed to disperse the target in two or more pieces, with the largest one having a mass $M/2$. Similarly Q_S^* is the threshold for shattering the body. More work is required to disperse a body than to shatter it, so evidently $Q_D^* > Q_S^*$. It is worth keeping in mind that in detail the outcome of a particular collision will depend upon many factors, including the mass ratio between the

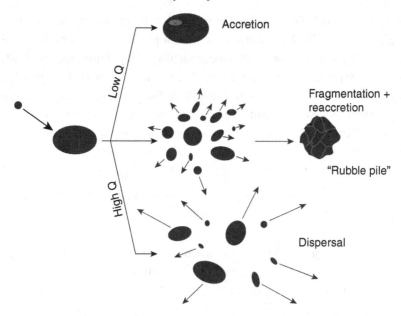

Figure 5.4 Possible outcomes of collisions between bodies.

target and the impactor, the angle of impact, and the shape and rotation rate of the bodies involved. Quoted values of Q_D^* are often averaged over impact angles, but even when this is done the parameterization of collision outcomes in terms of Q is only an approximation.

The estimated values of Q_D^* for a target of a particular size vary by more than an order of magnitude depending upon the composition of the body, which can broadly be categorized into solid or shattered rock, and solid or porous ice. For any particular type of body, however, two distinct regimes can be identified:

- **Strength dominated regime**. The ability of small bodies to withstand impact without being disrupted depends upon the material strength of the object. In general the material strength of bodies declines with increasing size, owing to the greater prevalence of defects that lead to cracks. In the strength dominated regime Q_D^* decreases with increasing size.

- **Gravity dominated regime**. Large bodies are held together primarily by gravitational forces. In this regime Q_D^* must at the very least exceed the specific binding energy of the target, which scales with mass M and radius s as $Q_B \propto GM/s \propto \rho_m s^2$. In practice it requires a great deal more than this minimum amount of energy to disrupt the target – so Q_B is *not* a good estimate of Q_D^* – but nonetheless Q_D^* does increase with increasing size.

Although the transition between these regimes is reasonably sharp there is *some* influence of the material properties (in particular the shear strength) on the catastrophic disruption threshold for smaller bodies within the gravity dominated regime.

Values of Q_S^* and Q_D^* can be determined experimentally for small targets (e.g. Arakawa *et al.*, 2002). Experiments are not possible in the gravity dominated regime, but Q_D^* can be estimated theoretically using numerical hydrodynamics (Benz & Asphaug, 1999; Leinhardt & Stewart, 2009) or (for rubble piles) rigid body dynamics simulations (Leinhardt & Richardson, 2002; Korycansky & Asphaug, 2006). A simple parameterization of the numerical results is as a broken power-law that includes terms representing the strength and gravity regimes:

$$Q_D^* = q_s \left(\frac{s}{1\,\text{cm}} \right)^a + q_g \rho_m \left(\frac{s}{1\,\text{cm}} \right)^b. \tag{5.17}$$

Often (but not always) Q_D^* is averaged over impact geometry, and q_s, q_g, a, and b are all constants whose values are derived by fitting to the results of numerical simulations.

Benz and Asphaug (1999) and Leinhardt and Stewart (2009) determined the values of the fitting parameters in Eq. (5.17) from the results of an ensemble of simulations of impacts into icy or rocky targets. Their results are given in Table 5.1 and plotted as a function of target size in Fig. 5.5. One observes immediately that the results for a particular target material vary with the impact velocity, and hence that Q_D^* is *not* the sole determinant of the outcome of collisions. There is, however, a clear transition between the strength and gravity dominated regimes, with the weakest bodies being those whose size is comparable to the cross-over point. The most vulnerable bodies are generally those with radii in the 100 m to 1 km range. Just how vulnerable such bodies are to catastrophic disruption depends sensitively on their makeup, and it would be unwise to place too much trust in precise numbers. As a rough guide, however, the weakest icy bodies have minimum $Q_D^* \sim 10^5$ erg g^{-1}, while the strongest conceivable planetesimals (unfractured rocky bodies) have minimum $Q_D^* > 10^6$ erg g^{-1}.

As a reality check, we may note that asteroids in the main belt with $e \simeq 0.1$ would be expected to collide today with typical velocities of the order of 2 km s^{-1}.

Table 5.1 *Parameters for the catastrophic disruption threshold fitting formula (Eq. 5.17), which describes how Q_D^* scales with the size of the target body. The quoted values were derived by Benz and Asphaug (1999) and Leinhardt and Stewart (2009) using numerical hydrodynamics simulations of collisions, which are supplemented in the strength dominated regime by experimental results.*

	v (km s^{-1})	q_s (erg g^{-1})	q_g (erg cm^3 g^{-2})	a	b
Ice (weak)	1.0	1.3×10^6	0.09	-0.40	1.30
Ice (strong)	0.5	7.0×10^7	2.1	-0.45	1.19
Ice (strong)	3.0	1.6×10^7	1.2	-0.39	1.26
Basalt (strong)	3.0	3.5×10^7	0.3	-0.38	1.36
Basalt (strong)	5.0	9.0×10^7	0.5	-0.36	1.36

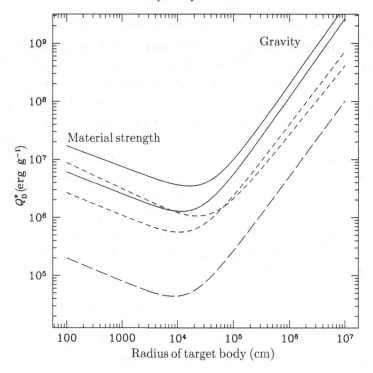

Figure 5.5 The specific energy Q_D^* for catastrophic disruption of solid bodies is plotted as a function of the body's radius. The solid and short-dashed curves show results obtained using fits to theoretical calculations for impacts into "strong" targets by Benz and Asphaug (1999). The long-dashed curve shows the recommended curve for impacts into "weak" targets from Leinhardt and Stewart (2009), derived from a combination of impact experiments and numerical simulations. In detail the solid curves show results for basalt at impact velocities of $5\,\mathrm{km\,s^{-1}}$ (upper curve) and $3\,\mathrm{km\,s^{-1}}$ (lower curve). The short-dashed curves show results for water ice at $3\,\mathrm{km\,s^{-1}}$ (the lower curve for small target sizes) and $0.5\,\mathrm{km\,s^{-1}}$ (upper curve for small target sizes). The long-dashed curve shows results for normal impacts into weak water ice targets at $1\,\mathrm{km\,s^{-1}}$.

For a mass ratio $m/M = 0.1$ the specific energy of the collision is then around $Q = 2 \times 10^9\,\mathrm{erg\,g^{-1}}$, which from Fig. 5.5 is sufficient to destroy even quite large solid bodies with $s \simeq 100\,\mathrm{km}$. This is consistent with the observation of asteroid families, and the interpretation of such families as collisional debris. Evidently the random velocities that characterize collisions must have been *much* smaller during the epoch of planet formation if we are to build large planets successfully out of initially km-scale planetesimals.

The collision model described above is quite basic. Subsequent work shows that momentum, rather than specific energy, controls the outcome in many gravity dominated collisions (Leinhardt & Stewart, 2012). The use of the most up-to-date collision models is highly recommended for any quantitative applications.

5.2 Statistical Models of Planetary Growth

The basic assumption underlying statistical models of planetary growth is that the number of bodies is large enough that the population can be described by one or more probability distributions that encode the probability that a body has a specified mass, inclination, and eccentricity. The details of actual orbits are assumed to be unimportant, and accordingly the distribution of the longitude of pericenter and the longitude of the ascending node is taken to be uniform. For bodies of a given mass the population is then described by the distribution of Keplerian orbital elements $f(e,i)$, which can be regarded as the analog of the Maxwellian distribution of velocities for particles in a gas. Numerical experiments show that the appropriate distribution that is set up as a result of gravitational interactions among a population of bodies is a Rayleigh distribution. Specifically, for a population of planetesimals of mass m and local surface density Σ_p, the probability distribution of eccentricity and inclination takes the form

$$f(e,i) = 4 \frac{\Sigma_p}{m} \frac{ei}{\langle e^2 \rangle \langle i^2 \rangle} \exp\left[-\frac{e^2}{\langle e^2 \rangle} - \frac{i^2}{\langle i^2 \rangle} \right], \tag{5.18}$$

where $\langle e^2 \rangle$ and $\langle i^2 \rangle$ are the mean square values of the eccentricity and inclination, respectively (Lissauer, 1993). This distribution is equivalent to a Gaussian distribution of the *random* components (v_r, v_z, and $\delta v_\phi = v_\phi - v_K$) of planetesimal velocities

$$f(z, \mathbf{v}) = \frac{\Omega \Sigma_p}{2\pi^2 m \sigma_r^2 \sigma_z^2} \exp\left[-\frac{(v_r^2 + 4\delta v_\phi^2)}{2\sigma_r^2} - \frac{(v_z^2 + \Omega^2 z^2)}{2\sigma_z^2} \right], \tag{5.19}$$

where the velocity dispersions σ_r and σ_z in the radial and vertical direction are related to the mean square eccentricities and inclinations via

$$\sigma_r^2 = \frac{1}{2} \langle e^2 \rangle v_K^2, \tag{5.20}$$

$$\sigma_z^2 = \frac{1}{2} \langle i^2 \rangle v_K^2. \tag{5.21}$$

One should note that the exact conversion between orbital elements and the random components of the velocity depends upon the problem under study. The definitions given in Eq. (5.21) are appropriate for converting between mean square orbital elements and velocity dispersions in the planetesimal distribution function, but for other applications different formulae are needed. For a single planetesimal with eccentricity e and inclination i at least three slightly different definitions of the "random velocity" σ can be useful (Lissauer & Stewart, 1993):

(1) The planetesimal velocity relative to a circular orbit with $i = 0$ with the same semi-major axis

$$\sigma = (e^2 + i^2)^{1/2} v_K. \tag{5.22}$$

(2) The planetesimal velocity relative to the local circular orbit with $i = 0$

$$\sigma = \left(\frac{5}{8}e^2 + \frac{1}{2}i^2\right)^{1/2} v_K. \tag{5.23}$$

(3) The planetesimal velocity relative to other planetesimals

$$\sigma = \left(\frac{5}{4}e^2 + i^2\right)^{1/2} v_K. \tag{5.24}$$

Yet more variations on these expressions arise due to the need to average over the planetesimal distribution when calculating, for example, the collision rates of planetesimals with protoplanets.

Very often the physics of the excitation and damping of e and i is such that the velocity distribution closely approximates an isotropic form. In this limit

$$\langle e^2 \rangle^{1/2} = 2\langle i^2 \rangle^{1/2}, \tag{5.25}$$

and the problem has one fewer degree of freedom. Whether or not we make this assumption, however, our task is to calculate, for a given distribution of e and i, the rate of collisions while accounting properly for the effects of both gravitational focusing and three-body dynamics. Having done this we then need to consider the mechanisms that determine appropriate values for $\langle e^2 \rangle$ and $\langle i^2 \rangle$.

5.2.1 Approximate Treatment

Simple results for the growth rate of protoplanets can readily be derived for the dispersion dominated regime. Let us assume that a relatively massive body of mass M, physical radius R_s, and surface escape speed v_{esc} is embedded within a swarm of smaller planetesimals. The planetesimal swarm has a local surface density Σ_p and a velocity dispersion σ (assumed here to be isotropic so that no distinction between σ_r and σ_z is needed). The vertical scale-height of the swarm is then $h_p \simeq \sigma/\Omega$ and its volume density is

$$\rho_{sw} \simeq \frac{\Sigma_p}{2h_p}. \tag{5.26}$$

The statistical (or "particle in a box") estimate for the growth rate of the massive body is then simply the product of the density of the planetesimal swarm, the encounter velocity at infinity, and the collision cross-section Γ (Eq. 4.63):

$$\frac{dM}{dt} = \rho_{sw}\sigma\pi R_s^2 \left(1 + \frac{v_{esc}^2}{\sigma^2}\right). \tag{5.27}$$

Substituting for ρ_{sw} and h_p we find that

$$\frac{dM}{dt} = \frac{\sqrt{3}}{2}\Sigma_p\Omega\pi R_s^2 \left(1 + \frac{v_{esc}^2}{\sigma^2}\right), \tag{5.28}$$

where the numerical pre-factor, which we have not derived, is correct for an isotropic velocity dispersion (Lissauer, 1993). The inverse scaling of the density with the velocity dispersion results in a cancellation of the direct effect of the velocity dispersion, which enters only via the gravitational focusing term.

Two simple solutions to this equation give insight into the properties of more sophisticated models of planetary growth. First, we assume that the gravitational focusing factor,

$$F_g = \left(1 + \frac{v_{esc}^2}{\sigma^2}\right), \tag{5.29}$$

is a constant, and that the surface density of the planetesimal swarm does not change with time. In this limit the equation for the growth of the protoplanet's radius,

$$\frac{dR_s}{dt} = \frac{\sqrt{3}}{8} \frac{\Sigma_p \Omega}{\rho_m} F_g, \tag{5.30}$$

where ρ_m is the density of the protoplanet, can be trivially integrated. We find that

$$R_s \propto t, \tag{5.31}$$

and the radius grows at a linear rate. For an icy body with $\rho_m = 1\,\mathrm{g\,cm^{-1}}$ at 5.2 AU (the current orbital radius of Jupiter),

$$\frac{dR_s}{dt} \approx 1 \left(\frac{\Sigma_p}{10\,\mathrm{g\,cm^{-2}}}\right) F_g\,\mathrm{cm\,yr^{-1}}. \tag{5.32}$$

Unless gravitational focusing is dominant this is a very slow growth rate. At this rate it would take 10 Myr – as long as or longer than the lifetime of typical protoplanetary disks – to grow a solid body up to a size of 100 km. We conclude immediately that to build large bodies in the outer regions of the protoplanetary disk large gravitational focusing enhancements to the cross-section are unavoidable.

Given the necessity of gravitational focusing, the other obvious limit to consider is that in which $v_{esc} \gg \sigma$. In this regime

$$F_g \simeq \frac{v_{esc}^2}{\sigma^2} = \frac{2GM}{\sigma^2 R_s}, \tag{5.33}$$

and the rate of growth of the protoplanet mass becomes

$$\frac{dM}{dt} = \frac{\sqrt{3}\pi\,G\Sigma_p\Omega}{\sigma^2} MR_s = kM^{4/3}, \tag{5.34}$$

where k is a constant if the properties of the planetesimal swarm are fixed (i.e. if the growing protoplanet neither excites the velocity dispersion nor consumes a significant fraction of the planetesimals). If under these conditions we consider the growth of *two* bodies with masses M_1 and M_2 such that $M_1 > M_2$ we find that

$$\frac{dM_1/M_1}{dM_2/M_2} = \frac{R_1}{R_2} > 1. \tag{5.35}$$

The initially more massive body grows faster than its less massive cousin, both absolutely and as measured by the ratio of masses M_1/M_2. This phenomenon is called *runaway growth*, and it allows for much more rapid formation of large bodies. Indeed, if we formally integrate Eq. (5.34) for a fixed planetesimal velocity dispersion we obtain

$$M(t) = \frac{1}{(M_0^{-1/3} - k't)^3},\tag{5.36}$$

where k' is a constant. The planet attains an infinite mass at a finite time. In reality this singularity is avoided because the feedback from the growing planet results in an increase of the planetesimal velocity dispersion, which slows growth. The physical lesson, however, is correct – gravitational focusing can drive runaway growth and rapid formation of large bodies.

5.2.2 Shear and Dispersion Dominated Limits

In the dispersion dominated limit the approximate treatment leading to Eq. (5.28) suffices to expose most of the important physics. The shear dominated limit is trickier to analyze, and in general there is no alternative but to use the results of three-body numerical experiments or fitting formulae derived from such experiments. To gain some analytic insight into the important processes we content ourselves here with an approximate treatment that is a simplified version of that given by Greenberg *et al.* (1991). The calculation has two parts. First, we estimate the width of the annulus (described as the "feeding zone") surrounding the protoplanet within which planetesimals can be diverted on to trajectories that enter the Hill sphere. Second, we assess the fraction of incoming planetesimals that actually impacts the protoplanet.

To derive the width of the feeding zone we consider the limit in which the growing protoplanet, of mass M, has a circular orbit. We assume that the planetesimals likewise have very small eccentricities, so that we are firmly in the shear dominated regime. A planetesimal whose orbit differs from that of the protoplanet by Δa must then have its eccentricity excited to

$$e \approx \frac{\Delta a}{a}\tag{5.37}$$

if it is to be diverted on to a trajectory that would approach the planet.

Let us now evaluate the impulse that the planetesimal receives as it passes by the protoplanet. The strongest perturbations occur while the two bodies are separated by an azimuthal distance of Δa or less, which persists for a time interval $\delta t = 4/\Omega$. The impulse imparted to the planetesimal is then

$$\delta v \approx \frac{4}{\Omega} \frac{GM}{(\Delta a)^2},\tag{5.38}$$

which corresponds to an eccentricity of

$$e \approx \frac{4}{\Omega^2 a} \frac{GM}{(\Delta a)^2}. \tag{5.39}$$

Equating this value of the eccentricity to that *required* in order to yield an approach trajectory (Eq. 5.37) we find that the outer edge of the feeding zone is delimited by a half-width[2]

$$\Delta a \simeq \left(\frac{4M}{M_*}\right)^{1/3} a = 2.3 \, r_{\mathrm{H}}, \tag{5.40}$$

where the Hill sphere radius r_{H} is defined by Eq. (5.15).

Planetesimals with orbits as far away from the protoplanet as $(a - \Delta a)$ and $(a + \Delta a)$ can be deflecting on to approach trajectories. However, *not all* orbits within the feeding zone permit close approaches. As illustrated in Fig. 5.3, those planetesimals whose orbits are too *close* to that of the protoplanet describe horseshoe orbits that never encounter the Hill sphere. Roughly speaking, those planetesimals with semi-major axes between $(a - \Delta a/2)$ and $(a + \Delta a/2)$ are protected from encounters. This reduces the effective width of the feeding zone from $2\Delta a$ to Δa. The typical planetesimal that can take part in the feeding flow is then at a radial distance of $0.75\Delta a$ from the protoplanet. In the frame rotating with the protoplanet, the average relative velocity of approach for a planetesimal that will enter the Hill sphere is then

$$v_{\mathrm{shear}} = a \left| \frac{d\Omega}{da} \right| \frac{3}{4} \Delta a = \frac{9}{8} \Omega \Delta a. \tag{5.41}$$

Combining the results for the effective width of the feeding zone and the approach velocity we obtain

$$\frac{dM_{\mathrm{H}}}{dt} = \frac{9}{8} \Omega \Delta a \Sigma_{\mathrm{p}} \Delta a, \tag{5.42}$$

as the rate at which mass flows toward the protoplanet from the planetesimal disk in the shear dominated regime. Although we have not proved it here, this mass flux that is diverted toward the planet is to a good approximation also the mass flux that enters the Hill sphere around the protoplanet.

The planetesimals that flow into the Hill sphere do so with typical encounter velocities that are (using Eq. 5.37) of the order of

$$v_{\mathrm{enc}} \sim ev_{\mathrm{K}} \sim \Omega \Delta a. \tag{5.43}$$

Since Δa scales as the Hill sphere radius the mass ratio of the protoplanet to the star enters this expression only weakly, as the one-third power. Once the planetesimals

---

[2] This is an approximate derivation that happens to give an answer that is quite close to that derived from numerical experiments. Equally valid approximate treatments will all yield slightly different numerical factors.

have entered the Hill sphere the collision cross-section with the protoplanet can be evaluated using two-body dynamics. Provided that the half-thickness of the incoming planetesimal flow exceeds the gravitational capture radius

$$ai > R_{\text{capture}}, \tag{5.44}$$

where

$$R_{\text{capture}} = R_{\text{s}} \left(1 + \frac{v_{\text{esc}}^2}{v_{\text{enc}}^2} \right)^{1/2}, \tag{5.45}$$

the collision cross-section takes the same form as in the dispersion dominated limit:

$$\Gamma = \pi R_{\text{s}}^2 \left(1 + \frac{v_{\text{esc}}^2}{v_{\text{enc}}^2} \right). \tag{5.46}$$

Simple geometry, depicted in Fig. 5.6, then allows us to determine the *fraction* of planetesimals entering the Hill sphere that go on to impact the protoplanet. This fraction is

$$f \approx \frac{\Gamma}{(2r_{\text{H}})(2ai)}. \tag{5.47}$$

Collecting together results, the rate at which the protoplanet grows via accretion of planetesimals in the shear dominated regime is the product of this fraction with the mass flow rate into the Hill sphere (Eq. 5.42):

$$\frac{dM}{dt} = \frac{9}{32} \frac{(\Delta a)^2}{air_{\text{H}}} \Sigma_{\text{p}} \Omega \pi R_{\text{s}}^2 \left(1 + \frac{v_{\text{esc}}^2}{v_{\text{enc}}^2} \right). \tag{5.48}$$

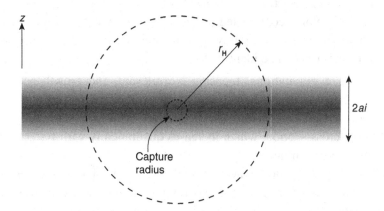

Figure 5.6 An edge-on view showing the flow of planetesimals into the protoplanet's Hill sphere in the shear dominated regime. In the limit depicted here, the vertical thickness of the planetesimal disk exceeds the capture radius of the protoplanet and the three-dimensional formula for gravitational focusing describes the cross-section. If the disk is *very* thin a different regime of essentially two-dimensional accretion applies.

Although this expression is superficially similar to the result that we deduced in the dispersion dominated regime (Eq. 5.28) there are a number of important differences. First, the gravitational focusing term within the parentheses no longer depends upon the random velocities of the planetesimals, but rather on the ratio between the escape velocity and the speed at which planetesimals enter the Hill sphere. Since $v_{enc}^2 \propto r_H^2 \propto M^{2/3}$, while $v_{esc}^2 \propto M$, there is a partial cancellation of the mass dependence of the gravitational focusing term. Second, there is an explicit dependence on the vertical thickness of the planetesimal swarm, $dM/dt \propto 1/i$. The eccentricity is formally irrelevant, although it will often be the case that the eccentricities and inclinations of the planetesimals continue to obey the relation $e \sim 2i$.

The above analysis was predicated on the assumption that $ai > R_{capture}$. If the planetesimal disk is extremely cold and thin this assumption can be violated, and there is a further transition from the three-dimensional accretion geometry discussed above to an essentially two-dimensional planar flow. In the very thin disk limit the rate at which planetesimals enter the Hill sphere remains unaltered, but the fraction of those planetesimals that are accreted becomes

$$f \approx \frac{R_{capture}}{r_H}. \tag{5.49}$$

The protoplanet growth rate in this regime contains no dependence upon either e or i.

The parameters that control the growth rate of the protoplanet are different in the shear and dispersion dominated limits, and hence there is no universality to the transition between the different regimes. In particular, the switch from dispersion to shear domination does not occur at a universal value of the gravitational focusing factor. Figure 5.7 shows the dependence of the collision cross-section (normalized to the geometric cross-section) as a function of the ratio v_{esc}/σ (after Greenberg *et al.*, 1991). This plot is constructed for specific protoplanet parameters assuming that v_{esc} remains fixed, while σ is varied maintaining $e = 2i$. In this case, and typically, four regimes can be identified:

- For $v_{esc}/\sigma < 1$ gravitational focusing is irrelevant, and growth occurs in the dispersion dominated regime according to the geometric cross-section.
- Once $v_{esc}/\sigma > 1$ gravitational focusing becomes significant and rapidly comes to dominate the cross-section, which increases quadratically with v_{esc}/σ whilst we remain dispersion dominated.
- Exiting the dispersion dominated regime the growth of the cross-section first steepens, before increasing more slowly (as $1/\sigma$) in the shear dominated regime.
- Finally, the thickness of the disk falls below the scale of the capture radius $R_{capture}$. In this thin disk regime the effective cross-section is constant.

Computations of the growth rate based on numerical integrations of test particles in the gravitational field of the protoplanet and the star show, not unexpectedly, that the transitions between these regimes occur quite smoothly.

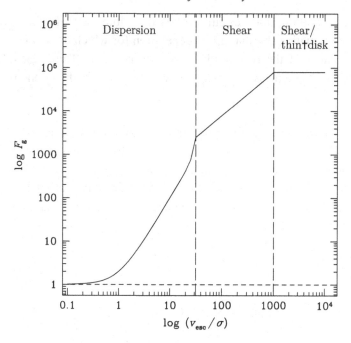

Figure 5.7 How the gravitational enhancement factor (the ratio of the true collision cross-section to the geometric cross-section) varies with v_{esc}/σ, the ratio of the escape speed from the protoplanet to the velocity dispersion of the planetesimal swarm. In constructing this plot it is assumed that the protoplanet mass is fixed and that the velocity dispersion of the planetesimals is the quantity being varied.

5.2.3 Isolation Mass

The growth of a protoplanet within the planetesimal disk is eventually limited by the feedback of the protoplanet on the properties of the planetesimals. The possible types of feedback include excitation of the random velocities of the planetesimals (which will slow growth by reducing the magnitude of gravitational focusing) and simple depletion of the local planetesimal surface density due to accretion. We can estimate the maximum mass that a protoplanet can attain during runaway growth by finding the mass of a single body that has consumed all of the planetesimals in its vicinity. This is known as the *isolation mass*. We work in the shear dominated regime, since for a massive body to have grown rapidly the planetesimal disk within which it is embedded must have had a low velocity dispersion.

To determine the isolation mass we make use of the fact that in the shear dominated regime a body of mass M can accrete only those planetesimals whose orbits lie within the feeding zone. The radial extent of the feeding zone Δa_{max} scales with the radius of the Hill sphere:

$$\Delta a_{max} = C r_{H}. \tag{5.50}$$

For planetesimals on initially circular orbits the width of the feeding zone can be estimated by evaluating the maximum separation for which collisions are possible within the context of the restricted three-body problem. This yields $C = 2\sqrt{3}$ (Lissauer, 1993).[3] The mass of planetesimals within the feeding zone is

$$2\pi a \cdot 2\Delta a_{\max} \cdot \Sigma_{\mathrm{p}} \propto M^{1/3}. \tag{5.51}$$

Note the one-third power of the planet mass, which arises from the mass dependence of the Hill radius. As a planet grows its feeding zone expands, but the mass of new planetesimals within the expanded feeding zone rises more slowly than linearly. We thus obtain the isolation mass by setting the protoplanet mass equal to the mass of the planetesimals in the feeding zone of the original disk:

$$M_{\mathrm{iso}} = 4\pi a \cdot C \left(\frac{M_{\mathrm{iso}}}{3M_*} \right)^{1/3} a \cdot \Sigma_{\mathrm{p}}, \tag{5.52}$$

which gives

$$M_{\mathrm{iso}} = \frac{8}{\sqrt{3}} \pi^{3/2} C^{3/2} M_*^{-1/2} \Sigma_{\mathrm{p}}^{3/2} a^3. \tag{5.53}$$

Evaluating this expression in the terrestrial planet region, taking $a = 1$ AU, $\Sigma_{\mathrm{p}} = 10$ gcm^{-2}, $M_* = M_{\odot}$, and $C = 2\sqrt{3}$ we obtain

$$M_{\mathrm{iso}} \simeq 0.07 \, M_{\oplus}. \tag{5.54}$$

Isolation is therefore likely to occur late in the formation of the terrestrial planets. Repeating the estimate for the conditions appropriate to the formation of Jupiter's core, using $\Sigma_{\mathrm{p}} = 10 \, \mathrm{g\,cm}^{-2}$ as adopted by Pollack *et al.* (1996) gives

$$M_{\mathrm{iso}} \simeq 9 \, M_{\oplus}. \tag{5.55}$$

This estimate is comparable to or larger than the current best determinations for the mass of the Jovian core. Full isolation may or may not be relevant to the formation of Jupiter and the other giant planets, depending upon the adopted disk model.

5.3 Velocity Dispersion

The random velocities of bodies during planet formation are determined by a competition between excitation and damping processes. Several processes can be important:

- **Viscous stirring**. The random motions of a population of planetesimals that are initially in a cold disk will be excited by the cumulative effect of weak gravitational encounters. The increase in the random energy of the bodies comes

---

[3] This is larger than the value ($C \simeq 2.3$) that we derived in Section 5.2.2 because here we consider all planetesimals that are not dynamically forbidden from eventually encountering the protoplanet, whereas previously we counted only those that would be deflecting into the Hill sphere on an orbital time scale.

ultimately from the orbital energy. This is the only excitation process that operates within a disk composed of equal mass bodies.

- **Dynamical friction**. Gravitational scattering behaves somewhat differently in the case where there is a spectrum of masses. The tendency for the system to seek equipartition of energy between the particles results in a transfer of energy from massive bodies to less massive ones, and the development of a mass-dependent velocity dispersion.

- **Gas drag**. Aerodynamic drag, although much weaker for planetesimals than for meter-scale bodies, continues to damp both e and i.

- **Inelastic collisions**. Physical collisions between bodies that result in energy dissipation also damp e and i.

- **Turbulent stirring**. Excitation of the random motions of bodies due to gravitational perturbations from a turbulent gas disk.

The most general formulation of the problem of planetary growth requires writing time-dependent equations for de^2/dt and di^2/dt that take account of these processes. The eccentricities and inclinations (or equivalently the velocity dispersion in the radial and vertical directions) depend upon mass, and are coupled to the equations describing the growth of protoplanets. If the time scales for growth and damping are short enough, the problem can be simplified by solving for the equilibrium values of e and i.

5.3.1 Viscous Stirring

We can estimate how quickly an initially cold disk of equal mass planetesimals will be heated by gravitational interactions by considering the amount of energy that is converted from ordered to disordered motion during a single encounter, and then summing over all encounters. We will work in the dispersion dominated regime in which two-body dynamics suffices to describe the encounters, and consider distant (or weak) encounters during which the perturbation to the initial trajectories is small. An individual flyby of two planetesimals with mass m, relative velocity σ, and impact parameter b then has the geometry shown in Fig. 5.8. The component of the gravitational force that is perpendicular to the trajectory is then

$$F_\perp = \frac{Gm^2}{d^2} \cos\theta, \tag{5.56}$$

where d is the instantaneous separation between the bodies. If we define $t = 0$ to coincide with the moment of closest approach, then the distance along the trajectory from the point of closest approach is just $x = \sigma t$ and we can write the time-dependent perpendicular component of the force as

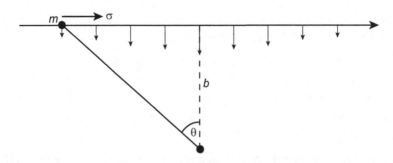

Figure 5.8 A body of mass m passes by a second body with an impact parameter (i.e. a distance of closest approach) b and a relative velocity σ. In the regime of a weak encounter the deflection angle is small and the impulse can be calculated assuming that the trajectory is unperturbed.

$$F_\perp = \frac{Gm^2}{b^2}\left[1 + \left(\frac{\sigma t}{b}\right)^2\right]^{-3/2}. \tag{5.57}$$

Integrating this force over the duration of the encounter yields the impulse felt by the planetesimals

$$|\delta v_\perp| = \int_{-\infty}^{\infty} \frac{F_\perp}{m} dt = \frac{2Gm}{b\sigma}, \tag{5.58}$$

and the change in kinetic energy

$$\delta E = \frac{1}{2}m|\delta v_\perp|^2 = \frac{2G^2m^3}{b^2\sigma^2}. \tag{5.59}$$

For consistency with the assumption that the trajectory remains almost unperturbed, we require that $|\delta v_\perp| \ll \sigma$, and this condition defines

$$b_{\min} = \frac{2Gm}{\sigma^2} \tag{5.60}$$

as the value of the impact parameter that delineates the boundary between weak (small deflection angle) encounters and strong (large angle) encounters.

To sum up the effect that many individual weak encounters have on the random velocity of one planetesimal, we note that the rate of encounters with an impact parameter in the range between b and $(b + db)$ is

$$\Gamma = 2\pi b \, db \, n_{sw} \sigma, \tag{5.61}$$

where n_{sw} is the number density of the planetesimal swarm. The rate of change of the kinetic energy is then

$$\frac{dE}{dt} = \frac{4\pi G^2 m^3 n_{sw}}{\sigma} \int_{b_{\min}}^{b_{\max}} \frac{db}{b}, \tag{5.62}$$

which is conventionally written in the form

$$\frac{\mathrm{d}E}{\mathrm{d}t} = \frac{4\pi G^2 m^3 n_{sw}}{\sigma} \ln \Lambda, \tag{5.63}$$

where $\ln \Lambda$ is known as the "Coulomb logarithm." The value of the Coulomb logarithm depends upon the size of the system (which determines the most distant encounters that need to be summed over) and may change with time, though given the weak (logarithmic) dependence it can often be approximated as a constant. We observe that for a given surface density of planetesimals viscous stirring will be more efficient if the planetesimal mass is large, since large masses imply larger individual kicks whose effects do not cancel out to the same extent as smaller more numerous scattering events.

To apply Eq. (5.63) to the problem of viscous stirring of a planetesimal disk we note that

$$n_{sw} = \frac{\rho_{sw}}{m} \simeq \frac{\Sigma_{sw}\Omega}{2m\sigma}, \tag{5.64}$$

and identify the energy E as the kinetic energy associated with the random component of the planetesimals' velocities. Equation (5.63) can then be written in the form

$$\frac{\mathrm{d}\sigma}{\mathrm{d}t} = \frac{2\pi G^2 m \Sigma_{sw}\Omega \ln \Lambda}{\sigma^3}, \tag{5.65}$$

where all the terms in the numerator on the right-hand-side are either exactly or (in the case of $\ln \Lambda$) approximately independent of time. Integrating, we predict that the random velocity of the planetesimals (and, hence, the vertical thickness of the planetesimal disk) ought to increase with time as

$$\sigma(t) \propto t^{1/4}. \tag{5.66}$$

Gravitational scattering is therefore an efficient mechanism for heating an initially cold thin disk, but the efficiency declines as the disk heats up and the encounter velocities increase. These properties are illustrated in Fig. 5.9, which shows the evolution of a single component disk under the action of viscous stirring calculated using N-body methods (Ohtsuki *et al.*, 2002). As noted previously, viscous stirring in this regime approximately maintains a ratio $e \simeq 2i$.

Our analysis is not sufficiently careful as to warrant any great faith in the accuracy of the pre-factor in Eq. (5.63). We can, however, estimate to an order of magnitude the time scale over which scattering will heat a planetesimal disk. To do so we adopt conditions that might be appropriate at 1 AU, where the surface density of 10 km radius planetesimals is $\Sigma_p = 10 \, \mathrm{g \, cm^{-2}}$. We assume that the planetesimals have a mass $m = 10^{19} \, \mathrm{g}$ and a random component of velocity $\sigma = 10^3 \, \mathrm{cm \, s^{-1}}$ (this is about an order of magnitude in excess of the Hill velocity). Taking b_{max} to be equal to the vertical thickness of the planetesimal disk yields an estimate for the

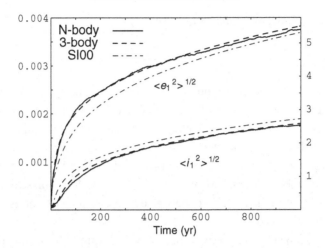

Figure 5.9 The evolution of the root mean square eccentricity and inclination of a disk composed of 10^3 equal mass bodies orbiting at or near 1 AU. The bodies have $m = 10^{24}$ g and lie in a disk with surface density $\Sigma_p = 10 \, \mathrm{g \, cm}^{-2}$. No damping processes are included. The solid lines show numerical N-body results, while the long-dashed lines labeled "3-body" and "S100" show different simplified results. Reproduced from Ohtsuki *et al.* (2002), with permission.

Coulomb logarithm $\ln \Lambda \simeq 9$, and we predict that the disk ought to be heated via gravitational scattering on a time scale

$$t_{\mathrm{VS}} = \frac{\sigma}{d\sigma/dt} \sim 6 \times 10^3 \, \mathrm{yr}. \tag{5.67}$$

This time scale is short enough that viscous stirring due to mutual gravitational perturbations will be an important source of heating for disks of planetesimals prior to the formation of any large bodies.

5.3.2 Dynamical Friction

Identical arguments can be applied to a two-component disk of planetesimals made up of bodies with masses m and M. As before, gravitational scattering *among* bodies of the same mass results in a steady increase of the random component of the planetesimal velocities, and a corresponding thickening of the disk. There also is a new effect, however, which derives from the fact that an encounter between a low mass body of mass m and a larger one of mass M gives a greater impulse to the lower mass object. The result of many such scatterings is that the system tries to attain a state of energy equipartition in which

$$\frac{1}{2} m \sigma_m^2 = \frac{1}{2} M \sigma_M^2, \tag{5.68}$$

where σ_m and σ_M are, respectively, the random velocities of the low and high mass bodies. This process is known as dynamical friction, and it leads to a mass

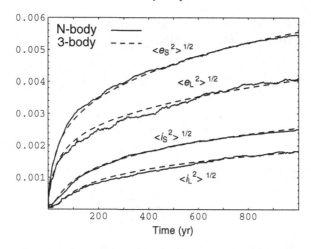

Figure 5.10 The evolution of the root mean square eccentricity and inclination of bodies within a two-component disk made up of 800 bodies with $m = 10^{24}$ g and 200 bodies with $m = 4 \times 10^{24}$ g. The summed surface density of the two components is $\Sigma_p = 10\,\mathrm{g\,cm}^{-2}$. The bodies are distributed randomly in a narrow annulus centered on 1 AU with initially small values of e and i. Viscous stirring heats both disks, but the eccentricity $\langle e_L^2 \rangle^{1/2}$ and inclination $\langle i_L^2 \rangle^{1/2}$ of the large bodies remains systematically lower than that of the smaller bodies due to dynamical friction. Reproduced from Ohtsuki *et al.* (2002), with permission.

dependence of the mean eccentricity and inclination of planetesimals and growing protoplanets. An example is shown in Fig. 5.10 from numerical results obtained by Ohtsuki *et al.* (2002). Although complete equipartition may not always be attained (Rafikov, 2003) this simple result is nonetheless of considerable import for studies of terrestrial planet formation. It implies that the relative velocity between a growing protoplanet and the surrounding planetesimals is *smaller* than that between two planetesimals. Lower velocities mean larger enhancements to the cross-section due to gravitational focusing, and hence dynamical friction tends to amplify the tendency toward runaway growth of the larger bodies.

5.3.3 Gas Drag

Damping of planetesimal eccentricity and inclination occurs as a consequence of gas drag. Due to the fact that the gas is partially supported against gravity by a pressure gradient, planetesimals experience aerodynamic drag even if $e = i = 0$ (cf. Section 2.4 and Eq. 4.31), and this modifies the damping rates for eccentric and/or inclined bodies. Defining the fractional difference between the Keplerian and gas disk velocities as

$$\eta' = \frac{v_K - v_{\phi,\mathrm{gas}}}{v_K}, \qquad (5.69)$$

Adachi *et al.* (1976) find that the damping rate of planetesimal eccentricity and inclination can be approximately described via

$$\frac{de}{dt} = -\frac{e}{t_{\text{drag}}} \left(\eta'^2 + \frac{5}{8}e^2 + \frac{1}{2}i^2 \right)^{1/2},$$ (5.70)

$$\frac{di}{dt} = -\frac{i}{2t_{\text{drag}}} \left(\eta'^2 + \frac{5}{8}e^2 + \frac{1}{2}i^2 \right)^{1/2},$$ (5.71)

where the characteristic time scale is defined as

$$t_{\text{drag}} = \frac{8}{3C_{\text{D}}} \frac{\rho_m s}{\rho v_{\text{K}}}.$$ (5.72)

Since η' is typically a few $\times 10^{-3}$ the term in the parentheses due to the background gas drag is dominant for very small values of e and i, and in this limit the damping time scale is independent of the actual values of the eccentricity and inclination. For low eccentricity planetesimals of mass 10^{19} g and radius 10 km orbiting within a gas disk of density $\rho = 5 \times 10^{-10}$ g cm^{-3} and $\eta' = 0.004$, for example, the damping time scale is of the order of 10^5 yr. This is a long time scale, but as we have noted it takes progressively longer to excite the random velocities of planetesimals by viscous stirring as their eccentricities and inclinations increase. It is therefore possible to reach an equilibrium state in which gas drag is sufficient to maintain the population of small bodies on almost circular orbits despite the heating due to gravitational scattering.

5.3.4 Inelastic Collisions

Sufficiently inelastic collisions between planetesimals can damp eccentricity and inclination by (ultimately) converting some of the random motion into heat within the bodies. To estimate the potential importance of this effect we first observe that if the gravitational focusing factor is large most "encounters" between planetesimals result in gravitational scattering (which is a heating process) rather than physical collisions. Inelastic collisions are only potentially significant when $\sigma > v_{\text{esc}}$, in which case the collision time scale is

$$t_{\text{inelastic}} = \frac{1}{n_{\text{sw}} \pi s^2 \sigma}.$$ (5.73)

Assuming an isotropic velocity dispersion, the mid-plane number density of planetesimals n_{sw} is inversely proportional to σ, which cancels as usual to yield an estimate

$$t_{\text{inelastic}} \simeq \frac{8\rho_m s}{3\Sigma_p \Omega}.$$ (5.74)

We can compare this time scale to the time scale for damping of planetesimal random motion by gas drag (Eq. 5.71), which in the limit of small e and i is

$$t_{\text{gas}} \simeq \frac{8\rho_m s}{3\eta' C_D \rho v_K}. \tag{5.75}$$

Equating $t_{\text{inelastic}}$ to t_{gas}, we find that up to factors of the order of unity the condition for inelastic collisions to dominate is

$$\Sigma_p \gtrsim \rho r \eta'. \tag{5.76}$$

This can be further simplified by noting that the deviation of the gas velocity from the Keplerian value $\eta' \sim (h/r)^2$. We find finally

$$\frac{\Sigma_p}{\Sigma} \gtrsim \frac{h}{r}, \tag{5.77}$$

as the condition for inelastic collisions to be more important than gas drag if, additionally, gravitational focusing is unimportant. Since Σ_p/Σ is assuredly smaller than the local dust to gas ratio – whose standard value of $f = 10^{-2}$ is in turn smaller than the typical $(h/r) \approx 0.05$ – this estimate justifies our neglect of inelastic collisions relative to aerodynamic drag in the preceding section. One can see, however, that the margin by which inelastic collisions are subdominant is not all that large, and as the gas disk evolves and dissipates it may be possible to encounter a regime in which inelastic collisions provide the main cooling mechanism.

Planetesimal excitation by gravitational coupling to turbulent fluctuations in the gas disk can also influence the velocity disperson. Many plausible sources of disk turbulence (including the magnetorotational instability) create spatial and temporal fluctuations in the gas surface density of the disk. Any solid body orbiting within the gas experiences stochastic forcing as a result of the gravitational forces from the nonaxisymmetric surface density fluctuations, and these randomly directed impulses act as an excitation mechanism for eccentricity and inclination. For terrestrial planet formation the main interest in this process derives from the possibility that excitation due to gravitational coupling to turbulence might dominate viscous stirring prior to the formation of large protoplanets, and thereby delay or (in some regions of the disk) even prevent collisional growth. Ormel and Okuzumi (2013) provide estimates of the importance of this effect. Since the mechanism or mechanisms that result in angular momentum transport within protoplanetary disks remain uncertain it will come as no surprise that there remain substantial uncertainties.

5.4 Regimes of Planetesimal-Driven Growth

The expressions derived above for the growth rate of protoplanets and for the velocity evolution of planetesimals are only suitable for order of magnitude estimates. More accurate semi-analytic formulae can be derived via two approaches: a kinetic formalism that is based on adding a collisional term to the collisionless Boltzmann equation describing the evolution of the phase-space density of bodies (Hornung *et al.*, 1985; Stewart & Wetherill, 1988), or a celestial mechanics treatment of three-body dynamics using Hill's equations (Ida, 1990). These approaches are equivalent (Stewart & Ida, 2000) and the approximate formulae that result can be further improved by reference to numerical integrations of the three-body problem. Inaba *et al.* (2001), Ohtsuki *et al.* (2002), and Chambers (2006) provide detailed expressions and fitting formulae that cover much of the parameter regime relevant for classical planetesimal-driven planet formation models.

Within their domain of validity these semi-analytic formulae for the excitation and damping of planetesimal eccentricity (supplemented where necessary with the similar expressions for dynamical friction and stirring by planetesimal–planetesimal scattering) are quite accurate. Figures 5.9 and 5.10 show, for example, the comparison between the semi-analytic formulae and N-body results for the cases of one- and two-component disks of bodies evolving under the action of viscous stirring and dynamical friction, and it is clear that the agreement is impressively good.

Returning to qualitative considerations, Fig. 5.11 is a sketch of the three possible regimes that can occur as planetesimals grow into planets. In *orderly* growth, the typical mass of bodies increases over time, but the dispersion in mass (in logarithmic units) does not. In *runaway growth*, a small number of bodies grow rapidly while the bulk of the population of bodies do not. Runaway growth is the norm in the early part of classical models of planetesimal-driven terrestrial planet formation (Greenberg *et al.*, 1978). At later times, ongoing growth occurs in a high velocity regime where the planetesimal disk is heated by viscous stirring from the

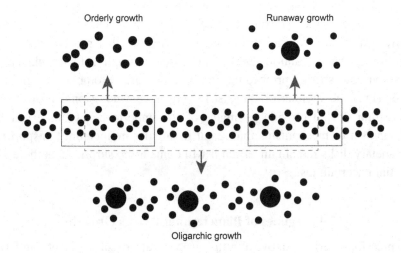

Figure 5.11 Illustration of the different regimes of planetesimal-driven growth.

protoplanets and cooled by gas drag (Ida & Makino, 1993). This regime, known as oligarchic growth, is partially self-limiting, since growth of any individual proto-planet increases the random velocities of planetesimals in its vicinity and decreases the gravitational focusing enhancement to the collision cross-section.

Although these regimes were originally defined in the context of models where planetesimals provided the only mechanism of growth, the same classification is useful in more general models that include the possibility of pebble accretion.

5.5 Coagulation Equation

For much of the preceding discussion we have assumed that the distribution of masses of the growing bodies can be partitioned into two groups: large planetary embryos with mass M and negligible random velocities, and much less massive planetesimals with mass m and significant random velocities. If runaway growth occurs this simple approximation is actually quite a good way to think about the problem, since each annulus of the disk *will* contain one body that is much larger than all the rest. One may rightly worry, however, that attempting to demonstrate the existence of runaway growth using a two-groups approximation involves a suspiciously circular logic. The right framework for approaching the problem of whether runaway growth occurs requires dropping the two-groups model and treat-ing the evolution of an arbitrary size distribution of bodies using the methods of coagulation theory (Smoluchowski, 1916). The application of this theory to planet formation was pioneered by Safronov (1969), and has subsequently been adopted by many authors. Clear descriptions of the basic method are given by Wether-ill (1990) and Kenyon and Luu (1998), whose nomenclature is largely adopted here.

Coagulation theory is based on solutions to the coagulation equation, which can be written equivalently in either integral (Eq. 4.74) or discrete form. In the discrete representation, we assume that at some time t there are n_k bodies within some fixed volume that have a mass km_1 that is an integral multiple of some small mass m_1. We treat n_k as a *continuous function*.[4] Ignoring radial migration and fragmentation, we can then write the discrete coagulation equation for an annulus of the disk in the form

$$\frac{dn_k}{dt} = \frac{1}{2} \sum_{i+j=k} A_{ij} n_i n_j - n_k \sum_{i=1}^{\infty} A_{ik} n_i, \qquad (5.78)$$

where the A_{ij}, known as the kernels, describe the probability of a collision that leads to accretion between bodies with masses im_1 and jm_1 per unit time. In general these will be nonlinear functions of the masses, random velocities, and physical properties (e.g. density) of the bodies involved. One observes that the number of

---

[4] This is reasonable provided that $n_k \gg 1$, but we must be cognizant that the statistical foundations of the method can break down if for some k of interest $n_k \sim 1$.

bodies in the kth mass bin changes due to two processes. First, the number of bodies of mass km_1 *increases* whenever there is a collision between any pair of bodies whose total mass $(i+j)m_1$ sums to km_1 (the factor of $1/2$ is present to avoid double counting these collisions). Second, the number of bodies of mass km_1 *decreases* whenever there is any collision with a body of any other mass.

There are no known analytic solutions to the coagulation equation for kernels that encode the critical physics of planet formation – gravitational focusing and mass-dependent velocity dispersion. Numerical solutions provide the only way to handle these complexities. We can, however, gain some insight into the nature of general solutions to the coagulation equation by studying three known solutions for cases with particularly simple kernels. If

$$A_{ij} = \alpha, \tag{5.79}$$

where α is a constant, and there are initially n_0 bodies with mass m_1, then at time t

$$n_k = n_0 f^2 (1 - f)^{k-1}, \tag{5.80}$$

$$f = \frac{1}{1 + \frac{1}{2}\alpha n_0 t}. \tag{5.81}$$

Inspection of the units shows that f is a dimensionless measure of the time. The physical interpretation of f is that it is the fraction of bodies in the smallest mass bin that have yet to collide with any other body.

The solution expressed in Eq. (5.81) is plotted in Fig. 5.12. This solution is an example of *orderly growth*. As time progresses the mean mass of the population increases, but the shape of the mass spectrum approaches an asymptotic form and its width (expressed in logarithmic units) does not increase. Qualitatively similar is the solution for the kernel

$$A_{ij} = \alpha(m_i + m_j), \tag{5.82}$$

which has the form

$$n_k = n_0 \frac{k^{k-1}}{k!} f(1 - f)^{k-1} \exp[-k(1 - f)], \tag{5.83}$$

$$f = \exp[-\alpha n_0 t]. \tag{5.84}$$

This solution also exhibits orderly growth.

The final known solution applies to the case of a kernel that is the product of the masses of the bodies,

$$A_{ij} = \alpha m_i m_j. \tag{5.85}$$

In this case,

$$n_k = n_0 \frac{(2k)^{k-1}}{k! \, k} \left(\frac{1}{2}\alpha n_0 t\right)^{k-1} \exp[-\alpha n_0 k t]. \tag{5.86}$$

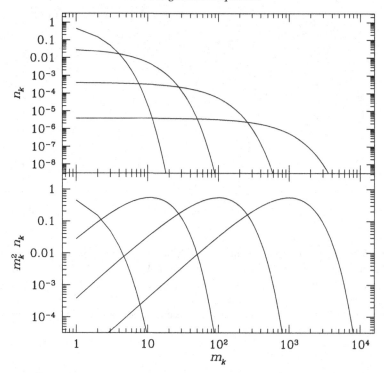

Figure 5.12 Solution to the coagulation equation for the simple case in which the kernel $A_{ij} = \alpha$ is a constant. Initially all bodies have mass m_1. The solution is plotted for scaled times $t' \equiv \alpha n_0 t$ equal to 1, 10, 100, and 10^3. The upper panel shows the evolution (on an arbitrary vertical scale) of the number of bodies n_k as a function of mass, while the lower panel shows the evolution of the mass distribution.

This solution is plotted in Fig. 5.13 together with a solution for the case of a constant kernel that has the same initial collision rate. Equation (5.86) represents a qualitatively different class of solution to the coagulation equation, in which *runaway growth* develops. The mass distribution develops a power-law tail toward high masses as a small number of bodies grow rapidly at the expense of all of the others. Solutions to the coagulation equation that display runaway growth generally apply only at early times, since once most of the mass accumulates into a single massive body the assumptions upon which the coagulation equation is based break down. Formally Eq. (5.86) is valid up until a time $\alpha n_0 t = 1$.

Realistic kernels for planet formation do not scale with mass in the same way as any of the analytically tractable forms. If gravitational focusing is unimportant, the collision cross-section scales with the geometric area and $A \propto R_s^2 \propto m^{2/3}$. This scaling is bracketed by the two analytic solutions that display orderly growth. Conversely, if gravitational focusing dominates then $A \propto m R_s \propto m^{4/3}$, which lies between analytic solutions exhibiting orderly and runaway growth. This observation does not allow us to draw any definitive conclusions, but numerical calculations

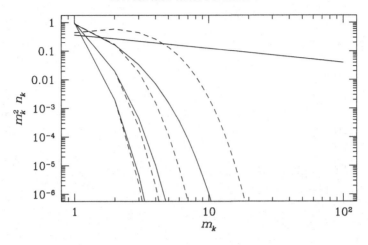

Figure 5.13 Comparison of the analytic solution to the coagulation equation with the kernel $A_{ij} = \alpha$ (dashed lines) to that with $A_{ij} = \alpha m_i m_j$ (solid lines). The solutions are plotted for scaled times $t' \equiv \alpha n_0 t$ equal to 10^{-3}, 10^{-2}, 0.1, and 1. The collision rate for $i = j = 1$ is identical for the two solutions, so the very early time behavior is the same. Large differences develop just prior to the onset of runaway growth at a time $\alpha n_0 t = 1$.

suggest that runaway growth can occur for kernels that describe collisions during the early phases of planet formation.

5.6 Pebble Accretion

The "classical" model of planet formation, in which growth occurs due to collisions between and among protoplanets and planetesimals, is the only possibility if the planetesimal formation process efficiently removes all (or almost all) small solid particles from the disk. Observationally, this is not what appears to happen. Dust and mm-sized grains are present in protoplanetary disks during all evolutionary phases, including at late times when planetesimals and larger bodies are very likely to have formed. This fact motivates us to consider how a growing planet would interact with solid bodies of arbitrary sizes during the period when gas is still present within the disk.

Figure 5.14 illustrates some of the main possibilities. The upper panels depict possible flows for bodies whose gravity is weak (think, for example, of a km-scale planetesimal). Very small dust particles, which are tightly coupled to the gas via aerodynamic forces, simply flow around the body. Larger objects, which experience negligible aerodynamic forces, collide and accrete only if their trajectories at large distance intersect the geometric cross-section of the body.

The lower panels of the figure illustrate what can happen for larger bodies whose gravity can significantly deflect the trajectories of passing objects. This leads to accretion cross-sections that exceed the geometric one. We have already discussed

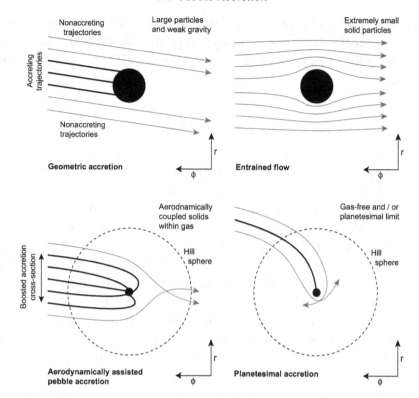

Figure 5.14 Illustration of some of the regimes of interaction between a growing planet and solid particles or larger planetesimals (after Ormel 2017). For relatively large bodies, the cross-section for the aerodynamically assisted accretion of particles (lower left panel) can be much larger than for planetesimals (lower right panel).

the enhancement due to gravitational focusing, which is relevant for planetesimals interacting with growing planets in a regime that is effectively gas-free. The cross-section is also boosted for solid particles whose size is such that they experience significant gravitational *and* aerodynamic forces during their encounter. This is the regime of aerodynamically assisted or *pebble* accretion (Ormel & Klahr, 2010; Lambrechts & Johansen, 2012). Pebble accretion is a qualitatively distinct growth mechanism for planets. Instead of depending upon the local density and velocity dispersion of planetesimals, as is the case for planetesimal-driven growth, the key factors are the local density and dimensionless friction time of small solid particles. Pebble accretion can be more efficient than planetesimal accretion, and its dependence on different factors allows for growth in a broader range of circumstances.

In this section we discuss the key physics of pebble accretion. Before doing so, a note on nomenclature is appropriate. For most of us, the word "pebble" conjures up an image of stones with sizes similar to those found, for example, on beaches. There is also a formal geological definition of pebbles as particles with sizes between

2 mm and 64 mm. Neither of these notions is relevant to pebble accretion. For our purposes, a "pebble" is a particle, of any composition or physical size, that has a dimensionless friction time in a range where aerodynamic forces lead to enhanced accretion.

5.6.1 Encounter Regimes

The rate of pebble accretion can be calculated by numerically integrating the trajectories of particles as they encounter protoplanets, including both gravitational and aerodynamic forces. Many of the formal results, however, can be approximately reproduced using an analytic model that is based on a simple time scale argument. We assume, throughout, that the protoplanet has an orbit with negligible eccentricity and inclination, that the effects of turbulence within the gas on particle accretion are small,[5] and that the temperatures are such that particles do not sublimate into vapor prior to being accreted. (These neglected effects can be significant for some applications.) We follow closely the pedagogical treatment of Ormel (2017).

Figure 5.15 shows the flow of pebbles (here assumed to be solid particles that are small enough to move at close to the same speed as the gas) in a frame co-rotating with a planet (or protoplanet) on a circular orbit. At the orbital radius of the planet, the gas and any aerodynamically coupled solids have a nonzero relative velocity due to the existence of pressure gradients within the disk (Section 2.4). For a gas disk with a mid-plane pressure profile

$$P \propto r^{-n}, \tag{5.87}$$

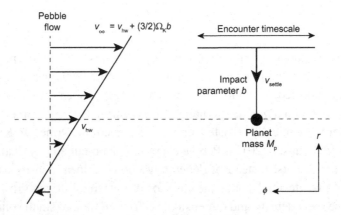

Figure 5.15 The flow of pebbles relative to a planet is a combination of a piece due to the headwind and a piece due to Keplerian shear. Pebbles are accreted if, within the encounter time (the time over which the particles feel the strongest gravity of the planet), they are coupled to the gas and are able to settle toward the planet.

---

[5] Turbulence was considered in early work on pebble accretion by Johansen and Lacerda (2010).

the resulting *headwind* velocity is

$$v_{hw} = \frac{1}{2}n\left(\frac{h}{r}\right)^2 v_K,$$ (5.88)

where (h/r) is the geometric thickness of the disk and $v_K = \sqrt{GM_*/r}$ is the Keplerian orbital speed. Numerically

$$v_{hw} = 35\left(\frac{n}{3}\right)\left(\frac{h/r}{0.05}\right)^2\left(\frac{M_*}{M_\odot}\right)^{1/2}\left(\frac{r}{10\ AU}\right)^{-1/2}\ m\ s^{-1}.$$ (5.89)

This is the speed with which a pebble, moving with the gas at the same orbital radius as the planet, would approach the planet.

The headwind speed depends upon the disk structure. For a disk with a power-law surface density profile and a mid-plane temperature profile, $T_c \propto r^{-1/2}$, n is a constant and $(h/r) \propto r^{1/4}$. In this special case the radial dependence arising from (h/r) and v_K cancels out in the above equation, and v_{hw} is just a constant throughout the disk.

Pebbles that encounter the planet with non-co-orbital trajectories do so with a relative velocity at large distance v_∞ that is the sum of the headwind speed and a piece due to Keplerian orbital shear. For a pebble with an impact parameter b a simple calculation gives

$$v_\infty = v_{hw} + \frac{3}{2}\Omega_K b.$$ (5.90)

If v_∞ is dominated by the first term on the right-hand-side, we say the encounter is in the headwind regime (also called the "Bondi" regime by some authors). Otherwise, the encounter is in the shear (or "Hill") regime. The dividing line between these regimes occurs for an impact parameter $b = (2/3)v_{hw}\Omega_K^{-1}$.

5.6.2 Pebble Accretion Conditions

Whether a pebble encountering a planet will be accreted can be estimated by considering the ordering of three time scales. We have already defined (in Section 4.2) the friction (or stopping) time scale t_{fric}, along with its dimensionless version $\tau_{fric} = t_{fric}\Omega_K$. In the Epstein drag regime the friction time is velocity independent,

$$t_{fric} = \frac{\rho_m}{\rho}\frac{s}{v_{th}},$$ (5.91)

and depends only on the particle size s, the particle material density ρ_m, the gas density ρ, and the gas thermal speed v_{th}.

The encounter with the planet introduces two new time scales. The first is the encounter time scale

$$t_{enc} = \frac{2b}{v_\infty},$$ (5.92)

which measures (in the impulse approximation) the duration of the period over which the pebble experiences the largest gravitational forces from the planet. The second is the settling time. This is the characteristic time scale on which a pebble, moving at its terminal velocity through gas at rest with respect to the planet, would settle. Equating the drag force, $|F_D| = mv/t_{fric}$, to the gravitational force, $F_{grav} = GM_p m/b^2$, for a pebble of mass m and initial distance b, we find

$$v_{settle} = \frac{GM_p}{b^2} t_{fric}.$$ (5.93)

The settling time is

$$t_{settle} = \frac{b^3}{GM_p} \frac{1}{t_{fric}}.$$ (5.94)

For pebble accretion in a laminar gas disk (i.e. ignoring the effects of turbulence, which would introduce an additional time scale) these three time scales are the key factors determining the dynamics of the encounter.

A pebble encountering a planet will be accreted if two conditions are met. First, the pebble needs to be sufficiently well-coupled to the gas that aerodynamic forces can modify its trajectory while the encounter is taking place. In terms of time scales, this means that

$$t_{fric} < t_{enc}.$$ (5.95)

Second, the gravity of the planet needs to be strong enough to pull the particle toward it, against the tendency of the disk flow to sweep it away. This translates to a condition

$$t_{settle} < t_{enc}.$$ (5.96)

Obviously both of these conditions are rather rough estimates. Rather surprisingly, however, imposing the requirement that they are simultaneously satisfied provides a reasonable approximation for when pebble encounters result in pebble accretion.

In the headwind regime, the encounter time scale $t_{enc} = 2b/v_{hw}$, and the settling time $t_{settle} = b^3/GM_p t_{fric}$. The condition that settling is faster than the time scale of the encounter is satisfied for $b \leq b_{hw}$, where the maximum impact parameter is

$$b_{hw} = \left(\frac{2GM_p}{v_{hw}} t_{fric} \right)^{1/2}.$$ (5.97)

The separate requirement that $t_{fric} < t_{enc}$ at $b = b_{hw}$ is satisfied if $M_p \geq M_{min}$, where

$$M_{min} = \frac{v_{hw}^3}{8G} t_{fric}.$$ (5.98)

Aerodynamically assisted accretion of pebbles, at rates significantly in excess of what would be predicted based on geometric grounds, sets in above a minimum mass that is linearly proportional to the stopping time (and hence to the particle size, in the Epstein regime).

In the shear regime, the encounter velocity $v_\infty \simeq (3/2)\Omega_K b$. The encounter time scale, $t_{enc} = 2b/v_\infty$, is then

$$t_{enc} \approx \Omega_K^{-1}. \tag{5.99}$$

The form of the encounter time scale in the shear regime guarantees that the condition that $t_{fric} < t_{enc}$ is satisfied for all particles with $\tau_{fric} \lesssim 1$. Equating t_{settle} with t_{enc}, we find that the maximum impact parameter for pebble accretion is

$$b_{sh} = \left(\frac{GM_p}{\Omega_K}t_{fric}\right)^{1/3}. \tag{5.100}$$

The maximum impact parameter scales more weakly with planet mass in the shear regime as compared to the headwind regime. The transition between the regimes occurs at a critical mass M_{crit}, which can be found by setting $b_{hw} = b_{sh}$. The result is

$$M_{crit} = \frac{1}{8}\frac{M_t}{\tau_{fric}}, \tag{5.101}$$

which we have written in this way to emphasize that the transition depends upon both the planet mass and the pebble size, defined in terms of its dimensionless stopping time. The characteristic mass M_t depends only on the headwind speed and on the angular velocity,

$$M_t \equiv \frac{v_{hw}^3}{G\Omega_K}. \tag{5.102}$$

The headwind regime is valid for low planet masses, while the shear regime applies for $M_p > M_{crit}$.

Unless τ_{fric} is very small the shear regime of pebble accretion is appropriate for all except the lowest protoplanet masses. Using the expression for v_{hw} (Eq. 5.88) the characteristic mass is

$$M_t = \frac{1}{8}n^3\left(\frac{h}{r}\right)^6 M_*. \tag{5.103}$$

The strong dependence on the disk thickness means that M_t is an increasing function of orbital radius (scaling as $r^{3/2}$ for simple irradiated disk models). Taking $n = 3$ and $M_* = M_\odot$, we obtain $M_t = 2.4 \times 10^{-4}\,M_\oplus$ for $(h/r) = 0.03$ and $M_t = 5.2 \times 10^{-3}\,M_\oplus$ for $(h/r) = 0.05$.

The potential importance of pebble accretion as a mechanism for planetary growth can be grasped by rewriting the maximum impact parameter for accretion in the shear regime (Eq. 5.100) in a different, but equivalent, form:

$$b_{sh} = \left(\frac{GM_p}{\Omega_K}t_{fric}\right)^{1/3} = \left(\frac{GM_p}{\Omega_K^2}\right)^{1/3}\tau_{fric}^{1/3} \approx \tau_{fric}^{1/3}r_H, \tag{5.104}$$

where r_H is the Hill radius of the planet. For particles with $\tau_{fric} \sim 1$ the cross-section for pebble accretion can be as large as the entire Hill sphere of the accreting planet. This is a very large cross-section indeed. Suppose we have a planet at orbital radius a with mass M_p, radius R_s, and density ρ_m. Comparing the geometric cross-section πR_s^2 to the maximum pebble accretion cross-section πr_H^2 gives a "pebble enhancement factor" that is loosely analogous to the gravitational focusing factor for planetesimal accretion. The maximum pebble enhancement factor is

$$F_{pebble} = \frac{r_H^2}{R_s^2} = \left(\frac{4\pi\rho_m}{9M_*}\right)^{2/3} a^2. \tag{5.105}$$

Taking $\rho_m = 1 \text{ g cm}^{-3}$ and $M_* = M_\oplus$,

$$F_{pebble} \sim 2 \times 10^4 \left(\frac{a}{\text{AU}}\right)^2. \tag{5.106}$$

At large orbital radii, especially, the peak cross-section for a planet to sweep up pebbles is very large. Realizing this theoretical limit requires, of course, that the pebbles in the vicinity of the planet have close to the optimal size, and obtaining a large accretion rate additionally requires a large enough density in pebbles. Because of these caveats, detailed models are needed to determine whether the pebble accretion rate exceeds that of planetesimal accretion for a given situation.

5.6.3 Pebble Accretion Rates

If the pebbles form a layer whose thickness exceeds the maximum impact parameter of the planet, the pebble accretion rate is simply

$$\dot{M}_p = \pi b^2 v_\infty \rho_{pebble}, \tag{5.107}$$

where ρ_{pebble} is the volume density of pebbles encountering the planet. Substituting for the impact parameter one finds

$$\dot{M}_p \approx 2\pi G M_p t_{fric} \rho_{pebbles}, \tag{5.108}$$

which is valid in both the headwind and shear regimes of pebble accretion. If, instead, the pebbles occupy a very thin layer with surface density Σ_{pebble}, the accretion rate is given by $\dot{M}_p = 2bv_\infty \Sigma_{pebble}$. Across the different regimes the scaling of the accretion rate with planet mass is at most linear. Accordingly, pebble accretion does not facilitate runaway growth.

5.6.4 Relative Importance of Pebble Accretion

There is no simple answer to the question of when pebble accretion dominates over planetesimal accretion. Among other factors, the balance depends upon:

- The size distribution of aerodynamically-relevant solids, and the gas disk structure, which together determine t_{fric}.
- The fraction of the total solid mass that forms planetesimals, rather than remaining in the pebble size range.
- The time-dependent drift of pebbles radially through the disk, and the extent to which protoplanets at larger radii (than the one being considered) sweep them up first.
- The thickness of the pebble layer, which will be set by turbulent stirring. (Note that we have ignored turbulence in our discussion, but that this is another relevant parameter.)
- The full suite of processes that determine the gravitational focusing factor for planetesimals.

The accretion of pebbles can be a self-limiting process. As the planet grows in mass, the surface density profile of the gas in the vicinity is modified by angular momentum exchange mediated by gravitational torques. If this leads to the formation of a local pressure maximum upstream of the planet, radial drift of pebbles toward the planet will be halted (see Section 4.3.2). This occurs at the *pebble isolation mass*, which has been estimated to be (Morbidelli & Nesvorny, 2012; Ormel, 2017)

$$M_{\text{p,pebble-iso}} \sim 40 M_{\oplus} \left(\frac{h/r}{0.05} \right)^3 . \tag{5.109}$$

Pebble isolation is closely related to the process of gap opening (discussed in Chapter 7), and it likewise involves complex and not fully understood aspects of disk physics. However, on order of magnitude grounds, it can clearly be an important phenomenon as planet masses approach those where runaway gas accretion occurs in core accretion models of giant planet formation.

5.7 Final Assembly

A fiducial model for the formation of terrestrial planets with properties (masses and orbital radii) similar to those in the Solar System, based on planetesimal-driven accretion only, can be developed based on the aforementioned physics. Starting from a large population of planetesimals the first two phases are:

- **Runaway growth**. Initially there are no large bodies, so the random velocity of planetesimals is set by a balance between viscous stirring among the planetesimals themselves and damping via gas drag. The combined influence of dynamical friction and gravitational focusing results in runaway growth of a small fraction of planetesimals.
- **Oligarchic growth**. Runaway growth ceases at the point when the rate of viscous stirring of the planetesimals by the largest bodies first exceeds the rate

of self-stirring among the planetesimals. The resulting boost in the strength of viscous stirring increases the equilibrium values of planetesimal eccentricity and inclination, partially limiting the gravitationally enhanced cross-section of the protoplanets. In this regime the growth of protoplanets continues to outrun that of planetesimals, but the dominance is local rather than global. Across the disk many *oligarchs* grow at similar rates by consuming planetesimals within their own largely independent feeding zones.

These initial stages of terrestrial planet formation are rapid (of the order of 0.01–1 Myr), and result in the formation of perhaps 10^2 to 10^3 large bodies across the terrestrial planet zone. These are massive objects (of the order of 10^{-2} $M_\oplus$ to 0.1 $M_\oplus$, so comparable to the mass of the Moon or Mercury) but they are not yet terrestrial planets.

The final assembly of terrestrial planets gets underway once the oligarchs have depleted the planetesimal disk to the point that dynamical friction can no longer

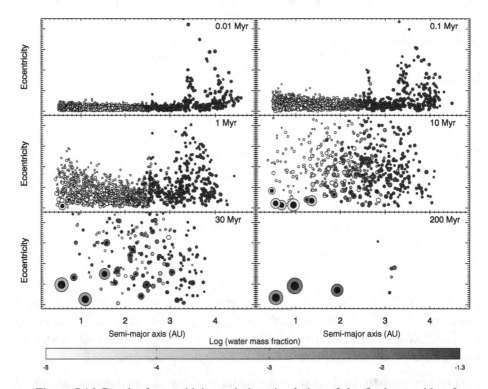

Figure 5.16 Results from a high resolution simulation of the final assembly of terrestrial planets (Raymond *et al.*, 2006). The initial conditions for this simulation are representative of the late stages of oligarchic growth, with approximately 2000 planetary embryos (shaded according to their water content) distributed in a disk with a surface density profile of $\Sigma_p \propto r^{-3/2}$ and a surface density at 1 AU of 10 g cm^{-2}. Jupiter is assumed to have a circular orbit at 5.5 AU during the duration of the final assembly phase, which takes around 100 Myr. Reproduced in modified form from Raymond *et al.* (2006), with permission.

maintain low eccentricities and inclinations of the oligarchs. Beyond this point the assumption that each oligarch grows in isolation breaks down, and the largest bodies start to interact strongly, collide, and scatter smaller bodies across a significant radial extent of the disk. Numerical N-body simulations (examples include Raymond *et al.*, 2006 and O'Brien *et al.*, 2006), an example of which is shown in Fig. 5.16, show that this final stage is by far the slowest, with large collisions continuing out to at least tens of Myr. The process is chaotic, and hence identical initial conditions can give rise to a range of outcomes that must be compared to the observed properties of the Solar System's terrestrial planets statistically. The level of agreement attained is by no means perfect, with the small mass of Mars (compared to that of the Earth and Venus) being the most important discrepancy. Viewed on a coarser level, however, reasonable values of the disk surface density in the inner Solar System do appear roughly to reproduce the observed properties of the terrestrial planets.

It is possible to extend the fiducial model for Solar System terrestrial planet formation, and consider how modest changes in the input parameters would affect the likely outcome. For example, both elementary stability arguments and numerical simulations predict that there ought to be a dependence on the surface density of the planetesimal disk, in the sense that higher Σ_p typically yields a smaller number of more massive planets (Wetherill, 1996; Kokubo *et al.*, 2006). This type of extrapolation has its limits, however, and when considering planets around solar-type stars in closer orbits, or habitable planets around lower-mass stars, qualitatively different formation histories may be involved. A significant role for pebble accretion is one possibility. Another is that migration may play a dominant role in determining the final architecture.

5.8 Further Reading

"Planet Formation by Coagulation: A focus on Uranus and Neptune," P. Goldreich, Y. Lithwick, & R. Sari (2004), *Annual Review of Astronomy & Astrophysics*, **42**, 549. This review includes useful "back of the envelope" estimates of planetary growth mechanisms.

"The Emerging Paradigm of Pebble Accretion," C. W. Ormel (2017), in *Formation, Evolution, and Dynamics of Young Solar Systems*, Astrophysics and Space Science Library, **445**, Cham: Springer, p. 197.

"Terrestrial Planet Formation at Home and Abroad," S. N. Raymond, *et al.* 2014, in *Protostars and Planets VI*, H. Beuther, R. S. Klessen, C. P. Dullemond, & T. Henning (eds.), Tucson: University of Arizona Press, p. 595.

6

Giant Planet Formation

Planets with masses above a few times the mass of the Earth typically have radii that are large enough to indicate the presence of a substantial gaseous envelope. In the Solar System the gas giants, Jupiter and Saturn, and the ice giants, Uranus and Neptune, fall into this class, with properties quite different from the terrestrial planets. In extrasolar planetary systems the masses of apparently rocky worlds appear to overlap with those of planets that have significant envelopes, but there is still a rough threshold mass above which envelopes are the norm.

The formation of giant planets within the core accretion model is understood as an extension of the processes that form terrestrial planets. A "core" of rocky and/or icy material grows, via a combination of pebble and planetesimal accretion, to a mass large enough to capture a significant gaseous envelope *before* the gas disk is dispersed. Initially the envelope is less massive than the core, and if the gas disk is dispersed while this is still the case the result is a planet qualitatively similar to a Solar System ice giant or an extrasolar mini-Neptune. If, instead, the envelope grows to become comparable in mass to the core, a phase of rapid gas accretion starts and the result is a fully formed gas giant. The time scale for giant planet formation in this model hinges on how quickly the core is assembled and on how rapidly the gas in the envelope can cool and accrete on to the core.

The goal in this chapter is to describe the physical principles behind core accretion and to provide a summary of some of the relevant observational constraints. We also discuss the conditions under which the gas disk can become gravitationally unstable and fragment into bound objects. Historically, gravitationally instability was studied primarily as an alternative mechanism for forming giant planets. More recent work suggests that the masses of objects formed as a result of gravitational instability would normally exceed those of most planets, but the process is still of interest as a way of rapidly forming substellar objects (including, perhaps, some planetary mass objects) at large orbital radii.

6.1 Core Accretion

The core accretion model for gas giant formation rests on one assumption: that a seed planet or *core* grows via pebble accretion or two-body collisions rapidly enough that it can exceed a certain critical mass prior to the dissipation of the gas disk. If this condition is satisfied, it can be shown (Perri & Cameron, 1974; Mizuno, 1980) that the core triggers a hydrodynamic instability that results in the onset of rapid gas accretion on to the core. Since the critical core mass is typically of the order of 5–10 $M_\oplus$, the end result is a largely gaseous but heavy element enriched planet that at least qualitatively resembles Jupiter or Saturn.

Figure 6.1 illustrates the four main phases in the formation of giant planets via core accretion:

- **Core formation**. A solid protoplanet (which henceforth we will rename a "core") grows via a combination of pebble accretion and planetesimal collisions

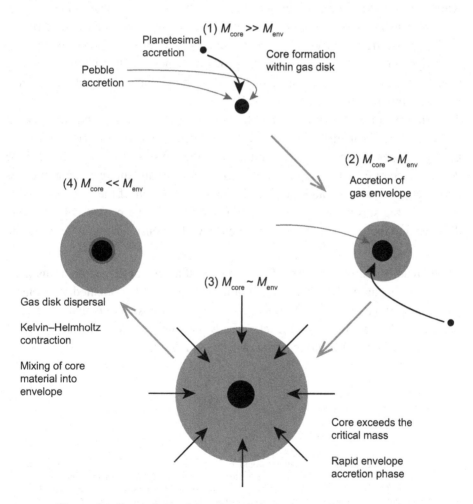

Figure 6.1 Stages in the formation of giant planets via core accretion.

until it becomes massive enough to retain a significant gaseous atmosphere or envelope. The physics during this initial phase is identical to that discussed in the context of terrestrial planet formation in Chapter 5, with the rate of growth being controlled by the aerodynamics of small solid bodies, the initial surface density of rocky and icy planetesimals, and by the extent of gravitational focusing.

- **Hydrostatic growth**. Initially the envelope surrounding the solid core is in hydrostatic equilibrium. Energy liberated by planetesimals impacting the core, together with gravitational potential energy released as the envelope itself contracts, must be transported through the envelope by radiative diffusion or convection before it is lost to the large gas reservoir of the protoplanetary disk. Over time both the core and the envelope grow until eventually the core exceeds a critical mass. The critical mass is not a constant but rather a calculable function of (primarily) the planetesimal accretion rate and opacity in the envelope.

- **Runaway growth**. Once the critical mass is exceeded a runaway phase of gas accretion ensues. The rate of growth is no longer demand limited (defined by the cooling properties of the envelope) but instead supply limited and defined by the hydrodynamic interaction between the growing planet and the disk. For massive planets the bulk of the planetary envelope is accreted during this phase, which is typically rather brief – of the order of 10^5 yr.

- **Termination of accretion**. Eventually the supply of gas is exhausted, either because of the dissipation of the entire protoplanetary disk or as a consequence of the planet opening up a local *gap* in the disk (the physics of gap opening will be discussed more fully in Chapter 7). Accretion tails off and the planet commences a long phase of cooling and quasi-hydrostatic contraction. If the core material is soluble in the surrounding envelope, some fraction of the primordial core may dissolve into the gas to form a fuzzy structure with a dense inner core and a dilute outer core.

We can readily estimate some of the masses that characterize the transitions between these phases. The weakest condition that must be satisfied if a planet embedded within a gas disk is to hold on to a bound atmosphere is that the escape speed v_{esc} at the surface of the planet exceeds the sound speed c_s within the gas.[1] A solid body of mass M_p and material density ρ_m has a radius

$$R_s = \left(\frac{3}{4\pi} \frac{M_p}{\rho_m} \right)^{1/3}, \tag{6.1}$$

---

[1] A planet that marginally satisfies this condition would lose its atmosphere on approximately the dynamical time scale in the absence of the gas disk, so this is *not* the same condition as for an isolated planet to be able to retain an atmosphere. To avoid atmospheric erosion via the process known as Jeans escape, an isolated planet must have a much higher surface escape speed that suffices to retain molecules far out in the Maxwellian tail of the particle velocity distribution.

and a surface escape speed

$$v_{\rm esc} = \sqrt{\frac{2GM_{\rm p}}{R_{\rm s}}}. \tag{6.2}$$

Noting that the sound speed in the protoplanetary disk can be written in terms of the disk thickness (h/r) and Keplerian velocity $v_{\rm K}$ via

$$c_{\rm s} = \left(\frac{h}{r}\right) v_{\rm K}, \tag{6.3}$$

the condition that $v_{\rm esc} > c_{\rm s}$ can be expressed as

$$M_{\rm p} > \left(\frac{3}{32\pi}\right)^{1/2} \left(\frac{h}{r}\right)^3 \frac{M_*^{3/2}}{\rho_{\rm m}^{1/2} a^{3/2}}, \tag{6.4}$$

where a is the orbital radius. This mass is very small. Substituting numbers appropriate for an icy body at 5 AU in a disk around a solar mass star with $(h/r) = 0.05$ we find that *some* atmosphere will be present provided that $M_{\rm p} \gtrsim 5 \times 10^{-4} \, M_\oplus$.

The existence of a tenuous wisp of an atmosphere will not have any dynamical importance. A more pertinent question is what is the minimum core mass able to maintain an envelope with a mass that is a non-negligible fraction of the mass of the core? To estimate this mass we assume that the envelope makes up a small fraction ϵ of the total planet mass, in which case the equation expressing hydrostatic equilibrium for the envelope can be written as

$$\frac{dP}{dr} = -\frac{GM_{\rm p}}{r^2}\rho, \tag{6.5}$$

where r is the distance from the center of the planet and $M_{\rm p}$ can be considered to be a constant. To keep the problem simple (this is after all only an estimate) we also assume that the envelope is isothermal, so that the pressure $P = \rho c_{\rm s}^2$. Equation (6.5) can then be integrated immediately to yield an expression for the radial density profile of the gas in the envelope:

$$\ln \rho = \frac{GM_{\rm p}}{c_{\rm s}^2} \frac{1}{r} + {\rm const.} \tag{6.6}$$

The constant of integration on the right-hand-side can be evaluated by matching the envelope density ρ to the density ρ_0 in the unperturbed disk at the radius $r_{\rm out}$, where the escape velocity from the planet matches the disk sound speed,

$$r_{\rm out} = \frac{2GM_{\rm p}}{c_{\rm s}^2}. \tag{6.7}$$

With the constant thereby determined, the envelope density profile is

$$\rho(r) = \rho_0 \exp\left[\frac{GM_{\rm p}}{c_{\rm s}^2}\frac{1}{r} - \frac{1}{2}\right]. \tag{6.8}$$

The density at the disk mid-plane ρ_0 can itself be written in terms of the surface density Σ and vertical scale-height h as $\rho_0 = (1/\sqrt{2\pi})(\Sigma/h)$ (Eq. 2.19).

Most of the mass of an envelope with the above density profile lies in a shell close to the surface of the solid core. We can therefore approximate the envelope mass as

$$M_{\rm env} \approx \frac{4}{3}\pi R_{\rm s}^3 \rho(R_{\rm s}), \tag{6.9}$$

where $\rho(R_{\rm s})$ is the envelope density evaluated at the surface of the core. Substituting for both the density profile and $R_{\rm s}$, the condition that the envelope makes up a non-negligible fraction of the total mass,

$$M_{\rm env} > \epsilon M_{\rm p}, \tag{6.10}$$

reduces to

$$M_{\rm p} \gtrsim \left(\frac{3}{4\pi\rho_{\rm m}}\right)^{1/2} \left(\frac{c_{\rm s}^2}{G}\right)^{3/2} \left[\ln\left(\frac{\epsilon\rho_{\rm m}}{\rho_0}\right)\right]^{3/2}. \tag{6.11}$$

The dominant dependence is on the disk sound speed, which is a decreasing function of radius. A growing protoplanet will therefore start to acquire a substantial gaseous envelope at a lower mass if it is located in the cool outer regions of the disk. To give a concrete example, we consider an icy body ($\rho_{\rm m} = 1\,{\rm g\,cm}^{-3}$) growing at 5 AU in a disk with $\rho_0 = 2 \times 10^{-11}\,{\rm g\,cm}^{-3}$ and sound speed $c_{\rm s} = 7 \times 10^4\,{\rm cm\,s}^{-1}$. Taking $\epsilon = 0.1$ to represent the threshold above which the envelope could be said to be significant, we predict that this will occur for planet masses

$$M_{\rm p} \gtrsim 0.2\,M_\oplus. \tag{6.12}$$

At 1 AU, on the other hand, where the disk parameters might be a density $\rho_0 = 6 \times 10^{-10}\,{\rm g\,cm}^{-3}$ and a sound speed $c_{\rm s} = 1.5 \times 10^5\,{\rm cm\,s}^{-1}$, a rocky protoplanet ($\rho_{\rm m} = 3\,{\rm g\,cm}^{-3}$) must grow to $M_{\rm p} \sim M_\oplus$ before we would expect to find it enshrouded in a dense massive envelope.

In light of the rather crude analysis that we have employed, Eq. (6.11) should only be trusted to an order of magnitude. Taking the estimate at face value, however, we can motivate the classical argument for why the Solar System's gas giant planets formed only in the cool outer regions of the protoplanetary disk. As we have seen, a protoplanet must grow to something like the mass specified in Eq. (6.11) before it starts to acquire a massive gaseous envelope, and this must occur *prior* to the dispersal of the protoplanetary disk. If we assume that the maximum protoplanet mass that can be attained prior to disk dispersal is comparable to the isolation mass (Eq. 5.53), we can determine for any particular disk model where in the disk planets will grow fast enough to capture envelopes. As an example, let us suppose that the radial profiles of the gas surface density and solid surface density are those specified by the minimum mass Solar Nebula model given in Section 1.2, and that the sound

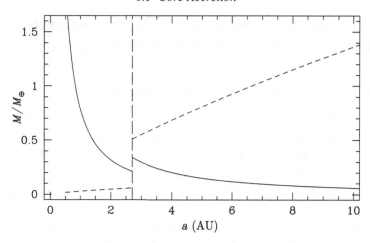

Figure 6.2 The radial dependence of the isolation mass (dashed curve) as compared to the minimum planet mass needed to sustain a massive envelope (solid curve). Both curves are computed for a solar mass star surrounded by a disk with $(h/r) = 0.05$. The surface density of the gas and solid component is taken to follow Hayashi's minimum mass Solar Nebula model, with a snow line at 2.7 AU. In computing the minimum mass necessary for envelope capture it has been assumed that interior to the snow line $\rho_m = 3 \, \mathrm{g \, cm}^{-3}$, while beyond the snow line $\rho_m = 1 \, \mathrm{g \, cm}^{-3}$.

speed is that appropriate to a disk with a constant geometric thickness $(h/r) = 0.05$.[2] Equations (5.53) and (6.11) then yield estimates for the isolation mass and the minimum mass necessary for envelope capture, and these estimates are plotted in Fig. 6.2. The result is suggestive. Interior to the snow line the isolation mass is *smaller* than the minimum mass required for envelope capture. In this region it is unlikely that protoplanets will grow fast enough to capture envelopes prior to disk dispersal, and the ultimate outcome of planet formation will instead be terrestrial planets. At orbital radii beyond the snow line, conversely, the isolation mass *exceeds* the minimum envelope capture mass. Giant planet formation is much more probable in this case, though as yet we cannot say whether the outcome is planets with a modest but still significant envelope (akin to Uranus and Neptune) or true gas giants whose mass is dominated by the contribution from the envelope.

It is fair to assume that in a counterfactual world, where the properties of extra-solar planetary systems were known but those of the Solar System were not, the above argument would never have surfaced. We now know that a large fraction of stars have planets with significant gas content that orbit within 1 AU, contrary to the prediction of the simple model. The basic factors that determine whether planets end up as rocky or gaseous worlds, however, remain valid. Rocky planets are the result if growth, whether from planetesimal or pebble accretion, is too slow,

---

[2] This value is adopted for consistency with most of the other estimates given in this section. It is not strictly consistent with the temperature profile usually assumed for the minimum mass Solar Nebula.

and there is a radial dependence to the ease with which a planet can capture an envelope and proceed to the runaway phase of accretion.

6.1.1 Core/Envelope Structure

The physical underpinning of the core accretion model for gas giant planet formation is the existence of a critical core mass – a core mass beyond which it is not possible to find a hydrostatic solution for the structure of a surrounding gaseous envelope. To understand the origin of the critical core mass, we modify some standard results from the theory of stellar structure to describe the structure of a giant planet envelope. We assume that the planet of total mass M_p has a well-defined solid core of mass M_{core} and an envelope mass M_{env}. Rotation is neglected and the envelope is taken to be in hydrostatic and thermal equilibrium. Conservation of mass and momentum then yield two differential equations for the structure of the envelope:

$$\frac{dM}{dr} = 4\pi r^2 \rho,$$
$$\frac{dP}{dr} = -\frac{GM}{r^2}\rho. \tag{6.13}$$

Here P is the pressure and $M = M(r)$ is the total mass enclosed within radius r. We have now dropped the simplifying assumption that $M \simeq M_{core}$ so the above equations are valid regardless of the mass of the envelope.

Although these equations only involve P and ρ, the temperature invariably enters as a third variable because the pressure $P = P(\rho, T)$ depends upon it. The temperature is determined by the requirement that the radial temperature *gradient* must be sufficient to transport the luminosity of the planet – generally produced at or near the surface of the core by the accretion of planetesimals – from the deep interior to the surface. This transport can occur via radiative diffusion or convection. In the radiative case the resulting temperature gradient is

$$\frac{dT}{dr} = -\frac{3\kappa_R \rho}{16\sigma T^3}\frac{L}{4\pi r^2}, \tag{6.14}$$

where κ_R is the Rosseland mean opacity (defined in terms of the frequency-dependent opacity via Eq. 2.68), σ is the Stefan–Boltzmann constant and L could in principle be a function of radius. The temperature gradient is directly proportional to the local flux $L/(4\pi r^2)$, so if the luminosity originates from planetesimal bombardment of the core we expect that a large flux and a correspondingly steep temperature gradient will be present for $r \sim R_s$. This immediately suggests that we need to be alert to the possibility that the radiative temperature gradient, which we conventionally define by combining Eq. (6.14) with the equation of hydrostatic equilibrium to yield

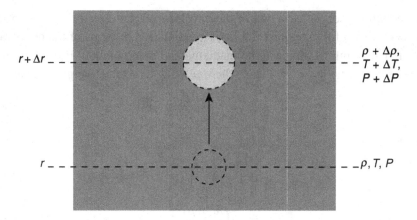

Figure 6.3 To assess the stability of the planetary envelope to convective instability, we imagine displacing a notional blob of fluid from r to $r + \Delta r$ under adiabatic conditions. Stability requires that the displaced fluid be denser than its new surroundings.

$$\nabla_{\text{rad}} \equiv \left(\frac{d \ln T}{d \ln P} \right)_{\text{rad}} = \frac{3 \kappa_R L P}{64 \pi \sigma G M T^4}, \tag{6.15}$$

may be unstable to the onset of convection.

We can establish the stability of the envelope to the onset of convection with the aid of a well-known thought experiment (e.g. Kippenhahn & Weigert, 1990, whose treatment we follow here) illustrated in Fig. 6.3. Convection is a buoyancy instability that occurs when the radial structure of an initially stationary atmosphere or envelope is prone to break up into upward-moving under-dense blobs and downward-moving over-dense streams. To assess whether this will happen, we imagine that the gas in the envelope is initially at rest with some specified gradient of density $(d\rho/dr)_{\text{env}}$, temperature $(dT/dr)_{\text{env}}$, and pressure $(dP/dr)_{\text{env}}$. We now displace a notional blob of fluid upward from r to $r + \Delta r$ *slowly* (so that the fluid within the blob remains in pressure equilibrium with the surrounding envelope) and *adiabatically* (so that no exchange of energy between the blob and the envelope occurs). It is immediately obvious that if the displaced blob finds itself denser than the envelope at $r + \Delta r$ it will tend to sink back down, whereas if it is less dense then buoyancy will cause it to rise and we conclude that the initial equilibrium is unstable. Mathematically the equilibrium is *stable* if

$$\left(\frac{d\rho}{dr} \right)_{\text{ad}} > \left(\frac{d\rho}{dr} \right)_{\text{env}}, \tag{6.16}$$

where the left-hand-side represents the density change within the blob under adiabatic conditions and the right-hand-side is the density gradient in the surrounding envelope. The fact that both gradients are negative quantities occasions a good deal of confusion, but the physical argument in terms of buoyancy is clear.

The equations for the structure of the envelope do not directly specify $(d\rho/dr)$ and hence it is useful to express the stability condition in terms of the temperature gradient. This requires only routine mathematical manipulation. For any equation of state of the form $\rho = \rho(P,T,\mu)$, where μ is the molecular weight, we can write

$$\frac{d\rho}{\rho} = \alpha\frac{dP}{P} - \delta\frac{dT}{T} + \varphi\frac{d\mu}{\mu}, \tag{6.17}$$

where the α, δ, and φ are defined via

$$\alpha \equiv \left(\frac{\partial \ln \rho}{\partial \ln P}\right)_{T,\mu}, \tag{6.18}$$

$$\delta \equiv -\left(\frac{\partial \ln \rho}{\partial \ln T}\right)_{P,\mu}, \tag{6.19}$$

$$\varphi \equiv \left(\frac{\partial \ln \rho}{\partial \ln \mu}\right)_{P,T}. \tag{6.20}$$

These parameters are specified by the equation of state. For an ideal gas one has that $\rho \propto P\mu/T$ and hence $\alpha = \delta = \varphi = 1$.

We can now convert the convective stability condition given by Eq. (6.16) into an equivalent condition based on temperature with the aid of Eq. (6.17). Noting that the molecular weight of the gas within the blob does not change as it is displaced, the condition for stability becomes

$$\left(\frac{\alpha}{P}\frac{dP}{dr}\right)_{ad} - \left(\frac{\delta}{T}\frac{dT}{dr}\right)_{ad} > \left(\frac{\alpha}{P}\frac{dP}{dr}\right)_{env} - \left(\frac{\delta}{T}\frac{dT}{dr}\right)_{env} + \left(\frac{\varphi}{\mu}\frac{d\mu}{dr}\right)_{env}. \tag{6.21}$$

This expression can be further simplified by noting that the two terms involving the pressure gradient vanish on account of the assumed pressure balance between the blob and its surroundings. Canceling these terms and multiplying through by $-P(dr/dP)$ yields the Ledoux criterion for the stability of the envelope against the onset of convection:

$$\left(\frac{d \ln T}{d \ln P}\right)_{env} < \left(\frac{d \ln T}{d \ln P}\right)_{ad} + \frac{\varphi}{\delta}\left(\frac{d \ln \mu}{d \ln P}\right)_{env}. \tag{6.22}$$

If the composition is uniform, the simpler Schwarzschild criterion is adequate:

$$\left(\frac{d \ln T}{d \ln P}\right)_{env} < \left(\frac{d \ln T}{d \ln P}\right)_{ad}. \tag{6.23}$$

We are now in a position to answer the question of when the envelope of our growing giant planet will develop convection. All we need to do is to compute the radiative gradient ∇_{rad} (Eq. 6.15) that would exist *in the absence* of convection and compare it to the adiabatic gradient defined by

$$\nabla_{ad} \equiv \left(\frac{d \ln T}{d \ln P}\right)_{ad}. \tag{6.24}$$

If (assuming for simplicity uniform composition)

$$\nabla_{\mathrm{rad}} < \nabla_{\mathrm{ad}}, \tag{6.25}$$

then the envelope is stable to convection and the radiative temperature gradient defined by Eq. (6.15) is self-consistent. If, conversely,

$$\nabla_{\mathrm{rad}} > \nabla_{\mathrm{ad}}, \tag{6.26}$$

then convection and bulk fluid motion will set in and at least some of the luminosity will be transported by convection. An approximate theory known as "mixing-length" theory exists, which can be used to calculate *how much* of the luminosity is carried convectively along with the value of the actual temperature gradient in convectively unstable regions. The details of this theory can be found in Kippenhahn and Weigert (1990) or in any other stellar structure textbook. In many circumstances, however, it turns out that convection, once it is established, is extremely efficient at transporting energy – so efficient in fact that it almost succeeds in erasing the unstable temperature gradient that set up convection in the first place! Simply replacing the radiative gradient ∇_{rad} with the adiabatic gradient ∇_{ad} in convectively unstable zones is thus often a good approximation that suffices for constructing simple envelope models.

Equations (6.13), along with the appropriate expression for the temperature gradient (either ∇_{rad} or ∇_{ad}) form a coupled set of differential equations that describe the envelope. They must be supplemented by expressions for the equation of state (i.e. the functional relationship $P = P(\rho, T, \mu)$), Rosseland mean opacity, and luminosity, and solved subject to appropriate boundary conditions. A simple case to consider is that in which the luminosity is entirely due to the collision of planetesimals with the core. In that case

$$L \simeq \frac{G M_{\mathrm{core}} \dot{M}_{\mathrm{core}}}{R_{\mathrm{s}}} \tag{6.27}$$

is a constant throughout the envelope. One inner boundary condition is obvious: at $r = R_{\mathrm{s}}$ we require that $M = M_{\mathrm{core}}$. The outer boundary conditions require slightly more thought. Physically we require that the envelope ought to match on smoothly to the disk at the "outer" radius of the planet, but how exactly should this outer radius be defined? There is no precise answer to this question (and indeed different authors use slightly different definitions) but there are two basic possibilities, the accretion radius

$$r_{\mathrm{acc}} = \frac{G M_{\mathrm{p}}}{c_{\mathrm{s}}^2}, \tag{6.28}$$

which is a measure of the maximum distance at which gas in the disk with sound speed c_s will be bound to the planet, and the Hill sphere radius

$$r_{\mathrm{H}} = \left(\frac{M_p}{3M_*}\right)^{1/3} a, \tag{6.29}$$

which is a measure of the distance beyond which shear in the Keplerian disk will unbind gas from the planetary envelope. For gas in the envelope to be bound to the planet, it must satisfy both $r < r_{\mathrm{acc}}$ and $r < r_{\mathrm{H}}$, so a logical choice for the location to specify the outer boundary conditions is at

$$r_{\mathrm{out}} = \min\left(r_{\mathrm{acc}}, r_{\mathrm{H}}\right). \tag{6.30}$$

At r_{out} we have that $M = M_p, P = P_{\mathrm{disk}}$, and $T \simeq T_{\mathrm{disk}}$.[3] Given these boundary conditions the envelope structure can be computed numerically using standard methods developed for stellar structure calculations.

6.1.2 Critical Core Mass

An intuitive argument for the existence of a critical core mass follows from consideration of the simplest situation: a hydrostatic envelope surrounding a core that is *not* being bombarded by planetesimals. In this limit, elementary thermodynamics tells us that the envelope – given enough time – must eventually lose any excess heat left over from its accretion and cool down to match the temperature of the gas in the neighboring protoplanetary disk. The resulting density profile for a low mass envelope is exponential (Eq. 6.8) with the pressure that supports the outer envelope being furnished by the high density close to the surface of the core. Now imagine slowly increasing the core mass. As we have already demonstrated, this causes the *fraction* of the total mass that is contained in the envelope to increase. At least initially this more massive envelope can be supported against gravity by simply increasing the density (and hence pressure) near the core. Once $M_{\mathrm{env}} \sim M_{\mathrm{core}}$, however, this equilibrium ceases to exist, since for still more massive envelopes increasing the base density also significantly increases the mass and hence the gravitational force. For higher core masses no hydrostatic solution exists, since any possible boost to the pressure by adding yet more gas to the envelope fails to compensate for the additional mass.

Real envelopes of giant planets are not isothermal, but the order of magnitude conclusion that hydrostatic equilibrium is possible only for $M_{\mathrm{env}} \lesssim M_{\mathrm{core}}$ carries over to physical models. In general, solutions to the equations for envelope structure given in Section 6.1.1 fall into two classes depending upon the energy transport mechanism that dominates near the outer boundary of the planet (for a detailed

---

[3] The approximate equality reflects the fact that the envelope temperature T will slightly exceed T_{disk} on account of the luminosity flowing through the planet (see e.g. Papaloizou & Terquem, 1999).

discussion see Rafikov (2006) and references therein). Both classes admit the existence of a critical core mass. One class of solution is fully convective, with an envelope entropy that is set by the external boundary conditions imposed by the protoplanetary disk. The second class has a radiative layer separating the inner envelope (which is usually convective) from the disk. The radiative layer decouples the structure of the planet from the disk, and as a consequence the critical core mass of planets described by these solutions is almost independent of the pressure and temperature in the surrounding disk. The critical core mass depends instead upon the luminosity of the planet and upon the opacity within the radiative layer. It is this second class of solution that probably describes planets forming in the outer regions of the protoplanetary disk, and accordingly we will focus exclusively on models of this class when describing the predicted sequence of giant planet formation in Section 6.1.3.

Accurate models of the structure of giant protoplanets are by necessity numerical, and it is not possible to calculate accurate values of the critical core mass analytically. The rather simple explanation that we gave above for the *existence* of a critical core mass suggests, however, that this ought to be a generic feature of any gaseous envelope in hydrostatic equilibrium around a solid core. This intuition is correct, and in fact we can gain considerable additional insight into the critical core mass by considering a simple model of a massive protoplanet with a purely radiative envelope. This model, unlike the fully realistic models that have both convective and radiative zones, is amenable to an approximate analytic treatment (Stevenson, 1982).

The goal of Stevenson's (1982) analysis is to obtain an approximate analytic form for the density profile within a radiative envelope, which can then be integrated to give an expression for the envelope mass. We first seek a relation between the temperature and the pressure. Starting from the equation of hydrostatic equilibrium (Eq. 6.13) and the expression for the radiative temperature gradient (Eq. 6.14), we eliminate the density by dividing one equation by the other:

$$\frac{dT}{dP} = \frac{3\kappa_R L}{64\pi\sigma GMT^3}.$$
(6.31)

We now integrate this equation inward from the outer boundary, making the approximation that $M(r) \approx M_p$ and taking L and κ_R to be constants:

$$\int_{T_{\text{disk}}}^{T} T^3 dT = \frac{3\kappa_R L}{64\pi\sigma GM_p} \int_{P_{\text{disk}}}^{P} dP.$$
(6.32)

Once we are well inside the planet we expect that $T^4 \gg T_{\text{disk}}^4$ and that $P \gg P_{\text{disk}}$, so the integral yields, approximately,

$$T^4 \simeq \frac{3}{16\pi} \frac{\kappa_R L}{\sigma GM_p} P.$$
(6.33)

Substituting P in this equation with an ideal gas equation of state,

$$P = \frac{k_B}{\mu m_p} \rho T,$$

(6.34)

we eliminate T^3 in favor of the expression involving dT/dr from Eq. (6.14) and integrate once more with respect to radius to obtain

$$T \simeq \left(\frac{\mu m_p}{k_B}\right) \frac{GM_p}{4r},$$

(6.35)

$$\rho \simeq \frac{64\pi\sigma}{3\kappa_R L} \left(\frac{\mu m_p GM_p}{4k_B}\right)^4 \frac{1}{r^3}.$$

(6.36)

Having derived the density profile, the mass of the envelope follows immediately:

$$\begin{aligned}
M_{\rm env} &= \int_{R_s}^{r_{\rm out}} 4\pi r^2 \rho(r)\, dr \\
&= \frac{256\pi^2\sigma}{3\kappa_R L} \left(\frac{\mu m_p GM_p}{4k_B}\right)^4 \ln\left(\frac{r_{\rm out}}{R_s}\right).
\end{aligned}$$

(6.37)

The right-hand-side of this equation has a strong dependence on the total planet mass M_p and a weaker dependence on the core mass $M_{\rm core}$ via the expression for the luminosity:

$$L = \frac{GM_{\rm core}\dot{M}_{\rm core}}{R_s} \propto M_{\rm core}^{2/3}\dot{M}_{\rm core}.$$

(6.38)

In principle there are further dependencies to consider since $r_{\rm out}$ is a function of M_p and R_s is a function of $M_{\rm core}$, but these enter only logarithmically and can be safely ignored. Noting that

$$M_{\rm core} = M_p - M_{\rm env},$$

(6.39)

we find that

$$M_{\rm core} = M_p - \left(\frac{C}{\kappa_R \dot{M}_{\rm core}}\right) \frac{M_p^4}{M_{\rm core}^{2/3}},$$

(6.40)

where we have shown explicitly the dependence on the envelope opacity and planetesimal accretion rate but have swept all the remaining constants (and near-constants) into a single constant C.

Solutions to Eq. (6.40) are plotted in Fig. 6.4. For fixed values of the core accretion rate $\dot{M}_{\rm core}$ and envelope opacity κ_R there is a maximum or critical core mass $M_{\rm crit}$ beyond which no solution is possible. The exact values of the critical core mass derived from this toy model should not be taken too seriously, but characteristic masses of the order of 10 $M_\oplus$ are obtained using plausible estimates of κ_R and $\dot{M}_{\rm core}$ (Stevenson, 1982). One also observes that $M_{\rm crit}$ is a function (though not a very rapidly varying one) of the product $\kappa_R \dot{M}_{\rm core}$ – reducing either the opacity or

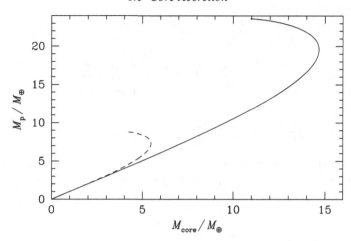

Figure 6.4 The total planet mass M_p is plotted as a function of the core mass M_{core} using Stevenson's (1982) approximate analytic solution for a fully radiative envelope. For any given choice of envelope opacity and planetesimal accretion rate there is a maximum core mass beyond which no hydrostatic solution exists. The solid curve is plotted for typical values of the opacity and accretion rate, while the dashed curve shows the lower critical core mass that results from reducing $\kappa_R \dot{M}_{core}$ by a factor of ten.

the accretion rate results in a lower value of the critical core mass. This property of the analytic model is also a feature of more complete numerical models of giant protoplanet structure.

6.1.3 Growth of Giant Planets

The equations describing hydrostatic envelope solutions can be readily modified to model the *time-dependent* growth of a giant planet up to the point where rapid accretion of the envelope begins. The basic assumption of time-dependent models is that the planet mass (and other time-variable quantities such as the core accretion rate) varies slowly enough that the envelope is always in hydrostatic equilibrium. Since hydrostatic equilibrium is established on a time scale that is of the order of the sound crossing time scale, this is a very good approximation which permits us to treat the growth of the planet as a slowly changing sequence of hydrostatic models. The fact that the envelope mass changes with time means that we must modify Eq. (6.38) to include the luminosity that derives from contraction of the envelope, but apart from this the mathematical description of time-dependent models is identical to that of hydrostatic ones. Taking the enclosed mass M as the dependent variable (as is often convenient) Ikoma *et al.* (2000), for example, employ the following set of equations:

$$\frac{\partial P}{\partial M} = -\frac{GM}{4\pi r^4},$$

(6.41)

$$\frac{\partial r}{\partial M} = \frac{1}{4\pi r^2 \rho}, \tag{6.42}$$

$$\frac{\partial T}{\partial M} = \begin{cases} -\frac{3\kappa_R}{64\pi\sigma r^2 T^3}\frac{L}{4\pi r^2} & \text{if radiative,} \\ \left(\frac{\partial T}{\partial P}\right)_S \left(\frac{\partial P}{\partial M}\right) & \text{if convective,} \end{cases} \tag{6.43}$$

$$\frac{\partial L}{\partial M} = \epsilon_{\text{acc}} - T\frac{dS}{dT}, \tag{6.44}$$

where S is the specific entropy of the gas in the envelope and the energy released by infalling planetesimals is assumed to be liberated at the core/envelope interface

$$\epsilon_{\text{acc}} = \frac{\delta(r - R_{\text{s}})}{4\pi r^2 \rho}\dot{M}_{\text{core}}\int_{R_{\text{s}}}^{r_{\text{out}}}\frac{GM}{r^2}dr. \tag{6.45}$$

As with the hydrostatic solutions, different authors adopt slight variations of these equations that differ, for example, in the assumed location of the outer boundary and in where within the envelope the energy of accreted planetesimals is released. These differences are rarely consequential. Of greater importance is the fact that the equations for the planetary structure must be supplemented with a model for how the planetesimal accretion rate $\dot{M}_{\text{core}}$ varies with time. The core accretion rate can be calculated using methods similar to those employed in the study of terrestrial planet growth (Section 5.2) provided that we allow for the fact that the planet mass must now include both the core and the envelope contributions, and that the cross-section for accretion may be modified by aerodynamic processes within the bound envelope. The mass of heavy elements within the planet will also increase due to pebble accretion. Any uncertainties in the magnitude and evolution with time of $\dot{M}_{\text{core}}$ (for example due to different assumptions as to the degree of gravitational focusing) propagate directly into the calculation of giant planet growth, and influence both the time scale and feasibility of planet formation at a particular location within the disk.

The first fully consistent time-dependent models of giant planet growth were published by Pollack *et al.* (1996), and their calculation, reproduced in slightly simplified form in Fig. 6.5, identified the main phases of the process. For their baseline model Pollack *et al.* (1996) considered the formation of Jupiter from a core at 5.2 AU in a disk with a solid surface density $\Sigma_{\text{p}} = 10\,\text{g cm}^{-2}$ and a gaseous surface density $\Sigma = 7 \times 10^2\,\text{g cm}^{-2}$. They adopted an opacity in the outer envelope that would be appropriate for a solar mixture of small grains that follow an interstellar size distribution. The model assumed that growth occurs exclusively through planetesimal accretion, with no contribution from pebbles. With these assumptions the growth of Jupiter proceeds through three well-separated phases:

- **Core formation**. Initially the core grows very rapidly due to runaway accretion of planetesimals within its feeding zone. The envelope mass remains small, and the growth of the total mass is dominated by the growth of the core. This phase is relatively brief (≈ 0.5 Myr) and comes to an end once the core approaches its isolation mass.

- **Hydrostatic growth.** This phase is characterized by the slow accretion of gas from the protoplanetary disk. The increasing mass of the planet drives a slow expansion of the feeding zone which suffices to maintain continued accretion of planetesimals, albeit at a lower rate than in the first phase. The hydrostatic phase continues for ≈ 7 Myr and ends only when the mass of the envelope starts to approach the mass of the core.

- **Runaway growth.** Once $M_{\text{env}} \gtrsim M_{\text{core}}$ the rate of accretion accelerates dramatically and runaway growth of the envelope sets in. The rate of accretion becomes limited by the rate at which the disk can supply gas, and is eventually shut off either by the onset of local tidal effects (to be discussed in more detail in Chapter 7) or by global disk dispersal.

As shown in Fig. 6.5 this model yields a formation time scale for Jupiter of 7–8 Myr and a primordial core mass of $M_{\text{core}} \simeq 20\ M_\oplus$.

The Pollack *et al.* (1996) model is a single member of a broad family of "core accretion" models that can form giant planets with a range of properties under different physical conditions. Let us initially retain two of the basic assumptions,

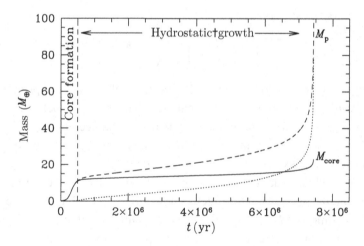

Figure 6.5 The evolution of the core mass (solid line), envelope mass (dotted line) and total mass (dashed line) from a time-dependent calculation of giant planet formation via core accretion (Rice & Armitage, 2003). In this illustrative calculation, which is based on a slightly simplified version of the physics described in Pollack *et al.* (1996), a core grows at a fixed radius of 5.2 AU in a disk with a solid surface density $\Sigma_p = 10$ g cm^{-2} and a gaseous surface density $\Sigma = 7 \times 10^2$ g cm^{-2}. With this choice of parameters and input physics one obtains a relatively short-lived phase of core formation that is followed by an extended period of slow coupled growth of the core and envelope. The critical core mass is exceeded and runaway growth starts after about 7–8 Myr, at which time the core mass is approximately $20\ M_\oplus$.

first that planetesimal accretion dominates over the pebble variety, and second that migration of the core as it grows can be neglected. Even in this limit, substantially different outcomes are possible given different histories of planetesimal accretion, or different envelope opacity. We can gain insight into the range of possibilities by considering the analytic fits to numerical models of hydrostatic planet envelopes calculated by Ikoma *et al.* (2000). For envelope opacities $\kappa_R \geq 10^{-2}\,\mathrm{cm^2\,g^{-1}}$ the critical core mass can be approximated by a power law in the planetesimal accretion rate $\dot{M}_{\mathrm{core}}$ and envelope opacity κ_R:

$$M_{\mathrm{crit}} \sim 7 \left(\frac{\dot{M}_{\mathrm{core}}}{10^{-7}\,M_\oplus\,\mathrm{yr}^{-1}} \right)^q \left(\frac{\kappa_R}{1\,\mathrm{cm^2\,g^{-1}}} \right)^s M_\oplus, \tag{6.46}$$

where the power-law indices q and s are both estimated to lie in the range 0.2–0.3. Ikoma *et al.* (2000) also calculated the time scale on which the envelope would grow in the complete absence of planetesimal accretion. Without ongoing core accretion the growth of the envelope is still limited by its ability to radiate away thermal energy as it contracts, which becomes more difficult for a lower mass core. Defining a growth time scale

$$\tau_{\mathrm{grow}} \equiv \left(\frac{1}{M_{\mathrm{env}}} \frac{\mathrm{d}M_{\mathrm{env}}}{\mathrm{d}t} \right)^{-1}, \tag{6.47}$$

Ikoma *et al.* (2000) estimate that

$$\tau_{\mathrm{grow}} \sim 10^8 \left(\frac{M_{\mathrm{core}}}{M_\oplus} \right)^{-2.5} \left(\frac{\kappa_R}{1\,\mathrm{cm^2\,g^{-1}}} \right) \mathrm{yr}. \tag{6.48}$$

Several important conclusions follow from these results. First, the primordial core mass that results from core accretion can vary substantially between otherwise identical models that vary in their assumed planetesimal accretion rate or envelope opacity. Core masses substantially larger than the fiducial 20 $M_\oplus$ could be formed within planetesimal disks whose higher surface density allowed for larger values of $\dot{M}_{\mathrm{core}}$. It is somewhat harder to tweak the model to yield lower core masses, because although reducing $\dot{M}_{\mathrm{core}}$ has the effect of lowering M_{crit} it has the side effect of increasing the time ($\sim M_{\mathrm{crit}}/\dot{M}_{\mathrm{core}}$) required to attain the critical mass. Lower core masses, however, arise naturally if the envelope opacity is smaller, as would be the case if grains coagulate (either in the disk or in the envelope itself) or settle. Lower opacities also result in substantially faster formation of giant planets via core accretion (Movshovitz *et al.*, 2010), greatly ameliorating older concerns that core accretion time scales were uncomfortably close to estimates of the disk lifetime.

Core accretion models that only include planetesimals fail to work beyond a threshold orbital radius that is of the order of 10 AU. Further out the *first* step in core accretion – the formation of the core itself – is slow because gravitational scattering of planetesimals becomes dominant over accretion at a low planet mass. Pebble accretion can be much more efficient at large orbital radii, allowing core accretion to form giant planets out to distances of the order of 50 AU (Lambrechts & Johansen, 2012). Indeed, if we allow ourselves the freedom to construct hybrid models that include both pebble and planetesimal accretion (which can still be important as a source of heat to slow down the onset of runaway growth), the range of orbital radii and time scales on which giant planets may in principle form is very wide. Theoretically determining what part of that range is actually populated is hard, because pebble variants of core accretion are intrinsically global problems in a way that older planetesimal-based models were not. In particular, core accretion models that include pebbles require a quantitative understanding of the abundance of protoplanetary seeds, the evolution of the radial pebble flux and its accretion on to those seeds, and the dynamical evolution of the cores themselves (Levison *et al.*, 2015). Improved observational knowledge of the exoplanet population at radii beyond 10 AU will likely be indispensable for an understanding of the physical processes responsible for the formation of wide-separation planetary systems.

The masses that we have been discussing for the pre-runaway phases of core accretion correspond closely to those for which orbital migration due to gravitational torques exerted by the disk on the planet is particularly rapid (discussed in Chapter 7). Migration is expected to alter the predictions of core accretion models in both their planetesimal and pebble-driven variants. In planetesimal-driven models, the relatively long time between the formation of a few $M_\oplus$ core and the onset of runaway gas accretion means that extensive orbital migration (potentially to very short-period orbits) is possible within many disk models. The faster growth made possible by pebbles helps, but the fact that pebble accretion is not a runaway process means that migration still cannot be ignored. The main logical possibilities are either that radial reshaping of forming giant planet systems due to migration is ubiquitous, or that giant planets form preferentially at planet traps where the disk structure supports convergent radial migration.

6.2 Constraints on the Interior Structure of Giant Planets

The best constraints on core accretion come from the Solar System. Spacecraft measurements of the gravitational field, when combined with knowledge of the equation of state and atmospheric composition, can be used to build models of the interior structure of the Solar System's giant planets. Individual measurements of extrasolar planets are much less powerful. The mass–radius relation of giant planets depends upon the total mass of heavy elements within the planet, but has little sensitivity to whether those elements are distributed uniformly within the planet or concentrated within a core. Modeling of the population of extrasolar planets is, in principle, more constraining. The observed correlation between host metallicity and the abundance of giant planets (Section 1.8.4), for example, is qualitatively consistent with core accretion expectations and can be investigated quantitatively within a full model for giant planet formation and migration.

6.2.1 Interior Structure from Gravity Field Measurements

The interior structures of Jupiter and Saturn can be constrained by comparing theoretical stucture models to measurements of their gravitational fields. The principle
of the method relies upon the fact that rotating giant planets (Jupiter has a rotation
period of approximately 10 hours) are axially rather than spherically symmetric
bodies. The gravitational field *outside* the planet then depends not just on the planet
mass and distance (as is the case for spherical bodies), but also on the distribution
of mass *within* the planet. This connection between the external potential and the
internal structure can be formalized by expanding the potential of a rotating planet
in Legendre polynomials (e.g. Guillot *et al.*, 2004):

$$\Phi(r,\theta) = \frac{GM_{\rm p}}{r}\left[1 - \sum_{i=1}^{\infty}\left(\frac{R_{\rm eq}}{r}\right)^{2i} J_{2i}\,P_{2i}(\cos\theta)\right]. \qquad (6.49)$$

Here $R_{\rm eq}$ is the planet's equatorial radius, θ is the angular distance measured from
the pole, and the P_{2i} are Legendre polynomials. The J_{2i} are called the gravitational
moments, and these can be measured by modeling the trajectories of spacecraft that
make flybys or enter orbit about the planet. In practice the series converges rapidly
and only the first few moments are required to yield an accurate description of the
gravitational field. For Jupiter, data from the *JUNO* mission (Iess *et al.*, 2018) have
been used to measure the even harmonics up to J_{10}. J_2 is determined to a precision
of about 1 part in 10^6, with the precision dropping steadily to about 1% for J_8.[4]
Data from the *Cassini* mission yield similarly precise measurements of Saturn's
external gravity field.

   To make use of the measured moments to constrain the internal structure, we note
that they can be written as a volume integral over the planet's density distribution
(e.g. Zharkov & Trubitsyn, 1974):

$$J_{2i} = -\frac{1}{M_{\rm p}R_{\rm eq}^{2i}}\int \rho r^{2i}\,P_{2i}(\cos\theta)\mathrm{d}V. \qquad (6.50)$$

A set of measured moments then yields multiple integral constraints on the density
distribution that must be satisfied by any physical planet model. The weighting
function inside the integral means that the higher order moments become increasingly sensitive to structure near the surface of the planet rather than to the central
regions.

   Even for Jupiter, where the gravitational field data are of exquisite quality, the
number of observational constraints that bear on the interior structure is quite
limited. We have the total mass, radius, and surface rotation rate, together with
four well-measured even harmonics of the gravity field and knowledge of the

---

[4] *JUNO* data have also been used to measure the *odd* harmonics. These measurements do not further constrain
   the static interior structure, but instead probe the depth to which dynamic features such as zonal flows extend
   within the planet.

atmospheric composition. This reality forces us to exercise careful judgement in determining what type of theoretical model to compare the data against. There is no hope of measuring the detailed properties of Jupiter's core in a model-independent way. Rather, the goal is to construct and constrain models that incorporate our best estimates of the equation of state at the relevant interior pressures, with physically reasonable transitions in the structure at different radii.

6.2.2 Internal Structure of Jupiter

Figure 6.6 shows an example interior structure for Jupiter that is consistent with the available observational constraints (Wahl *et al.*, 2017). The model structure has four layers. The envelope has a compositional transition, with a helium-poor upper envelope, in which hydrogen is molecular, overlying a helium-rich lower envelope, in which hydrogen is metallic. The core also has two layers. A dense inner core, made up of icy and rocky material, blends into a dilute outer core containing a mixture of metallic hydrogen and heavy elements.

Compared to earlier generations of models, current determinations of Jupiter's internal structure benefit from greatly improved measurements of the gravity field and from smaller theoretical uncertainties in the equation of state at high pressure. The mass of heavy elements in the core region is estimated at 7–25 $M_\oplus$ (Wahl *et al.*, 2017), with a preference for a dilute core structure of the type previously suggested theoretically (Stevenson, 1982; Lissauer & Stevenson, 2007). The large

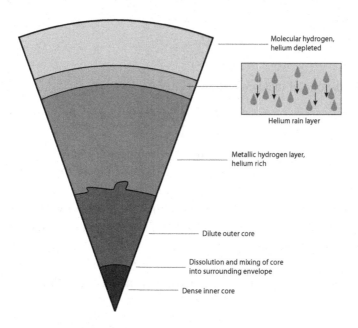

Figure 6.6 An example of an interior structure model that is compatible with *JUNO* measurements of Jupiter's gravitational field (based on Wahl *et al*, 2017).

error bar reflects degeneracies in the predictions of different theoretical models, and the fundamental difficulty of determining the deep interior structure of a planet from only exterior data, without benefit of the powerful seismic data that are available for the Earth and the Sun. The estimated core mass is consistent with the expectations of core accretion theory, but spans too wide a range to discriminate between different physically reasonable variants.

6.3 Disk Instability

Gaseous protoplanetary disks may undergo a phase in which they are massive enough to be unstable to instabilities arising from their own self-gravity. If the outcome of these instabilities is fragmentation, disk instability could be a channel for the formation of massive planets, brown dwarfs, or binary stellar systems. There is a long history of suggestions that disk instability could form planets, dating back to models by Kuiper (1951) and Cameron (1978) that predate any serious work on core accretion. More recent work by Boss (1997) was the foundation for renewed interest in disk instability models. As a possible planet formation process, the fundamental difference between the disk instability model and the core accretion model arises from the fact that in the disk instability model the solid component of the disk is a bystander that plays only an indirect role (via its contribution to the opacity) in the process of planet formation.

We have already derived the conditions needed for a protoplanetary gas disk to become unstable to its own self-gravity in Sections 3.4 and 4.6.1. Globally the disk mass must roughly satisfy

$$\frac{M_{\rm disk}}{M_*} \gtrsim \frac{h}{r}, \tag{6.51}$$

while the well-defined local condition is

$$Q \equiv \frac{c_s \Omega}{\pi G \Sigma} < Q_{\rm crit}, \tag{6.52}$$

where $Q_{\rm crit}$ is a dimensionless measure of the threshold below which instability sets in. It lies in the range $1 < Q_{\rm crit} < 2$. If we compare these requirements to observational determinations of protoplanetary disk properties the global condition suggests that widespread gravitational instability (i.e. instability that extends across a large range of disk radii) must be limited to disks at the upper end of the observed range, with masses of around a tenth of the stellar mass. Such massive disks may be commonly present early in the evolution of pre-main-sequence stars (e.g. Eisner *et al.*, 2005). For a disk around a solar mass star the local condition can be written in the form

$$\Sigma \gtrsim 3.8 \times 10^3 \left(\frac{Q_{\rm crit}}{1.5}\right)^{-1} \left(\frac{h/r}{0.05}\right) \left(\frac{r}{5\,{\rm AU}}\right)^{-2} {\rm g\,cm}^{-2}. \tag{6.53}$$

One immediately observes that the surface densities required for gravitational instability are large, more than an order of magnitude in excess of the minimum mass Solar Nebula value at 5 AU. Such high surface densities are, however, neither observationally excluded nor unreasonable on theoretical grounds. High surface densities are likely at early epochs when the disk accretion rate is large, especially if angular momentum transport within the disk is rather inefficient, and these conditions provide the most fertile ground for the development of gravitational disk instabilities.

Assuming for the time being that gravitational instability results in fragmentation, we can estimate the masses of the objects that would be formed using an identical argument to that given for planetesimals forming out of unstable *particle* layers in Section 4.6.2. Noting that the most unstable scale in a gravitationally unstable disk is $\lambda \sim 2c_s^2/(G\Sigma)$, we expect that fragmentation will result in objects whose characteristic mass is of the order of $M_p \sim \pi\lambda^2\Sigma$. For a disk around a solar mass star with $(h/r) = 0.05$ and $Q_{crit} = 1.5$ this characteristic mass is independent of orbital radius and equal to $M_p \approx 8\ M_J$, where M_J is the mass of Jupiter. This estimate is very crude, and evidently on the high side for Jupiter and for the majority of known extrasolar planets. It suffices, however, to establish that disk instability could result in the formation of substellar objects (massive planets or brown dwarfs).

6.3.1 Outcome of Gravitational Instability

A disk will become gravitationally unstable if $Q < Q_{crit}$, but satisfying this condition is not sufficient to conclude that the result of the instability will be fragmentation. The first linearly unstable modes in a gravitationally unstable disk are generally nonaxisymmetric ones, which develop into a pattern of spiral structure that is able to transport angular momentum outward via gravitational torques. The outward flow of angular momentum implies an inward flow of mass, which heats up the disk due to the dissipation of accretion energy. A self-gravitating disk thus has a tendency to increase its sound speed and decrease its surface density, both of which boost Q and act to stabilize the disk. These effects mean that stable angular momentum transport is a possible outcome of gravitational instability. In this regime heating of the accretion flow due to the dissipation of gravitational potential energy balances radiative cooling in such a way that the disk attains a quasi-steady state (Paczynski, 1978).

Fragmentation occurs if the stabilizing feedback loop described above fails. A disk that cools too quickly, or that gains mass on too short a time scale, becomes violently gravitationally unstable as Q is pushed lower. The associated density fluctuations can become large enough that clumps form and collapse into bound objects within the disk.

6.3.2 Cooling-Driven Fragmentation

We can express the cooling criterion for fragmentation in two equivalent ways, which, taken together, expose the important physics. The essential argument is based upon the fact that, if cooling is very rapid, gravitational instabilities must be rather violent to generate heating that is sufficient to mitigate these losses. Quantitatively we can define a local cooling time scale via

$$t_{\rm cool} \equiv \frac{U}{|dU/dt|}, \tag{6.54}$$

where U is the thermal energy content of the disk per unit area, and ask what is the equivalent value of α that would be required to generate enough offsetting heating. As we showed in Section 3.4.2 the answer is

$$\alpha = \frac{4}{9\gamma(\gamma - 1)\Omega t_{\rm cool}}, \tag{6.55}$$

where γ is the two-dimensional adiabatic index (Gammie, 2001). The equivalent α (a measure of the strength of the "self-gravitating turbulence" within the disk) is thus inversely proportional to the cooling time. We should caution that this simple result relies on the assumption that turbulence and transport within the disk is local. Since gravitational torques can act over long ranges one should worry about this, but in practice subsequent work shows that the formally local result is approximately valid for many conditions of interest for protoplanetary disks.

Noting that when $\alpha \sim 1$ the velocity perturbations within the disk become supersonic, we might guess that this would be when transient clumps would become dense enough to collapse and trigger fragmentation. This intuition is roughly correct. Early numerical simulations by Gammie (2001) showed that for a particular equation of state there is a critical cooling time scale below which fragmentation occurs. For $\gamma = 2$,

$$t_{\rm cool,\ crit} \simeq 3\Omega^{-1}. \tag{6.56}$$

Subsequent work identified subtleties in obtaining a numerical converged estimate of $t_{\rm cool,\ crit}$, but later determinations are broadly in agreement with the original value of $t_{\rm cool,\ crit} \simeq 3\Omega^{-1}$ (Deng *et al.*, 2017; Baehr *et al.*, 2017). The critical cooling time corresponds to a critical value of the equivalent α that divides the regimes of stable angular momentum transport from fragmentation:

$$\alpha_{\rm crit} \simeq 0.1. \tag{6.57}$$

A locally self-gravitating disk cannot sustain a larger α without fragmenting (Rice *et al.*, 2005).

The disk mass above which a disk becomes gravitationally unstable is equal (up to numerical factors) to the mass above which self-gravity provides the dominant

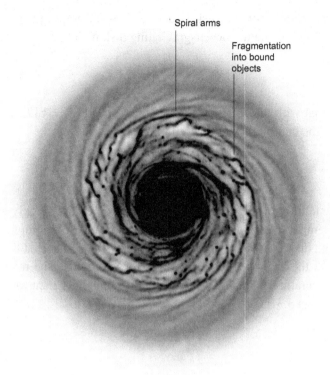

Figure 6.7 An image of the surface density of a simulated protoplanetary disk that is subject to gravitational instability (adapted from Rice *et al.* 2003). The non-linear outcome of gravitational instability in disks is generally a nonaxisymmetric pattern of transient spiral arms. If the cooling time of the gas within the disk is short enough – as in this example – the disk fragments into dense clumps. Provided that these clumps are able to survive and contract further they may form substellar objects (massive planets or brown dwarfs).

contribution to the vertical acceleration. It follows immediately that *if* fragmentation occurs, the collapse of the disk into dense clumps takes place on the orbital time scale. This result is supported by numerical simulations. Figure 6.7 depicts the onset of fragmentation in a disk that has been set up (rather artificially) with a cooling time that is everywhere slightly shorter than the critical cooling time needed to allow fragmentation. Fragmentation occurs promptly as soon as the disk has cooled to the point where $Q \lesssim Q_{\text{crit}}$, forming a number of clumps which, if they can survive and contract further, could be progenitors of massive planets or brown dwarfs.

6.3.3 Disk Cooling Time Scale

We can derive the radial dependence of the cooling time scale, or equivalently the α that would be needed to keep a self-gravitating disk in a stable thermal equilibrium, by combining a model for the disk structure with the requirement that $Q \sim 1$ (Levin, 2007). We assume that a self-gravitating disk maintains

$$Q = \frac{c_s \Omega}{\pi G \Sigma} = Q_0, \tag{6.58}$$

and combine that with the equation for the sound speed $c_s^2 = k_B T_c / \mu m_p$ to get an expression for the central disk temperature:

$$T_c = \pi^2 Q_0^2 G^2 \left(\frac{\mu m_p}{k_B} \right) \frac{\Sigma^2}{\Omega^2}. \tag{6.59}$$

To assess the stability of the disk we need to determine how quickly a disk with a given central temperature and surface density would cool. For a simple example, we assume that the disk is at least moderately optically thick, and that the form of the opacity law has the T_c^2 dependence expected for ice grains. The relation between the central temperature T_c and the effective temperature T_{disk} then follows from the equations (Section 3.3.3)

$$\frac{T_c^4}{T_{disk}^4} = \frac{3}{4} \tau, \tag{6.60}$$

$$\tau = \frac{1}{2} \kappa_R \Sigma, \tag{6.61}$$

$$\kappa_R = \kappa_0 T_c^2. \tag{6.62}$$

The cooling rate is then

$$\frac{dU}{dt} = 2\sigma T_{disk}^4 = \frac{16\sigma}{3\kappa_0 \Sigma} T_c^2. \tag{6.63}$$

Noting that the thermal energy content of the disk per unit area is

$$U = \frac{c_s^2 \Sigma}{\gamma(\gamma - 1)}, \tag{6.64}$$

Eqs. (6.54) and (6.55) allow us to find how α scales with the disk parameters:

$$\alpha = \frac{64}{27} \left(\frac{\mu m_p}{k_B} \right)^2 \frac{\pi^2 Q_0^2 \sigma G^2}{\kappa_0} \Omega^{-3}. \tag{6.65}$$

For this opacity law the only parameter that matters for the stress is the angular velocity Ω.

Substituting some reasonable numbers, $\kappa_0 = 2 \times 10^{-4}$ cm^2 g^{-1} K^{-2} and $Q_0 = 1.5$, the result for a solar mass star is

$$\alpha \sim 0.1 \left(\frac{r}{40 \text{ AU}} \right)^{9/2}. \tag{6.66}$$

Comparing this result to the estimate of the maximum stress that a self-gravitating disk can stably support (Eq. 6.57) the conclusion is that fragmentation is not possible in the inner disk, but may occur at radii between about 40 AU and 100 AU (Clarke, 2009; Rafikov, 2009).

The radii where fragmentation may be possible coincide with those where stellar irradiation, rather than heating due to accretion, normally controls the thermal state of the disk gas. This requires a revision to the argument given above, most obviously because if Σ is too small stellar irradiation may be able to keep $Q \gg 1$ and prevent the disk becoming self-gravitating in the first place. More generally, the possibility of significant stellar irradiation needs to be considered in most circumstances where protoplanetary disks are gravitationally unstable (Rice *et al.*, 2011; Hirose & Shi, 2019).

6.3.4 Infall-Driven Fragmentation

Our analysis of thermally driven fragmentation implicitly assumed the type of relatively isolated disk that is found in Class 2 YSOs. Common sense, however, suggests that the massive disks that might be gravitationally unstable are more likely to be present in the youngest Class 0 or Class 1 YSOs. In these systems the disks are not isolated, but instead are being actively fed with mass that is falling in from an envelope. This feeding, if it is too rapid, provides an alternative way to force a gravitationally unstable disk to fragment. Mass addition to the disk on a fast time scale can exceed the ability of self-gravity to transport mass inward, leaving fragmentation as the only other possibility.

Numerical results on the conditions for infall-driven fragmentation can be motivated by thinking about two distinct senses in which infall might be "too rapid." For a system with a total mass $M_{\text{total}} = M_* + M_{\text{disk}}$, a characteristic global time scale for the infalling gas is Ω^{-1}, where the angular frequency is measured at the circularization radius given the mean specific angular momentum of the envelope. A dimensionless parameter expressing the strength of infall is therefore

$$\Gamma \equiv \frac{\dot{M}_{\text{infall}}}{M_{\text{total}}} \Omega^{-1}. \tag{6.67}$$

A second relevant parameter can be defined by thinking about the ability of a self-gravitating disk to attain a steady-state by transporting the infalling gas away quickly enough. Setting $Q = Q_0$ as before, making use of the α prescription ($\nu = \alpha c_s^2 / \Omega$), and noting that in a steady-state disk $\nu \Sigma = \dot{M}/(3\pi)$, the transport rate through a self-gravitating disk is

$$\dot{M} = \frac{3\pi\alpha}{Q_0} \frac{c_s^3}{G}. \tag{6.68}$$

Based on this, we can can express the "strength" of the self-gravitating transport that is needed to stably remove mass being added at a rate $\dot{M}_{\text{infall}}$ via a parameter

$$\xi \equiv \frac{G\dot{M}_{\text{infall}}}{c_{\text{s}}^3}. \tag{6.69}$$

For a thin disk $\xi \propto \alpha$, and we could use the same line of reasoning as for thermally unstable disks to argue that fragmentation ought to occur when α exceeds some threshold value. We can still define ξ, however, in more general situations (for example when the disk mass exceeds the stellar mass) where the approximations involved in treating self-gravitating disks as pseudo-viscous structures break down.

Numerical simulations of isothermal disks undergoing infall show that the fragmentation condition can be expressed in terms of Γ and ξ. Kratter *et al.* (2010) find that fragmentation occurs when

$$\Gamma \lesssim 1.2 \times 10^{-3} \xi^{2.5}. \tag{6.70}$$

Kratter and Lodato (2016), and references therein, provide additional details on this regime of fragmentation.

6.3.5 Outcome of Disk Fragmentation

The range of outcomes that result from the fragmentation of gravitationally unstable disks is not well known. Observations of L1448 IRS3B show that the three protostars in the system co-exist with a disk showing spiral structure indicative of gravitational instability (Tobin *et al.*, 2016). This observation, together with less direct arguments based on larger samples of stars, suggests that disk fragmentation is a mechanism for forming binaries and other multiple stellar systems. It is consistent with the basic theoretical expectation that fragments, forming in a massive and highly turbulent disk, would often accrete rapidly from the remaining disk material.

Substellar companions (brown dwarfs or massive planets) may also be outcomes of disk fragmentation, especially if it takes place at a somewhat later epoch when the disk is less massive. Accretion on to clumps formed from fragmentation will be suppressed if the time scale for them to cool and contract is long, and may be further reduced (or even reversed) if the clumps migrate inward and lose their outer envelopes to tidal shear (Nayakshin, 2010). From an observational perspective, no extrasolar planetary systems are known whose properties unambiguously point to an origin from disk instability. Systems of one or more very massive planets, orbiting at the radii between about 50 AU and 200 AU characteristic of fragmentation, are the most obvious candidates, and a number of examples of these have been discovered.

6.4 Further Reading

"Formation of Giant Planets," G. D'Angelo & J. J. Lissauer (2018), in *Handbook of Exoplanets*. Cham: Springer, id. 140.
"Gravitational Instabilities in Circumstellar Disks," K. Kratter & G. Lodato (2016), *Annual Review of Astronomy and Astrophysics*, **54**, 271.

7

Early Evolution of Planetary Systems

The classical theory of giant planet formation described in the preceding chapter predicts that massive planets ought to form on approximately circular orbits, with a strong preference for formation in the outer disk at a few AU or beyond. Most currently known extrasolar planets have orbits that are grossly inconsistent with one or the other of these predictions. Even within the Solar System the existence of a large resonant population of Kuiper Belt Objects, including Pluto, is inconsistent with a model in which the giant planets formed in the same orbits that they currently inhabit.

In this chapter we describe a set of physical mechanisms – often generically described as *migration* processes – that can lead to the modification of planetary orbits. Any process that allows a planet to exchange energy and/or angular momentum with another component of the system will lead to migration, and there are several contenders. The other party can be the gas, planetesimal, or particle disk, another planet, the host star, or a binary companion. If substantial enough, migration can make the final architecture of the system unrecognizable from its state immediately after planet formation.

The study of planetary migration involves two sets of challenges. The first is to understand the physics of individual processes well enough that theoretical uncertainties are not the dominant limitation when building models of observed systems. To take an example, determining the rate of change of the semi-major axis of a $3\ M_\oplus$ planet embedded in a protoplanetary gas disk of given properties (and with specified physics) is a well-posed problem. Recent years have seen substantial progress in solving these types of problems. The second challenge is harder. Which processes dominated or contributed substantially to the observed architecture of typical planetary systems, or to specific examples such as the Solar System? This is harder because a model that includes several physical processes is only as reliable as its weakest link, and because, with imperfect knowledge of the initial conditions, we are forced to rely on late-time observations that may not contain enough information to uniquely reconstruct the earlier history. Many of the observations that motivate study of migration can, as a result, be fit with more than one model.

7.1 Migration in Gaseous Disks

Gas disk migration refers to a change in the semi-major axis of a planet that is caused by exchange of angular momentum between the planet and the surrounding gaseous protoplanetary disk. The exchange of angular momentum is mediated by gravitational torques between the planet and the disk. No torque is exerted on a planet by an axisymmetric disk, so gas disk migration can only take place if the planet excites nonaxisymmetric structure. In addition to angular momentum, energy is also exchanged between the planet and the disk, and depending upon the details of these exchanges the net result may be changes not just in a but also in e and i.

Migration is potentially important whenever a fully formed planet (very roughly of the mass of Mars or above – smaller bodies do not interact strongly with the disk) co-exists with a gaseous disk. Such co-existence is inevitable for the cores of giant planets forming via core accretion, and for gas giants themselves, and the possibility of large-scale gas disk migration can never be ignored when considering the formation of these objects. For terrestrial planets (including super-Earths in extrasolar planetary systems) the likely importance of gas disk migration is model dependent. In the classical scenario for terrestrial planet formation in the Solar System, the time scale for the final assembly of the terrestrial planets substantially exceeds the typical gas disk lifetime. It is then possible that the progenitors of our terrestrial planets remained of sufficiently low mass prior to disk dispersal that migration can be consistently ignored. Similar or more massive planets forming at smaller orbital radii, where the growth time scale is shorter, would instead fall into the regime where planet–disk interactions need to be considered.

7.1.1 Planet–Disk Torque in the Impulse Approximation

The detailed physics of the gravitational interaction between a planet and a sur-rounding gas disk is subtle, and several important aspects remain poorly under-stood. To gain some insight into the interaction it is useful to begin by considering a gaseous analog of the two-body dynamics discussed in the context of terrestrial planet formation in Section 5.3.1. We assume that a particle in the gas disk is initially on an unperturbed circular orbit, and calculate the angular momentum change that occurs as the particle is impulsively deflected during a close approach to the planet. The deflection is calculated as if the particle in the gas disk were a freely moving test particle, and hydrodynamics is ignored entirely except for the implicit assumption that the disk is able to "smooth out" the trajectories so that particles resume unperturbed orbits prior to making their next encounter with the planet. Despite these simplifications, calculation of the planet–disk interaction via the so-called *impulse approximation* (Lin & Papaloizou, 1979) yields correct scal-ings and better than order of magnitude estimates of the rate of angular momentum transport.

Working in a frame of reference moving with the planet, we consider the gravitational interaction between the planet and gas flowing past with relative velocity Δv and impact parameter b. We have already derived (in the two-body, free particle limit) the change to the perpendicular velocity that occurs during the encounter. It is (Eq. 5.58)

$$|\delta v_{\perp}| = \frac{2GM_p}{b\Delta v}, \tag{7.1}$$

where M_p is the planet mass. This velocity is directed radially, and hence does not correspond to any angular momentum change. The interaction in the two-body problem is however conservative, so the increase in the perpendicular velocity implies a reduction (in this frame) of the parallel component. Equating the kinetic energy of the gas particle well before and well after the interaction has taken place we have that

$$\Delta v^2 = |\delta v_{\perp}|^2 + (\Delta v - \delta v_{\parallel})^2, \tag{7.2}$$

which implies (for small deflection angles)

$$\delta v_{\parallel} \simeq \frac{1}{2\Delta v} \left(\frac{2GM_p}{b\Delta v} \right)^2. \tag{7.3}$$

If the planet has a semi-major axis a the implied angular momentum change per unit mass of the gas is

$$\Delta j = \frac{2G^2 M_p^2 a}{b^2 \Delta v^3}. \tag{7.4}$$

It is worth pausing at this juncture to fix the *sign* of the angular momentum change experienced by the gas and by the planet firmly in our minds. Gas exterior to the planet's orbit orbits the star more slowly than the planet, and is therefore "overtaken" by the planet. The decrease in the parallel component of the relative velocity of the gas therefore corresponds to an *increase* in the angular momentum of the gas exterior to the planet. Since the gravitational torque must be equal and opposite for the planet the sign is such that:

- Interaction between the planet and gas exterior to the orbit increases the angular momentum of the gas, and decreases the angular momentum of the planet. The planet will tend to migrate inward, and the gas will be repelled from the planet.
- Interaction with gas interior to the orbit decreases the angular momentum of the gas and increases that of the planet. The interior gas is also repelled, but the planet tends to migrate outward.

In the common circumstance where there is gas both interior and exterior to the orbit of the planet, the net torque (and sense of migration) will evidently depend upon which of the above effects dominates.

The total torque on the planet due to its gravitational interaction with the disk can be estimated by integrating the single particle torque over all the gas in the disk. For an annulus close to but exterior to the planet, the mass in the disk between b and $b + db$ is

$$dm \approx 2\pi a \Sigma db, \tag{7.5}$$

where Σ (assumed to be a constant) is some characteristic value of the gas surface density. If the gas in the annulus has angular velocity Ω and the planet has angular velocity Ω_p, all of the gas within the annulus will encounter the planet in a time interval

$$\Delta t = \frac{2\pi}{|\Omega - \Omega_p|}. \tag{7.6}$$

Approximating $|\Omega - \Omega_p|$ as

$$|\Omega - \Omega_p| \simeq \frac{3\Omega_p}{2a} b, \tag{7.7}$$

which is valid provided that $b \ll a$, we obtain the total torque on the planet due to its interaction with gas outside the orbit by multiplying Δj by dm, dividing by Δt, and integrating over impact parameters. Formally we would have that

$$\frac{dJ}{dt} = -\int_0^\infty \frac{8G^2 M_p^2 a \Sigma}{9\Omega_p^2} \frac{db}{b^4}, \tag{7.8}$$

but this integral is clearly divergent – there is what must be an unphysically infinite contribution from gas passing very close to the planet. Sidestepping this problem for now, we replace the lower limit with a minimum impact parameter b_{min} and integrate. The result is

$$\frac{dJ}{dt} = -\frac{8}{27} \frac{G^2 M_p^2 a \Sigma}{\Omega_p^2 b_{min}^3}. \tag{7.9}$$

It is possible to tidy up this calculation somewhat, for example by taking proper account of the rotation of the planet frame around the star, and if this is done the result is that the expression derived above must be multiplied by a correction factor (see e.g. Papaloizou & Terquem, 2006, and references therein).

Aside from the sign of the torque, the most important result that we can glean from this calculation is that the torque on the planet due to its interaction with the gas scales as the *square* of the planet mass. This scaling can be compared to the orbital angular momentum of the planet, which is of course linear in the planet mass. We conclude that if all other factors are equal – in particular if neither Σ in the vicinity of the planet nor b_{min} varies with M_p – the time scale for the planet to change its orbital angular momentum significantly will scale inversely with the planet mass. The migration velocity in this limit will then be proportional to M_p – more massive planets will migrate faster.

Finally, we can estimate the magnitude of the torque for parameters appropriate to different stages of giant planet formation. First, let us consider a rather low mass core ($M_p = M_\oplus$) growing at 5 AU in a gas disk of surface density $\Sigma = 10^2$ g cm^{-2} around a solar mass star. Our treatment of the interaction has assumed that the disk can be treated as a two-dimensional sheet, so it is arguably natural to take as a minimum impact parameter $b_{min} = h \approx 0.05a$. Using these numbers we find that the exterior torque would drive inward migration on a time scale

$$\tau = \frac{J}{|dJ/dt|} \sim 1 \text{ Myr}. \tag{7.10}$$

Of course this will be partly offset by the interior torque – which has the opposite sign – but unless there is some physical reason why these two torques should cancel to high precision, we would predict changes to the semi-major axis on a time scale of the order of a Myr. This is already rapid enough to be a potentially important effect during giant planet formation via core accretion. Second, we can evaluate the torque for a fully formed gas giant. A Jupiter mass planet has a Hill sphere $r_H > h$, so it seems reasonable to argue that the value of b_{min} that we adopted for an Earth mass core is too small in this case. Picking a modestly larger value $b_{min} = 0.2a$ results in a time scale,[1]

$$\tau \sim 2 \times 10^5 \text{ yr}, \tag{7.11}$$

that is an order of magnitude shorter than typical protoplanetary disk lifetimes. If these estimates can be trusted to even an order of magnitude the inescapable conclusion is that gas disk migration ought to be an important process for the early evolution of the orbits of giant planets.

7.1.2 Physics of Gas Disk Torques

The impulse approximation provides valuable intuition about planet–disk interactions, but does not capture all of the physical effects that can lead to migration torques. The total torque is the sum of at least three distinct processes, depicted in Figs. 7.1 and 7.2, torques exerted at Lindblad resonances, co-orbital torques, and additional torques from the vicinity of the planet due to thermal effects.

The torque component that is easiest to compute accurately arises from waves that are excited by the planet's gravitational perturbation at the location of *Lindblad*

---

[1] Physically the need to pick a larger value of b_{min} for a massive planet comes about precisely because the interaction close to the planet is strong – strong enough in fact to repel all the nearby gas so that there is nothing left for the planet to interact with. We will quantify this effect later, but for now the precise value of b_{min} can be considered either as an arbitrary but plausible guess, or as having been chosen with the bogus wisdom that comes from knowing the "right" answer so as to yield sensible estimates.

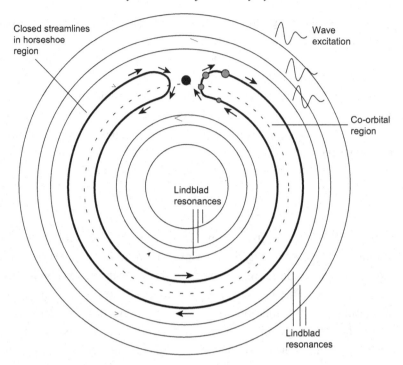

Figure 7.1 Illustration (not to scale) of disk locations where the purely gravitational perturbation of a planet leads to gas disk migration torques. The excitation of waves at exterior Lindblad resonances removes angular momentum from the planet, while waves launched at interior Lindblad resonances add angular momentum. In the co-rotating frame gas close to the planet follows closed horseshoe streamlines. Asymmetries in this region contribute additional co-rotation torques.

resonances. A Lindblad resonance occurs at a radius where gas in the disk is excited at a frequency that matches that of radial (or epicyclic) oscillations. If the planet has a circular orbit with angular frequency Ω_p, and the gas disk has orbital frequency $\Omega(r)$ and epicyclic frequency $\kappa(r)$, the resonant condition is

$$m[\Omega(r) - \Omega_p] = \pm\kappa(r), \tag{7.12}$$

where m is an integer. For a Keplerian disk the epicyclic and orbital frequencies are identical and the nominal locations of Lindblad resonances are given by

$$r_L = \left(1 \pm \frac{1}{m}\right)^{2/3} a, \tag{7.13}$$

where a is the semi-major axis of the planetary orbit. This simple expression is valid if the effects of pressure and self-gravity in the gas disk are negligible. The Lindblad resonances resemble a "comb" and become closely spaced (both to the inside and to the outside) near to the location of the planet.

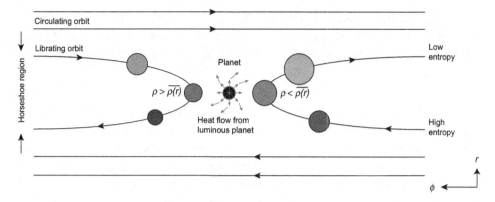

Figure 7.2 Illustration of two of the processes that can lead to torque from the vicinity of the co-orbital region. Gas librating on horseshoe orbits may execute horseshoe turns on a time scale that is faster than the thermal equilibration time with the background disk. This can result in an azimuthal density asymmetry close to the planet, and a net torque. Heat from a luminous planet will diffuse nonspherically due to shear in the disk. The resulting density perturbation can also be asymmetric, resulting in a thermal torque.

The Lindblad torque was computed in the context of interactions between satellites and planetary rings by Goldreich and Tremaine (1979, 1980), and the calculation has been revisited and refined many times since. In outline, the calculation has two steps. First, the perturbation to the gravitational potential caused by the planet is decomposed into Fourier modes that vary azimuthally as $\exp[im(\phi - \Omega_p t)]$, where m is an azimuthal wavenumber. Second, the response of the gas disk to the perturbations is calculated from the linearized hydrodynamic equations, from which can be derived the torque.[2] The key characteristics of the Lindblad torque are:

- The total torque on the planet is the signed sum of the torque exerted at each resonant location. The planet gains angular momentum from the interaction with the inner Lindblad resonances, and loses angular momentum to the outer ones, so there is a partial cancellation. When the effects of pressure in the gas disk are included the torque is cut off within a distance $\Delta r \approx 2h/3$ of the planet, removing the apparent divergence that was present in the impulse approximation.
- The planetary perturbation excites waves at the resonant locations, which propagate radially and deposit their angular momentum elsewhere in the disk. The propagating waves generate the spiral pattern that is seen in numerical simulations of planet–disk interactions. How the waves propagate and dissipate depends upon details of the disk structure, but the torque itself is not sensitive to these details.
- The torque does depend upon how far the inner and outer Lindblad resonances are from the planet, and on the surface density at each resonant location. As a

[2] Although the principles are easy to state, the actual calculation is quite intricate. For the reader interested in the details the derivation in Meyer-Vernet and Sicardy (1987) is reasonably compact and self-contained.

consequence the net torque has a dependence on the gradients (or more generally, the profiles) of surface density and temperature. However, for almost all physical disk models the outer resonances exert a greater torque on the planet than the inner ones.

Numerical simulations confirm that linear theory yields a good estimate of the net torque resulting from planet–disk interactions at Lindblad resonances. The strength of the torque depends upon how the disk responds to the exchange of angular momentum with the planet, which in turn depends upon the angular momentum transport and loss processes acting within the disk. We have noted previously that these aspects of disk physics are hard to calculate or measure observationally. However, for planets whose mass is low enough that the disk remains almost unperturbed, the robust expectation is that the Lindblad torque component results in inward migration (Ward, 1997).

A second component of the torque arises from gravitational perturbations to the gas disk in the co-orbital (or co-rotation) region. A part of the co-rotation torque can be calculated in an analogous manner to the Lindblad torque, by considering the linear response of the gas disk at the location of the co-rotation resonance where $\Omega(r) = \Omega_p$. The linear co-rotation torque depends upon the gradient of vortensity $(\nabla \times \mathbf{v})/\Sigma$ in the disk, and happens to vanish for the minimum-mass Solar Nebula case where $\Sigma(r) \propto r^{-3/2}$.

The linear co-rotation torque is not the entirety of the torque exerted in the co-orbital region, even for planet masses low enough to leave the disk surface density unaffected. A second part arises from the interaction of gas on closed *horseshoe orbits* with the planet (Ward, 1991). Horseshoe orbits, illustrated in Fig. 7.1, arise in the restricted three-body problem but can also be present in fluid disks. It is clear that fluid elements executing horseshoe orbits must continually exchange angular momentum with the planet. A parcel of gas that catches up with the planet on a slightly interior orbit must gain angular momentum to move to the exterior orbit, and the reverse process occurs in the switch from the exterior to interior part of the horseshoe trajectory. In the purely gravitational case of a near-co-orbital particle the gains and losses sum to zero, and there is no overall torque on the planet. In the case of a fluid disk, however, it is possible to generate a persistent asymmetry such that the gain in angular momentum during one of the horseshoe turns differs in magnitude from the loss during the other. Any such asymmetry leads to a net torque on the planet.

A specific example of physics that can generate a horseshoe asymmetry is shown in Fig. 7.2. Suppose that the disk in the vicinity of the planet has a radial gradient of entropy that differs from the adiabatic gradient, and that the time scale for a fluid element to come into thermal equilibrium is similar to or longer than the time it takes to complete a horseshoe turn. Under these conditions, the expansion of fluid moving from the inner portion of the horseshoe outward is *not* symmetric with the compression that occurs moving in the other direction. An azimuthal density perturbation is set up near the planet, which leads to a torque.

Entropy gradients are not the only cause of a nonzero horseshoe drag. Vortensity gradients are also important, so the total torque depends upon the surface density gradient, the temperature gradient, and thermodynamic properties of the fluid (including its adiabatic index and ability to equilibrate via radiative or diffusive processes).

Evaluating the co-orbital torques using the unperturbed disk structure and thermodynamic properties as a background yields what is described as the *unsaturated* torque. This is the instantaneous torque that would be experienced by a planet laid down within a specified disk model. Over time, however, the torque would significantly modify the angular momentum of fluid on the horseshoe orbits, which are closed and do not connect to other orbits within the disk. Numerical simulations show that in the absence of diffusion and viscosity, the back-reaction on the co-orbital region leads to the torque *saturating*, and eventually being driven to zero. This phenomenon introduces an implicit dependence of the corotation torque on the strength of angular momentum transport and loss processes within the disk.

To summarize, a planet experiences torque due to the gravitational perturbation it exerts on near-co-orbital gas as a consequence of several effects. There is a linear co-rotation torque, which depends upon the radial gradient of vortensity, and nonlinear horseshoe drag torques, which vary according to the gradients of surface density and temperature. These torques can only be persistent if there is sufficient diffusion (or loss) of heat and angular momentum to keep the co-orbital region unsaturated.

A minimal accounting of planetary migration requires consideration of the sign and magnitude of the combined Lindblad and co-rotation torques. There can also be other contributions. A luminous planet generates a thermal perturbation in the surrounding disk, which sources a corresponding density perturbation. If the density perturbation is asymmetric, the result is a *thermal torque*. For a planet that is already migrating, the flow of gas through the co-orbital region generates a dynamical torque that differs from that experienced by a stationary planet.

7.1.3 Torque Formulae

The results of analytic calculations and simulations of planet–disk interactions can be represented via formulae that give the magnitude and dependence of the migration torque on the planet. For a disk with a power-law scaling of surface density with radius,

$$\Sigma(r) \propto r^{-\alpha}, \tag{7.14}$$

the net Lindblad torque on a planet of mass M_p in a circular orbit at distance a from the star is given in linear theory as (Tanaka *et al.*, 2002)

$$\Gamma_{\text{LR}} = -(2.34 - 0.1\alpha)\Gamma_0, \tag{7.15}$$

where Γ_0, the reference torque, is,

$$\Gamma_0 = \left(\frac{M_p}{M_*}\right)^2 \left(\frac{h}{r}\right)^{-2} \Sigma a^4 \Omega_K^2. \tag{7.16}$$

Comparing the scalings here to those derived via the impulse approximation, we find that the more sophisticated calculation of the linear theory torque yields a dependence on the disk thickness that scales as $(h/r)^{-2}$ rather than $(h/r)^{-3}$. The reason for this difference is that the intrinsic asymmetry between the inner and outer Lindblad torques is itself an increasing function of the disk thickness. The remaining scalings with planet mass, disk surface density, and orbital radius, are identical to those deduced from elementary arguments.

The linear part of the co-rotation torque also scales with the disk and planet properties as a numerical multiple of the reference torque Γ_0. We do not quote the corresponding formulae, because it is rarely appropriate to consider the linear co-rotation torque without simultaneously accounting for the multiple components of the horseshoe drag. Fitting formulae for the latter, and for functions that represent the transition between unsaturated and saturated co-orbital torques, have been derived numerically, but they are more complex and remain subject to possible revision as simulations improve. The reader who needs accurate estimates of planetary migration torques is advised to start with current (at the time of writing) papers and reviews by Paardekooper *et al.* (2011), Kley and Nelson (2012), and Jiménez and Masset (2017), and work forward from there.

The physics of planet–disk interactions at co-rotation is undeniably messy, and hard to calculate to high precision. This complexity does not, however, preclude us from making a simple estimate of when gas disk migration torques are likely to be important for planet formation and planetary system evolution. To do so, we note that the *Lindblad* contribution to the torque is both physically independent of whatever happens at corotation, and amenable to a reliable linear calculation. If we estimate a migration time scale based solely on the Lindblad torque, and find it to be short, then that conclusion is highly unlikely to be generically reversed by the addition of independent corotation torques, whatever their strength and sign. (As we will discuss later, exact cancellation of torques can occur at special locations within the disk, known as *planet traps*.)

Proceeding in this spirit, we define a reference migration time scale for a planet with orbital angular momentum L as

$$\tau_0 \equiv \frac{L}{\Gamma_0} \propto M_*^{3/2} M_p^{-1} \left(\frac{h}{r}\right)^2 \Sigma^{-1} a^{-1/2}. \tag{7.17}$$

Substituting numbers roughly appropriate for super-Earths at 0.1 AU, planetary embryos in the Solar System at 1 AU, and giant planet cores at 10 AU, we find

$$\tau_0 \approx 2.0 \times 10^5 \left(\frac{M_p}{3M_\oplus}\right)^{-1} \left(\frac{h/r}{0.02}\right)^2 \left(\frac{\Sigma}{10^3 \, \text{g cm}^{-2}}\right)^{-1} \left(\frac{a}{0.1 \, \text{AU}}\right)^{-1/2} \text{yr,}$$

$$\approx 4.2 \times 10^6 \left(\frac{M_p}{0.1M_\oplus}\right)^{-1} \left(\frac{h/r}{0.03}\right)^2 \left(\frac{\Sigma}{10^3 \, \text{g cm}^{-2}}\right)^{-1} \left(\frac{a}{\text{AU}}\right)^{-1/2} \text{yr,}$$

$$\approx 7.4 \times 10^5 \left(\frac{M_p}{5M_\oplus}\right)^{-1} \left(\frac{h/r}{0.05}\right)^2 \left(\frac{\Sigma}{10^2 \, \text{g cm}^{-2}}\right)^{-1} \left(\frac{a}{10 \, \text{AU}}\right)^{-1/2} \text{yr.} \quad (7.18)$$

These estimates assume a solar mass star. We should note that there is substantial uncertainty in protoplanetary disk models, especially on sub-AU scales, and the fiducial values adopted above are not much more than guesses. With that caveat in mind, the reference migration time scales that we have estimated paint a clear physical picture. Gas disk migration ought to occur on time scales of a Myr or less, and thus be an important physical effect, for planets of Earth mass and above that orbit close to their stars ($a \sim 0.1$ AU), and for giant planet cores beyond the snow line ($a \sim 3$–10 AU). Migration is a weaker effect, which may be ignorable, within the classical model of Solar System terrestrial planet formation, provided that the progenitors of the terrestrial planets have masses that stay below $\sim 0.1 \, M_\oplus$ prior to gas disk dispersal.

7.1.4 Gas Disk Migration Regimes

As a planet grows the effect of the angular momentum that it deposits or extracts from the disk, at Lindblad resonances and around co-rotation, becomes more important. Eventually it results in a wholesale modification of the disk structure in the vicinity of the planet. It is conventional to distinguish two regimes:

- **Type 1** migration occurs for low mass planets whose interaction with the disk is weak enough as to leave the disk structure almost unperturbed. This will certainly be true if the local exchange of angular momentum between the planet and the disk is negligible compared to the redistribution of angular momentum due to disk viscosity. The planet remains fully embedded within the gas disk and material is present at all of the Lindblad resonances and in the co-orbital region.
- **Type 2** migration occurs for higher mass planets whose gravitational torques locally dominate angular momentum transport within the disk. As we have already noted, gravitational torques from the planet act to repel disk gas away from the orbit of the planet, so in this regime the planet opens an annular gap within which the disk surface density is reduced from its unperturbed value. Resonances close to the planet are severely depleted of material and contribute little or nothing to the total torque.

A visual impression of the two regimes is shown in Fig. 7.3, whose two panels show the simulated surface density structure of a disk interacting with a planet in the Type 1 and Type 2 regimes. That the interaction is concentrated at resonances

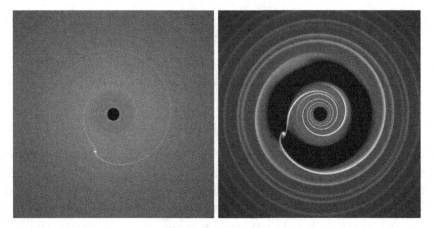

Figure 7.3 Two-dimensional hydrodynamic simulations depicting the interaction between a planet and a viscous protoplanetary disk in (left panel) the Type 1 regime appropriate to low mass planets and (right panel) the Type 2 regime relevant to giant planets. In both cases angular momentum exchange is the result of gravitational interaction with spiral waves set up within the disk as a consequence of the planetary perturbation. In the Type 1 regime the interaction is weak enough that the local surface density is approximately unperturbed, while in the Type 2 regime a strong interaction repels gas from the vicinity of the planet producing an annular gap.

cannot easily be discerned. Rather, one sees that the interaction results in the formation of a wake of enhanced surface density that trails the planet outside its orbit and leads inside. It is the gravitational back-reaction of this wake on the planet that produces the torque and leads (together with the co-orbital asymmetry, which is not obvious in this visualization) to migration.

7.1.5 Gap Opening and Gap Depth

The mass above which a planet is able to open a gap within the gas can be estimated by considering the competing effects that try to open and close annular gaps within the disk. The planetary torque *always* acts in such a way as to try and open a gap, but this tendency is opposed by viscosity (internal angular momentum transport processes that would be present in the absence of a planet) which acts diffusively to smooth out sharp radial gradients in the disk surface density. This competition is illustrated in Fig. 7.4.

The threshold mass above which a planet succeeds in opening a gap can be estimated using a simple approach that draws on our analysis of the torque in the impulse approximation. We assume that the minimum gap size is of the order of h, first because Lindblad resonances are at their most effective at about this distance from the planet, and second because gaps whose radial width was much smaller than the vertical thickness of the disk might well be unstable. The criterion for gap

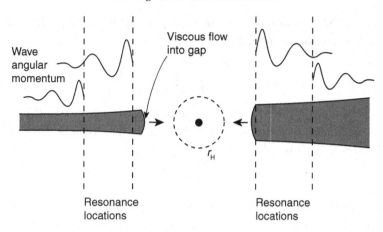

Figure 7.4 The balance of torques that determines (in part) whether a planet is able to open a gap within the disk. Waves excited at resonant locations act to remove angular momentum from the disk interior to the planet, and add angular momentum to the disk outside, thereby opening a gap. Viscous flow counteracts this tendency.

opening can then be estimated by equating the time scale for opening such a gap to the time scale on which viscosity would close it.

To estimate the gap opening time scale t_{open} we note that the amount of angular momentum that must be added to evacuate all of the gas between a and $(a + h)$ out of the annulus is

$$\Delta J = 2\pi a h \Sigma \cdot \left. \frac{dl}{dr} \right|_a \cdot h, \qquad (7.19)$$

where $l = \sqrt{GM_* r}$ is the specific angular momentum of gas in a Keplerian orbit. The gap opening time scale can then be estimated as

$$t_{\text{open}} = \frac{\Delta J}{|dJ/dt|}, \qquad (7.20)$$

with dJ/dt given by the impulse equation formula (Eq. 7.9) with $b_{\text{min}} = h$. The expression that this yields is not terribly enlightening. To make progress, we compare the gap opening time scale to the time scale on which viscosity would act to close a gap of characteristic scale h within the gas disk. The gap closing time scale is (Eq. 3.11)

$$t_{\text{close}} = \frac{h^2}{\nu}, \qquad (7.21)$$

where ν is the kinematic viscosity. Making use of the Shakura–Sunyaev α-prescription, $\nu = \alpha c_s h$ (Eq. 3.46), we then equate t_{open} and t_{close}. The result is an estimate for the critical mass ratio $q \equiv M_p/M_*$ between the planet and the star above which a gap can be opened:

$$q_{\text{crit}} \simeq \left(\frac{27\pi}{8}\right)^{1/2} \left(\frac{h}{r}\right)^{5/2} \alpha^{1/2}. \qquad (7.22)$$

In this expression (h/r) is to be interpreted as the geometric thickness that the disk would have in the absence of the planet. For typical disk parameters $(\alpha = 10^{-3}, h/r = 0.05)$ we obtain

$$q_{\text{crit}} \simeq 6 \times 10^{-5}, \qquad (7.23)$$

which suggests that planets of Saturn mass orbiting solar mass stars should be able to open gaps. The gap opening mass has no explicit dependence on the disk surface density, but varies strongly with h/r.

An additional consideration is whether a two-dimensional treatment of planet–disk interaction is adequate to study gap opening, or whether three-dimensional studies are necessary. (The argument above is essentially two-dimensional.) A two-dimensional analysis can be justified if the Hill sphere (Eq. 5.15) of the planet is larger than the scale-height of the disk,

$$r_{\text{H}} \gtrsim h. \qquad (7.24)$$

This will be true if the mass ratio exceeds

$$q_{\text{crit}} \gtrsim 3 \left(\frac{h}{r}\right)^3, \qquad (7.25)$$

which evaluates to $q_{\text{crit}} \approx 4 \times 10^{-4}$ for fiducial disk parameters. The similarity of this number to the threshold gap-opening mass implies that although two-dimensional models are useful, three-dimensional simulations are often necessary for accurate studies of the gap-opening process.

An analytic estimate for the *depth* of the gap created by a planet can be derived using an approach quite similar to that used to determine the gap-opening criterion. Following Kanagawa *et al.* (2015), we first note that the radial flux of angular momentum through the disk can be written as the sum of an advective and viscous piece (Eq. 3.5),

$$F_J(r) = -r^2 \Omega \dot{M} - 2\pi r^3 \nu \Sigma \frac{d\Omega}{dr}. \qquad (7.26)$$

We assume that the planet orbits at $r = a$, and that the gap extends out to $r = r_+$. If the gap is wide enough that essentially all of the one-side Lindblad torque is exerted on gas within the gap, then the angular momentum flux at the outer gap edge is equal to the sum of that passing through the disk at the location of the planet plus the planetary torque:

$$F_J(r_+) = F_J(a) + \Gamma_{\text{LB},1}. \qquad (7.27)$$

The impulse approximation yields the scalings of the one-sided planetary torque, but for a numerical estimate we use a better calculation, based on a sum over exterior Lindblad resonances, which gives (Tanaka *et al.*, 2002; Kanagawa *et al.*, 2015)

$$\Gamma_{\text{LB},1} = 0.12\pi \left(\frac{M_p}{M_*}\right)^2 \left(\frac{h}{r}\right)^{-3} a^4 \Sigma_{\text{gap}} \Omega_K^2. \tag{7.28}$$

We now assume that although the gap is wide enough to absorb all of the one-sided torque, it is narrow enough that $r_+ \approx a$ and $\Omega_K(r_+) \approx \Omega_K(a)$. In that case, $F_J(r_+)$ and $F_J(a)$ differ only via the second term in Eq. (7.26). Noting that $d\Omega/dr = -(3/2)(\Omega_K/r)$ for a Keplerian disk, and using the alpha prescription to write $\nu = \alpha h^2 \Omega_K$, Eq. (7.26) becomes

$$3\pi a^4 \alpha \left(\frac{h}{r}\right)^2 \Omega_K^2 \Sigma_{\text{disk}} = 3\pi a^4 \alpha \left(\frac{h}{r}\right)^2 \Omega_K^2 \Sigma_{\text{gap}}$$

$$+ 0.12\pi \left(\frac{M_p}{M_*}\right)^2 \left(\frac{h}{r}\right)^{-3} a^4 \Sigma_{\text{gap}} \Omega_K^2. \tag{7.29}$$

Simplifying and rearranging we get an expression for the gap depth,

$$\frac{\Sigma_{\text{gap}}}{\Sigma_{\text{disk}}} = \frac{1}{1 + 0.04K}, \tag{7.30}$$

where the power of a planet to open a gap is expressed via the parameter,

$$K \equiv \left(\frac{M_p}{M_*}\right)^2 \left(\frac{h}{r}\right)^{-5} \alpha^{-1}. \tag{7.31}$$

Making a somewhat arbitrary definition that a gap "opens" when $\Sigma_{\text{gap}}/\Sigma_{\text{disk}} \leq 0.5$, we find that the threshold mass ratio needed to open a gap has the same scaling derived before, $q_{\text{crit}} \propto (h/r)^{5/2} \alpha^{1/2}$. This is to be expected as the physical argument is almost the same. The numerical value of q_{crit} is slightly larger. The more refined analysis, however, does give us something new – the predicted depth of the gap over the transition between the Type 1 and Type 2 regimes. Numerical simulations collated by Kanagawa *et al.* (2015) show reasonable agreement with the analytic prediction.

7.1.6 Coupled Planet–Disk Evolution

Determining the migration history of a planet due to gas disk torques requires a well-specified model for the protoplanetary disk, which includes at a minimum the profiles of surface density, temperature, viscosity, and thermal diffusivity (together with any time dependence of these quantities). In the Type I regime, torque formulae can then be used to compute the rate of migration as a function of semi-major axis and planet mass within the given disk model. Examples of such *migration maps* are given by Bitsch *et al.* (2015). The results are quite sensitive to the adopted disk model, and as such are more reliable at orbital radii where the disk properties can be observationally constrained. Typically, there is a mass M_{min} below which migration is inward at all orbital radii, and a mass M_{max} above which the same is true. For intermediate masses there are alternating radial zones of outward migration (driven by co-orbital torques) and inward migration. These are separated by migration null points, known as planet traps, where the predicted instantaneous migration rate vanishes. Planets or planetary cores trapped at these special locations would experience slow drift, as the position of the traps themselves changes as the disk evolves.

In the Type II regime, determining how a planet migrates requires both a good model for the unperturbed disk and an understanding of how its structure is altered by the planet. A limited amount of insight can be obtained by considering a toy model of a massive planet that opens a clean gap (Fig. 7.3), so that the only significant torques are those exerted at a small number of low-m Lindblad resonances. The edges of the gap are defined by the location of the lowest m resonances that are just able to hold back the viscous inflow of the disk into the gap region. A conceptual model to understand how the disk–planet system behaves in this limit is to imagine that the planet is surrounded by "brick walls" that define the inner and outer edges of the gap. At the edge of the gap interior to the planet, the wall removes exactly enough angular momentum from gas to prevent the disk overflowing the wall, whereas at the outer edge of the gap the wall (or, physically, the planetary torque) adds the appropriate amount of angular momentum. If we now assume that in the region near the planet most of the angular momentum resides in the *gas rather than in the planet* it is fairly clear how the coupled system of the planet plus the disk will evolve. As gas from the outer disk flows inward due to ordinary viscous evolution it runs up against the barrier imposed by the planetary torque at the outer edge of the gap. To prevent the gas encroaching into the gap the planet must give angular momentum to the outer disk, which must be balanced by a loss of angular momentum from the inner disk. If the inner disk loses angular momentum, however, its gas must move inward and away from the inner wall. To maintain the edges of the gap at the resonant locations where torque is transmitted the planet itself must move inward.

This thought experiment suggests that the role of a relatively low mass planet migrating in the Type 2 regime is simply that of a catalyst whose torques bridge the gap in the disk surface density without otherwise altering the angular momentum flux that would be present in the disk in the absence of the planet. The gas then flows inward at the same rate as in a planet-free disk, and the planet migrates at the same rate as the local disk gas in order to remain at the center of the gap. We therefore define the nominal Type 2 migration rate as being equal to the velocity of gas inflow in a steady disk:

$$v_{nominal} = -\frac{3v}{2r}. \tag{7.32}$$

Using the α-prescription once more we obtain

$$v_{\text{nominal}} = -\frac{3}{2}\alpha \left(\frac{h}{r}\right)^2 v_{\text{K}}, \tag{7.33}$$

where v_{K} is the Keplerian orbital velocity. The speed of Type 2 migration depends upon poorly known disk parameters (in particular, it depends linearly on the efficiency of angular momentum transport as parameterized via α) but it is generically rather rapid. The coupled evolution of a massive planet embedded within an evolving disk can be approximately modeled with straightforward generalizations of the one-dimensional disk evolution equation discussed in Section 3.2 (Lin & Papaloizou, 1986). Figure 7.5 shows an example of such a calculation, which predicts migration time scales from 5 AU of the order of 10^5 yr. One should note that although migration is always inward in a steady disk of large radial extent, the argument actually implies only that the planet ought to track the

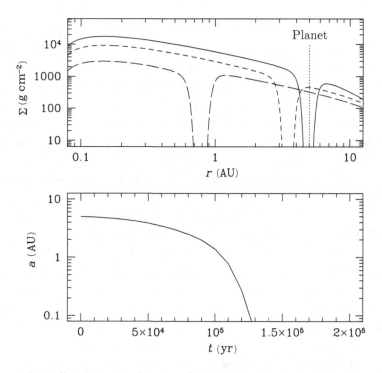

Figure 7.5 Predicted Type 2 migration of a 1 M_{J} planet formed within an evolving protoplanetary disk at 5 AU, based on an approximate one-dimensional treatment of the interaction (Armitage *et al.*, 2002). The calculation assumes that there is no mass flow across the gap, and that no mass is accreted by the planet. The upper panel shows the disk surface density as a function of radius at three different epochs, while the lower panel shows the evolution of the planetary semi-major axis. The model assumes that the planet formed at an early epoch while the gas disk was still massive. The planet migrates inward within a gap on a short time scale, reaching small radii after approximately 10^5 yr.

local motion of the surrounding gas. A planet that forms near the outer edge of a disk that is *expanding* viscously would therefore migrate with the gas to larger orbital radii (Veras & Armitage, 2004).

The actual Type 2 migration is expected to differ from the nominal rate due to at least two effects. Migration will slow if the local disk mass ($\sim \pi a^2 \Sigma$) is small compared to the planet mass, as a low mass disk is unable to transport away angular momentum at the high rate that would be needed to sustain the nominal speed. More generally, however, the rate of Type 2 migration is modified because simulations show that the edges of planet-carved gaps are rather porous dams. Gas can flow across the gap, violating the assumption of no flow that anchored the estimate of the nominal migration rate, and can be accreted by the planet, directly altering its angular momentum. Papers by Duffell *et al.* (2014), Dürmann and Kley (2015), Kanagawa *et al.* (2018), and Robert *et al.* (2018) report simulation results that quantify the magnitude of the discrepancy.

7.1.7 Eccentricity Evolution

Up to now we have focused on the evolution of the semi-major axes of planets experiencing gas disk migration. An important question is whether eccentricity can be excited as a consequence of planet–disk interactions during Type 2 migration, as this could be a mechanism able to explain the observed non-zero eccentricities of many giant extrasolar planets. Eccentricity evolution can be investigated within the same resonant framework used to compute the net torque, by evaluating the net damping or excitation of planetary eccentricity that results from the interaction of a slightly eccentric planet with the disk at all significant resonant locations (Goldreich & Tremaine, 1980). Table 7.1 lists the nine most important resonances,[3] along with the sign of the eccentricity change they induce on an embedded planet (Goldreich & Sari, 2003; Ogilvie & Lubow, 2003; Masset & Ogilvie, 2004). The interaction at some resonances excites eccentricity, while that at others damps it. For a massive planet that clears a gap (so that the three co-orbital resonances listed in the table play no role) the balance between excitation and damping at the remaining resonances is delicate, with damping winning by only a narrow margin. This conclusion could be reversed in a number of ways, most of which involve suppressing the damping effect of the first-order co-rotation resonances by at least a small amount. This suppression could occur if the planet clears a *very* wide gap (as happens for binary stars, Artymowicz *et al.*, 1991), or if the co-rotation resonances are intrinsically slightly weaker due to saturation at less than the nominal strength (Goldreich & Sari, 2003; Ogilvie & Lubow, 2003).

---

[3] The nomenclature of "fast" and "slow" denotes resonances that occur where the pattern speed is either faster or slower than the planet's orbital frequency.

Table 7.1 *List of the principal and first-order Lindblad and co-rotation resonances, together with their effect on the eccentricity of an embedded giant planet (after Masset & Ogilvie, 2004).*

Resonance	$(r/a)^{-2/3}$	Effect on e
Principal OLR	$m/(m+1)$	excite
Principal ILR	$m/(m-1)$	damp
Co-orbital CR	1	unclear
Fast first-order OLR	1	damp
Fast first-order ILR	$(m+1)/(m-1)$	excite
Fast first-order CR	$(m+1)/m$	damp
Slow first-order OLR	$(m-1)/(m+1)$	excite
Slow first-order ILR	1	damp
Slow first-order CR	$(m-1)/m$	damp

Numerical simulations confirm the analytic prediction that substantial eccentricities can be excited for very massive planets or brown dwarfs (with masses $M_p \gtrsim 5$–$20\ M_J$) that create deep and wide gaps. In these circumstances eccentricity is excited due to the dominant influence of the 3:1 exterior Lindblad resonance (Papaloizou *et al.*, 2001). Lower mass giant planets can also experience eccentricity growth, but the maximum eccentricity is limited. Simulations suggest that the limit may be of the order of $e \sim (h/r)$, and certainly cannot exceed a limit set by the requirement that the apocenter of the orbit remains within the gap $e \sim 0.1$ (D'Angelo *et al.*, 2006; Duffell & Chiang, 2015; Ragusa *et al*, 2017).

7.2 Secular and Resonant Evolution

In our survey of the Solar System (Section 1.1) we noted that a resonance occurs when there is a near-exact commensurability among the characteristic frequencies of one or more bodies. The use of the term "characteristic frequency" is deliberately vague, since it must encompass a wide variety of possibilities that include the orbital frequency, the spin frequency of a planet's rotation, the precession frequency of the orbit, and more. We can start, however, by considering the simplest case where the commensurability is between the orbital frequencies (or periods) of two planets on circular orbits. We have already noted (Eq. 1.3, here expressed in an equivalent form) that the condition for resonance can be written as

$$\frac{P_{\text{in}}}{P_{\text{out}}} \simeq \frac{p}{p+q}, \qquad (7.34)$$

where P_{in} and P_{out} are the orbital periods of the two planets and p and q are integers. The definition, of course, raises the immediate question of what is meant by the approximate equality sign. How close do the two planets have to be to the exact commensurability for them to be as a practical matter "in resonance"? To address this we first consider how gravitational perturbations affect a system that is close to an exact commensurability, and then sketch out how to mathematically define whether a given system is resonant or not.

7.2.1 Physics of an Eccentric Mean-Motion Resonance

The essential physics of mean-motion resonances resonances can be understood by considering the simple model system shown in Fig. 7.6 (Peale, 1976). A massive planet m_{in} with a circular orbit has a low mass outer companion m_{out} with a significantly eccentric orbit. We assume that the orbital periods of the planets (obtained by fitting Keplerian orbits to the instantaneous position and velocity data) are close to an integer ratio, and examine how the small perturbations the inner planet exerts on the outer affect the evolution.

At *conjunction* (when the two planets line up on one side of the star) and at *opposition* (when they line up with the star in between) the gravitational perturbation m_{in} exerts on m_{out} is radial, and does not act to change the angular momentum of the outer body's orbit. During the interval between opposition and conjunction the outer body is ahead of the inner one along its orbit, and the tangential component of the mutual gravitational force acts to remove angular momentum from m_{out}. Conversely, between conjunction and the next opposition the interactions works to add angular momentum to m_{out}. It is easy to see that if conjunctions occur at either

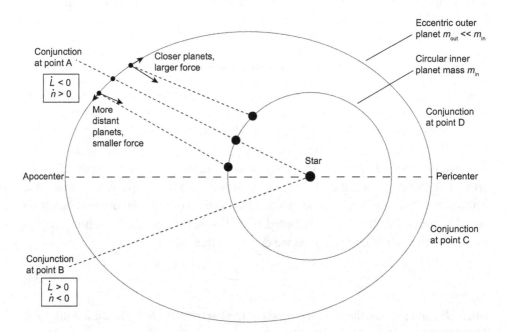

Figure 7.6 Illustration, after Peale (1976), of how gravitational perturbations act to keep two planets with nearly commensurate orbital periods in resonance. For simplicity, the model system consists of a massive inner body on a circular orbit together with a low mass outer body on an eccentric orbit.

pericenter or apocenter the angular momentum gaining and angular momentum losing phases are symmetric, and there is no net transfer of angular momentum between the bodies.

The symmetry is broken if conjunctions occur at a general point along the orbit. Suppose, for example, that a conjunction occurs at point A in Fig. 7.6. Because the trajectories of the planets are diverging as they pass through conjunction (due to the eccentric orbit of $m_{\rm out}$), the gravitational perturbation on $m_{\rm out}$ just prior to the conjunction is stronger than that just afterwards. Additionally, the angular velocity of $m_{\rm out}$ is slowing down as it passes through point A moving toward apocenter, so there is also more time for the perturbation to act prior to conjunction than there is afterwards. Both effects work in the same sense, such that the interaction (averaged between one conjunction and the next) acts to remove angular momentum L from $m_{\rm out}$. A similar argument shows that angular momentum is added to $m_{\rm out}$ if conjunctions occur instead at point B.

A planet that loses angular momentum moves to an orbit with a shorter period P and a larger mean motion $n \equiv 2\pi/P$. This means that a conjunction that occurs at point A will be followed by a conjunction that occurs closer to apocenter. The same is true for a conjunction at point B. We say that the point of conjunction *librates* (oscillates) stably about apocenter, maintaining the system in resonance even though the instantaneous orbital periods of the planets are not usually exactly commensurate. Stable libration remains possible even if we include the effect of orbital precession which is also driven by the mutual gravitational perturbations between the planets.

Exactly the same methods can be applied to analyze the stability of conjunctions that occur at points C and D, near pericenter. Both are driven away from pericenter, which is an unstable equilibrium point of the system.

The scenario that we have analyzed here applies to a number of Solar System resonances, including some between the satellites of Saturn and between Jupiter and non-co-orbital asteroids. It is not however general, and physically different outcomes are possible if different assumptions are adopted. For example, stable libration at pericenter as well as apocenter is possible if the eccentricity of the outer orbit is small. Peale (1976) provides an elementary explanation of this limit, following the same line of reasoning that we have applied to the high eccentricity case.

7.2.2 Example Definition of a Resonance

The physical description of resonance given above makes it fairly clear how to define the condition of being in resonance mathematically. Let us first assume that two planets are in exact resonance and rewrite Eq. (7.34) in terms of the mean motions of the planets:

$$\frac{n_{\rm out}}{n_{\rm in}} = \frac{p}{p+q}. \tag{7.35}$$

If we ignore any perturbations between the planets, the angle λ between the radius vector to one of the planets and a reference direction advances linearly with time. Defining $t = 0$ and $\lambda = 0$ to coincide with a moment when the two planets are in conjunction, we have that

$$\lambda_{\text{in}} = n_{\text{in}}t, \tag{7.36}$$
$$\lambda_{\text{out}} = n_{\text{out}}t, \tag{7.37}$$

and the resonance condition becomes

$$(p + q)\lambda_{\text{out}} = p\lambda_{\text{in}}. \tag{7.38}$$

Finally, we can define a *resonant argument*

$$\theta = (p + q)\lambda_{\text{out}} - p\lambda_{\text{in}}, \tag{7.39}$$

which will evidently remain zero for all time if the planets are in exact resonance. For planets on general circular orbits, on the other hand, λ_{in} and λ_{out} still advance linearly with time, but no small p and q can be found so that θ remains a constant. Sampled at random intervals, in fact, θ takes on all values in the range $[0, 2\pi]$. It is this basic distinction that furnishes a definition of what it means for two planets to be in resonance. A resonance occurs when one or more resonant arguments is bounded (though it need not be exactly constant), while there is no resonance if the argument takes on all possible values.

In more detail, we follow Murray and Dermott (1999) and analyze the conditions for resonance within the context of the circular restricted three-body problem. We first review some nomenclature for eccentric orbits. The *mean anomaly* is given by

$$M = n(t - t_{\text{peri}}), \tag{7.40}$$

where t_{peri} is the time of pericenter passage. The *mean longitude* is

$$\lambda = M + \varpi, \tag{7.41}$$

where ϖ is the longitude of pericenter. We now consider the condition for exact resonance between a planet on a circular orbit ($e_{\text{in}} = 0$) and a low mass body on an eccentric orbit ($e_{\text{out}} \neq 0$). The potential experienced by the low mass body is not quite Keplerian, but to a good approximation we can assume that the orbit is an instantaneous Keplerian orbit whose longitude of pericenter varies with time at some rate $\dot{\varpi}_{\text{out}} \neq 0$. Working in a frame that rotates along with the drift in the longitude of pericenter of the outer body, the condition for exact resonance is

$$\frac{n_{\text{out}} - \dot{\varpi}_{\text{out}}}{n_{\text{in}} - \dot{\varpi}_{\text{out}}} = \frac{p}{p + q}, \tag{7.42}$$

and the resonant argument is

$$\theta = (p + q)\lambda_{\text{out}} - p\lambda_{\text{in}} - q\varpi_{\text{out}}. \tag{7.43}$$

Both the physical argument given already and a mathematical treatment show that if the system is close to the exact resonance, perturbations between the bodies act such as to keep θ bounded even if the instantaneous periods are not exactly commensurable. We say that the system is:

- In resonance if θ *librates*, by which we mean that the resonant argument may be time dependent but varies only across some limited range of angles.
- Out of resonance if θ *circulates*, taking on all values between 0 and 2π.

In fact for this model problem it can be shown that the behavior of θ over time is remarkably simple, and the angle oscillates in a manner that is mathematically identical to that of a pendulum:

$$\frac{d^2\theta}{dt^2} + \omega_0^2 \sin\theta = 0. \tag{7.44}$$

The oscillation frequency ω_0 is known in this context as the libration frequency, and the corresponding time scale,

$$t_{\text{lib}} \equiv \frac{2\pi}{\omega_0}, \tag{7.45}$$

as the libration time scale. The final basic quantity of interest is the range of semi-major axis values for which θ librates rather than circulates. This quantity is called the width of the resonance. Although one cannot state any simple general formulae for these quantities, it is possible to derive analytic estimates of the widths and libration time scales for particular resonances that are of interest (see e.g. Holman & Murray, 1996 for a concise description of the method).

Our focus up to this point on mean-motion resonances should not blind the reader into forgetting that a planetary system has a large number of characteristic frequencies, and as a consequence the number of possible resonant arguments is very large. Other types of resonance that are of interest include:

- Secular resonances, where the commensurable frequencies involve the slow precession of the longitude of pericenter or the longitude of the ascending node due to mutual interactions between the bodies. A simple example of a secular resonant argument is $\theta = \varpi_{\text{in}} - \varpi_{\text{out}}$.
- Resonances that involve inclination.
- Three-body resonances. A Solar System example is the three-body resonance involving Jupiter, Saturn, and Uranus: $n_J - 7n_U = 5n_{\text{Sat}} - 2n_J$.
- So-called secondary resonances, which include cases where the libration frequency of two planets in resonance is itself commensurable with some other frequency of the system.

Although in the Solar System there are no simple mean-motion resonances among the planets, a plethora of these weaker resonances are nonetheless believed to have important dynamical effects. In particular, spatial overlap of different resonances is the accepted explanation for the apparent presence of chaos in the motion of the planets. The review by Lecar *et al.* (2001) provides a good starting point for the reader who wishes to explore this intricate subject in more detail.

7.2.3 Resonant Capture

Since resonances have finite widths, there is some probability that two planets in a randomly assembled planetary system will happen to find themselves in a mean-motion resonance. Even a cursory inspection of data on the Solar System, however, convinces one of the need for a causal rather than a probabilistic explanation for entry into resonance. Chiang *et al.* (2007), for example, find that in excess of 20% of Kuiper Belt Objects with well-determined orbits are in mean-motion resonances with Neptune, with the 3:2 resonance occupied by Pluto being the most heavily populated. Among satellites too there are numerous known resonances, of which the most striking is the 4:2:1 resonance that involves three of the Galilean satellites of Jupiter – Io, Europa, and Ganymede. Mean-motion resonances between planets themselves also appear to be common among extrasolar planetary systems. Resonances have been securely identified in approximately 20% of known multiple planet systems, and there are additional systems that are plausibly but not yet provably in resonance.[4] One well-known example is the GJ 876 system (Marcy *et al.*, 2001), in which two massive planets orbit in a 2:1 resonance with orbit periods of approximately one and two months. For this system both of the lowest-order mean motion resonant arguments,

$$\theta_1 = \lambda_{in} - 2\lambda_{out} + \varpi_{in}, \qquad (7.46)$$

$$\theta_2 = \lambda_{in} - 2\lambda_{out} + \varpi_{out}, \qquad (7.47)$$

and the secular resonant argument,

$$\theta_3 = \varpi_{in} - \varpi_{out}, \qquad (7.48)$$

librate about zero degrees (Lee & Peale, 2001).

The likely origin of most of these mean-motion resonances lies in the phenomenon of resonant capture, which was first studied for spin–orbit resonances (Goldreich & Peale, 1966) and subsequently developed to explain resonances among planetary satellites (Sinclair, 1972; Yoder, 1979; Borderies & Goldreich, 1984). Resonant capture is possible when two bodies are driven toward resonance (i.e. their orbits converge) by the application of a (generally weak) external force – which could arise due to tidal effects on planetary satellites or torques from the

---

[4] Identifying resonances in extrasolar planetary systems is generally difficult because in many cases the orbital solutions are of limited precision. Making an accurate statistical statement as to the true fraction of resonant planetary systems is even harder, because the selection function for detecting two planets of given masses will generally differ depending upon whether they are in a resonance or not.

gas disk in the case of extrasolar planets. As the bodies approach and then enter the resonance there is a secular exchange of angular momentum between them that acts so as to maintain the resonance despite the ongoing external forcing that would otherwise result in exit from the resonance. In the case of two planets whose orbits converge because the outer planet is subject to a torque causing inward migration, for example, the resonant coupling acts to remove angular momentum from the inner planet so that the two bodies move inward in lockstep. A heuristic (but still quite intricate) description of the dynamics that results in this angular momentum transfer is given by Peale (1976). An important general result of the theory is that capture can occur only for converging orbits. Two bodies whose orbits *diverge* such that they approach and enter a resonance still experience an increased strength of interaction – which may lead to significant excitation of eccentricity – but cannot be captured into resonance according to the classical theory.

Even when it is possible (in the case of converging orbits) capture is generally a probabilistic phenomenon whose likelihood depends (in part) upon the relationship between the libration time scale t_{lib} and the time scale on which the external forcing would result in resonance crossing if capture did not occur:

$$t_{cross} = \frac{\Delta a_{res}}{|\dot{a}_{in} - \dot{a}_{out}|}. \tag{7.49}$$

Here Δa_{res} represents the width of the resonance under consideration. If $t_{cross} \gg t_{lib}$ then capture is guaranteed provided that the eccentricity of the body being captured is below some limit, while for faster crossing the probability of capture will depend upon the eccentricity, the rate of convergence of the orbit, and the particular resonance being encountered. Explicit expressions for the maximum drift rates that permit capture can be found in Quillen (2006, and references therein).

Resonant capture is a particularly appealing explanation for resonant extrasolar planets in short-period orbits, whose small semi-major axes provide circumstantial evidence for the existence of migration torques that could have both shrunk the orbits and driven the planets into resonance. Figure 7.7 shows a numerical calculation of resonant capture in a coplanar system of two Jupiter mass planets, where the outer planet is driven toward the inner one under the action of an external torque. For this choice of parameters capture occurs into the 2:1 resonance, which is then maintained as the orbits continue to shrink. Once in resonance, further decay of the semi-major axes of the planets is accompanied by growth of the eccentricity, though this may be damped (at least partially) by the interaction of the planets with the gas disk. Although existing theory is not fully predictive – in particular planets migrating through a turbulent gas disk can diffuse out of the resonance due to random variability in the torques they experience (Murray-Clay & Chiang, 2006; Adams *et al.*, 2008) – resonant capture is at least qualitatively consistent with the observation of a significant population of extrasolar planets in mean-motion resonances.

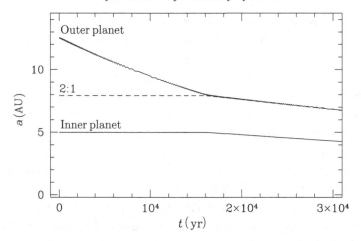

Figure 7.7 Numerical example of resonant capture in a coplanar system of two initially well-separated Jupiter mass planets. For illustrative purposes the outer planet is subjected to a fictitious drag force which results in steady decay of its semi-major axis. The inner planet is almost unperturbed until the planets encounter their mutual 2:1 mean-motion resonance, at which point capture occurs and the two planets move inward in lockstep.

7.2.4 Kozai–Lidov Dynamics

Kozai–Lidov dynamics is a quite different type of resonant behavior that is implicated in the evolution (and possibly formation) of some extrasolar planetary systems. The simplest system that exhibits the characteristic Kozai–Lidov behavior is a hierarchical triple system, in which a massive outer binary companion perturbs a low mass body on an inclined orbit closer in. The dynamics of this setup was analyzed by Mikhail Lidov, in the context of the motion of artificial satellites around the Earth under the perturbation of external bodies, and by Yoshihide Kozai, in the context of the motion of high inclination asteroids perturbed by Jupiter (Lidov, 1962; Kozai, 1962).[5] For the model system that consists of a low mass body orbiting well interior to a massive perturber on a highly inclined orbit the resonant argument is simply the argument of pericenter of the inner body itself.

The interest in this particular resonance derives from the very unusual evolution that occurs when the resonance is active. The simplest situation to analyze (and also the one most relevant for extrasolar planetary systems) is that shown in Fig. 7.8. A planet of mass M_p and semi-major axis a_{in} orbits the primary star (mass M_*) of a binary system. The secondary star of the binary system has mass M_s, semi-major axis a_s, and eccentricity e_s, and orbits in a plane inclined to that of the planet by an angle i. Triple systems of this kind are typically only stable if they are *hierarchical* (i.e. the outer binary has a much larger orbit than that of the inner binary, which is here the planet), so we consider the limit in which

---

[5] Both of these now celebrated papers languished in relative obscurity for more than 30 years.

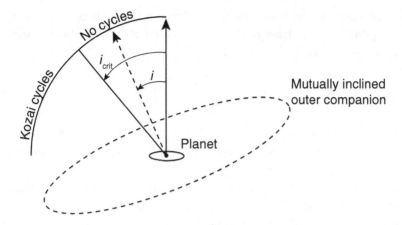

Figure 7.8 Kozai cycles are possible in hierarchical triple systems in which the orbit of the inner body (here a planet) does not lie in the same plane as that of the outer body (here a stellar companion). Cyclic exchange of angular momentum occurs if the mutual inclination $i > i_{crit}$.

$$a_s \gg a_{in}, \tag{7.50}$$

$$M_s \gg M_p. \tag{7.51}$$

The second of these conditions implies that essentially all of the angular momentum in the system resides in the orbit of the binary companion. Nothing that happens to the planet can affect the orbit of the binary, which thus remains fixed in space.

Given these approximations, Kozai (1962) and Lidov (1962) showed that the behavior of the orbit of the inner planet depends upon the magnitude of the mutual inclination i between the orbital planes. In particular, if i exceeds a critical value

$$i_{crit} = \cos^{-1}\left[(3/5)^{1/2}\right] \simeq 39.2°, \tag{7.52}$$

then the eccentricity e and inclination i of the planet describe cyclic oscillations known as Kozai–Lidov cycles. These oscillations can have a large amplitude. In the case where the planet has an initially circular orbit, the maximum value of the eccentricity during the cycle is

$$e_{max} = \left[1 - \frac{5}{3}\cos^2 i\right]^{1/2}, \tag{7.53}$$

where i is here the initial relative inclination. If i is large enough we find that $e_{max} = 1$ and the planet will be driven into collision with the star! Even more surprisingly *none of these results* depends upon the masses of the stars in the binary, the semi-major axis of the binary, or its eccentricity. At least in the model problem (where there are only three bodies, all of which can be described as Newtonian point masses) an extremely distant companion, whose perturbations one might assume would be negligible, can still excite dramatic evolution of the planetary orbit. In fact the only influence that the binary properties have on the cycle is to set its period. If the orbital period of the planet is P_{in}, the characteristic time scale of Kozai–Lidov cycles is of the order of

$$\tau_{KL} \sim P_{in} \left(\frac{M_*}{M_s}\right) \left(\frac{a_s}{a_{in}}\right)^3 \left(1 - e_s^2\right)^{3/2}. \tag{7.54}$$

Binary companions that cause weaker perturbations, either on account of their lower mass or their greater distance, increase the time scale for cycles but do not otherwise alter their properties.

As a secular resonance, the Kozai effect is vulnerable to being washed out if there are additional sources of secular precession in the system, for example due to the presence of other planetary companions or due to general relativistic effects. Approximately, the condition that must be fulfilled to allow Kozai cycles is that no other sources of precession are significant on the time scale of the cycle.

Figure 7.9 shows a numerically computed example of Kozai–Lidov dynamics for a Jupiter-mass planet orbiting a star that has a low mass binary companion. This specific system is similar to the model problem, and the numerical solution is in excellent qualitative and reasonable quantitative agreement with the analytic

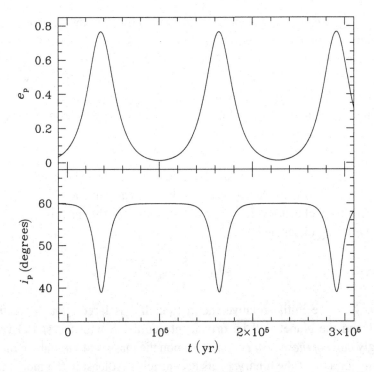

Figure 7.9 Numerical illustration of a Kozai–Lidov cycle in a marginally hierarchical system. The initial conditions assume that a Jupiter mass planet orbits a solar mass star in a circular orbit at 5 AU. The star has a low mass binary companion (0.1 $M_\odot$) in a circular but inclined orbit at 50 AU.

theory. It is possible to relax the restrictions of the original analytic theory, and consider situations where the orbiting bodies both have non-negligible mass, or where the outer perturber starts on an eccentric orbit. The dynamics in this more general case still permits large changes to the eccentricity and inclination of the inner body, but there are qualitative differences, including the possibility for the inclination of the inner body's orbit to switch from prograde to retrograde (for a review, see Naoz, 2016).

From a theoretical perspective one would have to be dismayed if such a beautiful dynamical phenomenon remained an academic curiosity devoid of widespread application in nature. Early applications of Kozai–Lidov dynamics to extrasolar planetary systems focused on systems in known binaries with properties that could plausibly drive Kozai–Lidov cycles. The planet around 16 Cyg B is an example, and a strong argument can be made for the probable effectiveness of the mechanism in such specific cases (Holman *et al.*, 1997).

Kozai–Lidov dynamics is also a channel that can potentially lead to the formation of hot Jupiters in orbits that are misaligned to the spin of their host stars, via a high-eccentricity tidal migration mechanism similar to that studied in the context of triple stellar systems (Eggleton & Kiseleva-Eggleton, 2001). Starting with a giant planet on a relatively wide orbit, Kozai–Lidov dynamics can generate large-amplitude excursions in eccentricity that lower the minimum pericenter distance $a_{in}(1 - e_{max})$ to the point where tidal interactions with the star can act to shrink the orbit (Wu & Murray, 2003; Fabrycky & Tremaine, 2007). (Note that Kozai–Lidov dynamics on its own does not alter the semi-major axis.) The basic prediction of this model – that a significant fraction of hot Jupiters ought to have misaligned orbits – is borne out by the observations of the Rossiter–McLaughlin effect discussed in Chapter 1. Population studies suggest that a large fraction of the hot Jupiters that show large projected obliquities could feasibly form as a consequence of high-eccentricity tidal migration (Naoz *et al.*, 2012; Petrovich & Tremaine, 2016).

7.2.5 Secular Dynamics

A planetary system in which no mean-motion resonances significantly affect the dynamics can be studied in a secular approximation. Averaging over the orbital motion, the gravitational potential of a planet on an eccentric orbit can be represented by a ring of matter whose mass per unit length is simply inversely proportional to the velocity of the planet at that point along its orbit. Two or more such rings, which may be eccentric and/or inclined with respect to each other, exert mutual gravitational torques, which cause the eccentricity and mutual inclination to evolve with time. The semi-major axis of each planet remains fixed under secular evolution.

The mathematical description of secular evolution, which in its simplest form is an approximation valid for small eccentricities, is known as Laplace–Lagrange theory (e.g. Murray & Dermott, 1999). Secular evolution conserves, to a good approximation, a quantity known as the *angular momentum deficit* (AMD). For a system of planets with masses M_j, semi-major axes a_j, eccentricities e_j, and inclinations with respect to the invariable plane i_j, the angular momentum deficit is defined as,

$$\text{AMD} = \sum_j \Lambda_j \left(1 - \cos i_j \sqrt{1 - e_j^2}\right), \tag{7.55}$$

where

$$\Lambda_j = \frac{M_* M_j}{M_* + M_j} \sqrt{G(M_* + M_j)a_j}. \tag{7.56}$$

The angular momentum deficit is the difference between the angular momentum of the actual system and a reference system in which the planets have the same masses and semi-major axes, but zero eccentricity and no mutual inclinations.

The conservation of AMD has two significant consequences. If the AMD is low enough, it may be possible to show that the orbits of planets in the system can never cross, *no matter how the AMD is apportioned among the planets*. This argument does not guarantee stability, because secular theory itself is only an approximation to the dynamics, but it provides a useful and easy to calculate stability metric (Laskar, 1997). Conversely, if the AMD is high enough (and the planet spacing large enough), it may be possible to place enough of the AMD in the innermost planet to drive $e \rightarrow 1$. Numerical simulations show that the secular evolution of systems of multiple planets can be chaotic, and if this is the case it is likely that the innermost planet *will* acquire most of the AMD of the whole system given only enough time. Secular chaos in the high AMD limit is another mechanism that can lead to high eccentricity tidal migration, and the formation of hot Jupiters (Wu & Lithwick, 2011).

7.3 Migration in Planetesimal Disks

A generic prediction of the core accretion model is that the time scale needed to assemble the core becomes longer in the outermost reaches of the protoplanetary disk. If giant planets form via core accretion we would therefore expect that outside a "giant planet zone" there ought to lie a region of debris that has been unable to form large bodies. This expectation is consistent with the fact that although planetesimals formed in the Solar Nebula out to a distance of ≈ 50 AU (as evidenced by the presence of the Kuiper Belt) there are no large bodies beyond Neptune in the Solar System. Our main goal in this section is to assess the possible dynamical influence that a disk of leftover debris (which we will dub a planetesimal disk, even though the bodies may have growth substantially by accretion or been destroyed by collisions) can exert on the orbits of neighboring giant planets.

Our first task is to estimate the amount of mass that might plausibly be present within a planetesimal disk in the outer region of a newly formed planetary system. This will depend upon the outer extent of the zone of giant planet formation (inside this radius the bulk of the planetesimals presumably end up within planets) and upon the assumed profile of the solid component of the protoplanetary disk. If we assume that the planetesimal disk tracks the surface density of the solid component of the minimum mass Solar Nebula[6] (Eq. 1.6), the integrated mass between radii of r_{in} and r_{out} is

$$M_{disk} = 4\pi \Sigma_0 \left(r_{out}^{1/2} - r_{in}^{1/2} \right), \tag{7.57}$$

where Σ_0 defines the normalization (the fiducial value for the outer minimum mass Solar Nebula corresponds to 30 g cm^{-2} at 1 AU). As we will note later, there is considerable evidence to suggest that the Solar System's ice giants formed considerably closer to the Sun than their current locations. We therefore assume that the zone of giant planet formation in the early Solar System extended out to 20 AU, so that the disk of leftover debris ranged between 20 AU and 50 AU. With these parameters we estimate that

$$M_{disk} \simeq 40 \, M_\oplus. \tag{7.58}$$

This estimate exceeds the *observed* mass of the present-day Kuiper Belt by more than two orders of magnitude. For it to be even remotely correct, dynamical processes must have removed almost all of the primordial material at some point during the lifetime of the Solar System.

The existence of a substantial debris disk in the outer reaches of a planetary system provides a long-lived (compared to the gas disk) reservoir of mass and angular momentum that can drive migration of any planets that are able to interact with it. Suppose, for example, that a planet of mass M_p orbiting interior to the disk scatters a mass of planetesimals δm into shorter period orbits at smaller radii. Elementary considerations suggest that the resulting change in the planetary semi-major axis ought to be of the order of

$$\frac{\delta a}{a} \sim \frac{\delta m}{M_p}. \tag{7.59}$$

Note that this cannot be the same process of small-angle scattering that we discussed in the context of gas disk migration (Section 7.1) since the sense of migration is reversed – we are imagining the planet to scatter planetesimals into lower

---

[6] Although we have previously warned of the dangers of placing too much trust in the minimum mass Solar Nebula profile, its use here is justifiable, since it is derived empirically from estimates of the mass of heavy elements within the planets. Given only weak assumptions (primarily that the planetesimal disk had a continuous surface density distribution in the outer Solar System) it is reasonable to use it to estimate the mass of planetesimals that *did not* form planets beyond the orbit of the last giant planet.

angular momentum orbits and hence to migrate outward toward the exterior disk. For significant migration to occur we therefore require, at a minimum, that

$$M_p \lesssim M_{disk}. \tag{7.60}$$

True gas giants with masses (typically) of hundreds of Earth masses are therefore relatively impervious to planetesimal-driven migration, whereas this estimate suggests that lower mass ice giants such as Uranus and Neptune could suffer substantial migration given the existence of planetesimal disks with plausible masses. There is, however, an obvious complication. Even if the total disk mass is large enough to permit migration, the mass of material that the planet is able to interact with at any one time is much smaller. A planet migrating through a planetesimal disk may stall (or at least slow down dramatically) if the disk surface density is too small, since the planet will then scatter all of the bodies it is able to perturb without moving far enough to start interacting with a fresh population. This argument – which is a close cousin of the reasoning behind the existence of the isolation mass (Section 5.2.3) – suggests that not just the integrated disk mass but also the local surface density are important parameters that determine the behavior of planetesimal-driven migration.

An approximate analytic model for migration within planetesimal disks can be developed by borrowing ideas from the more complete treatments given by Ida *et al.* (2000) and by Kirsh *et al.* (2009). We consider an asymmetric configuration in which a planet of mass M_p and orbital radius a lies just inside an exterior disk of planetesimals of surface density Σ_p. At $t = 0$ there are no planetesimals at $r < a$. We assume that the planet migrates outward into the disk as a consequence of scattering planetesimals inward on to lower angular momentum orbits, and that its migration is fast enough that any individual planetesimal is scattered inward and does not then interact further with the planet.

An analysis of this model is straightforward. We first note that the planet will be able to perturb planetesimals strongly only within a radial zone whose width Δr is of the order of the radius of the planet's Hill sphere (cf. Section 5.1.2):

$$\Delta r \approx \left(\frac{M_p}{3M_*} \right)^{1/3} a. \tag{7.61}$$

The mass of planetesimals within this zone is

$$\Delta m = 2\pi a \Sigma_p \Delta r. \tag{7.62}$$

We now need to estimate the average change in the specific angular momentum of a planetesimal that results from scattering. This can be calculated accurately, but for now we simply assume that it is of the order of the difference in specific angular momentum of circular orbits across the scattering zone. If this is the case, then once all of the planetesimals within the scattering zone have encountered the planet they will have collectively lost an amount of angular momentum

$$\Delta J \approx \Delta m \left.\frac{dl}{dr}\right|_a \Delta r, \tag{7.63}$$

where $l = \sqrt{GM_* r}$ is the specific angular momentum for a circular orbit at distance r. The angular momentum lost by the planetesimals is gained by the planet, which (if it stays on a circular orbit) migrates outward a distance

$$\Delta a \approx \frac{2\pi a \Sigma_p \Delta r^2}{M_p}. \tag{7.64}$$

For rapid migration to occur, Δa must be large enough to move the planet into a region of the disk stocked with as yet unperturbed planetesimals. Requiring that $\Delta a \gtrsim \Delta r$ we obtain a condition on the planet mass

$$M_p \lesssim 2\pi a \Sigma_p \Delta r. \tag{7.65}$$

Fast planetesimal-driven migration requires that the planet mass be *smaller* than the mass of planetesimals within a few Hill radii of the planet. One may observe again that this behavior is the opposite of that predicted for Type 1 migration in a gas disk, which becomes more rather than less efficient as the planet mass increases.

The rate of migration can be determined by estimating how long it takes for all of the planetesimals within the scattering zone to encounter the planet and be scattered inward. The relevant time scale is that set by the shear across the scattering zone (Eq. 7.6):

$$\Delta t \sim \frac{2}{3} \frac{a}{\Delta r} P, \tag{7.66}$$

where P is the planetary orbital period. Combining this with the expression for the distance that the planet moves (Eq. 7.64), we conclude that the migration rate in the fast regime will be

$$\frac{da}{dt} \sim \frac{a}{P} \frac{\pi a^2 \Sigma_p}{M_*}. \tag{7.67}$$

Up to numerical factors of the order of unity this expression matches that derived by Ida *et al.* (2000), and is in good agreement with numerical results. An interesting feature of the result is that the migration of a low mass planet through a massive planetesimal disk occurs at a rate that is independent of the planet mass. Slower but still substantial migration can occur for planets that are up to an order of magnitude more massive than the rough limit defined by Eq. (7.65). Kirsh *et al.* (2009) provide an empirical fit to numerical calculations of migration in the slow high-mass planet regime.

7.3.1 Application to Extrasolar Planetary Systems

Multiple giant planets and massive planetesimal disks are likely ubiquitous features of a fraction of young planetary systems, and hence it is reasonable to expect that dynamical interactions between them affect the final architecture of the system. Numerical experiments suggest that the range of different outcomes possible from the interaction of several planets with a disk of debris is substantial, and will depend at a minimum on the masses of the planets involved, the mass and extent of the disk, and the architecture of the system (for example, whether the most massive planet is generically closest to the star, as in the Solar System).

As an illustration of just one possibility, Fig. 7.10 shows the predicted evolutionary track of a system of three giant (but sub-Jupiter mass) planets that form in a compact configuration within an exterior planetesimal disk. The system expands

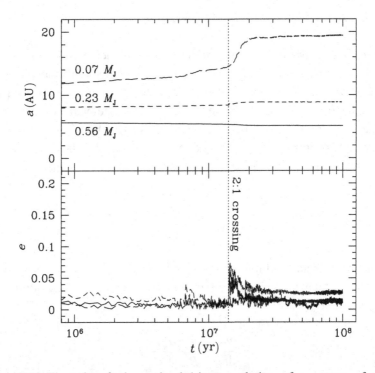

Figure 7.10 Example of planetesimal-driven evolution of a system of three relatively low mass giant planets, assumed here to form in close proximity to each other just interior to 10 AU. Beyond the planets lies a planetesimal disk containing 50 $M_\oplus$ of material distributed according to a $\Sigma_p \propto r^{-1}$ surface density profile between 10 AU and 20 AU. In this realization planetesimal scattering results in a slow expansion of the orbits of the outer two planets and an even more gradual contraction of the orbit of the innermost planet. Divergent crossing of the 2:1 resonance between the inner pair of planets briefly excites the eccentricities of the orbits, leading to enhanced scattering and more rapid outward migration of the outermost planet. The final values of the eccentricity are all small. Based on simulations by Sean Raymond.

slowly as the planetesimal disk is depleted via scattering before experiencing an accelerated phase of evolution that coincides with the eccentricity excitation accompanying a 2:1 resonance crossing. Once the resonance has been crossed, further interaction between the planets and the remaining disk material damps planetary eccentricities to small values.

7.4 Planetary System Stability

In the guise of the stability of the Solar System the problem of planetary system dynamics has a history dating back to Isaac Newton, who understood that the near-circular orbits of the planets were a special case whose survival was endangered by the planet–planet forces required by his law of gravitation. Those forces are small but they work over a long time – billions of years in a modern understanding – and it is not obvious whether their effects ultimately average to zero or rather build up until they destabilize the system. Newton himself feared the worst. In his *Opticks* he wrote that "blind fate could never make all the planets move one and the same way in concentric orbits, some inconsiderable irregularities excepted which may have arisen from the mutual actions of comets and planets upon one another, and which will be apt to increase, till this system wants a Reformation." This was guesswork and it has taken three centuries of mathematical development to show that Newton was largely, though not entirely, wrong.

In a colloquial sense the fact that the Earth has sustained life for a substantial fraction of the age of the Solar System provides compelling evidence for the basic stability of at least our planetary system. Similarly, it would be shocking to discover that any known extrasolar planetary system (most of which are also billions of years old) was about to suffer some catastrophic event such as a planetary collision or ejection. The interest in planetary system stability concerns not such gross questions, but rather two substantially subtler aspects. One is the mathematical question of whether (and if so when) it is possible to *prove* that an N-body system is stable for all time, especially in the practically important case that the motions of the bodies are chaotic and thus intrinsically unpredictable over sufficiently long time scales. The second arises from the fact that nothing we have discussed about the *formation* of planetary systems requires that the outcome will be a stable system. On the contrary, since planet formation plays out in a highly dissipative environment – where either gas or planetesimal disks can frustrate the development of slow-growing gravitational instabilities – it is plausible to imagine that the typical planetary system forms in what turns out to be a long-term unstable state. If this is correct, features such as the chaotic orbits of Solar System planets and the eccentric orbits of some extrasolar planets may be endpoints of the evolution of initially unstable planetary systems, and it is important to understand how unstable systems evolve.

7.4.1 Hill Stability

The trajectories of two point masses moving under Newtonian gravitational forces are stable and well known: elliptical orbits if the bodies are gravitationally bound, and parabolae or hyperbolae otherwise. For systems of three or more bodies, on the other hand, no general closed form analytic solutions to the motion of the bodies exist.[7] Nonetheless, it is sometimes possible to demonstrate the stability of a system of three bodies *despite our ignorance* of the actual trajectories, and although the stability criteria derived in this way are formally irrelevant for studies of more complex planetary systems they provide at least a framework for interpreting numerical results.

The only example of three-body motion for which it is easy to derive an analytic stability bound is the circular restricted three-body problem. We consider the motion of a test particle of negligible mass in the gravitational field of a circular star–planet system, and work in a frame that co-rotates with the orbital motion of the planet at angular velocity Ω. We have already derived the equations describing the motion of a test particle in this situation (the geometry is shown in Fig. 5.2 and discussed in Section 5.1.2). If the star and planet orbit in the (x, y) plane, then in a Cartesian coordinate system in which the star and planet lie along the $y = 0$ line at $x = -x_*$ and $x = x_p$ we have

$$\ddot{x} - 2\Omega\dot{y} - \Omega^2 x = -G\left[\frac{M_*(x + x_*)}{r_*^3} + \frac{M_p(x - x_p)}{r_p^3}\right], \qquad (7.68)$$

$$\ddot{y} + 2\Omega\dot{x} - \Omega^2 y = -G\left[\frac{M_*}{r_*^3} + \frac{M_p}{r_p^3}\right]y, \qquad (7.69)$$

$$\ddot{z} = -G\left[\frac{M_*}{r_*^3} + \frac{M_p}{r_p^3}\right]z. \qquad (7.70)$$

Here r_* and r_p are the instantaneous distances between the test particle and the star and planet. No assumptions have been made as to the masses of the star M_* and planet M_p.

The acceleration due to the centrifugal force can be subsumed into a pseudo-potential. Defining

$$U \equiv \frac{\Omega^2}{2}\left(x^2 + y^2\right) + \frac{GM_*}{r_*} + \frac{GM_p}{r_p}, \qquad (7.71)$$

we obtain

$$\ddot{x} - 2\Omega\dot{y} = \frac{\partial U}{\partial x}, \qquad (7.72)$$

---

[7] The exceptions to this statement include fascinating mathematical curiosities, such as the "figure of eight" orbit for three equal mass bodies that was discovered by Chenciner and Montgomery (2000).

$$\ddot{y} + 2\Omega\dot{x} = \frac{\partial U}{\partial y}, \tag{7.73}$$

$$\ddot{z} = \frac{\partial U}{\partial z}. \tag{7.74}$$

The Coriolis terms on the left-hand-side of the first two equations can be eliminated by multiplying through by $\dot{x}, \dot{y}$, and $\dot{z}$ and adding. We then obtain

$$\dot{x}\ddot{x} + \dot{y}\ddot{y} + \dot{z}\ddot{z} = \dot{x}\frac{\partial U}{\partial x} + \dot{y}\frac{\partial U}{\partial y} + \dot{z}\frac{\partial U}{\partial z}, \tag{7.75}$$

$$\frac{d}{dt}\left(\frac{1}{2}\dot{x}^2 + \frac{1}{2}\dot{y}^2 + \frac{1}{2}\dot{z}^2\right) = \frac{dU}{dt}. \tag{7.76}$$

This equation integrates immediately to give

$$\dot{x}^2 + \dot{y}^2 + \dot{z}^2 = 2U - C_J, \tag{7.77}$$

$$C_J = 2U - v^2, \tag{7.78}$$

where v is the velocity and C_J, called the *Jacobi constant*, is the arbitrary constant of integration. Note that C_J is an energy-like quantity that is a conserved quantity in the circular restricted three-body problem.

The existence of this integral of motion is important because it places limits on the range of motion possible for the test particle. For a particle with a given initial position and velocity we can use Eq. (7.78) to compute C_J, and hence to specify *zero-velocity surfaces*, defined via

$$2U = C_J, \tag{7.79}$$

which the particle can never cross. If the volume enclosed by one of the zero-velocity surfaces is finite, then a particle initially within that region is guaranteed to remain there for all time. By this argument one can prove stability of the system without needing to derive an explicit form for the motion of the test particle.

The topology of the zero-velocity surfaces in the restricted three-body problem varies according to the value of C_J. An example is shown in Fig. 7.11. In this instance the zero-velocity surfaces define three disjoint regions in the (x, y) plane, one corresponding to orbits around the star, one corresponding to orbits around the planet, and one corresponding to orbits around the star–planet binary. A particle in any one of these states is stuck there – it cannot cross the forbidden zone between the different regions to move into a different state.

Given these results a test particle's orbit is guaranteed to be stable – in the sense that it can never have a close encounter with the planet – if its Jacobi constant is such that a zero-velocity surface lies between it and the planet. This condition can be written in terms of a minimum orbital separation needed for stability. As an example, consider a test particle in a circular orbit of radius a_{out} in a system where

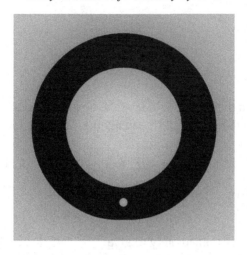

Figure 7.11 Forbidden zones (dark regions) in the (x, y) plane in an example of the restricted three-body problem. For this particular choice of the Jacobi constant C_{J}, particles can orbit the star at small radii, the planet in a tight orbit, or the star–planet binary as a whole. The existence of zero-velocity surfaces, however, means that particles cannot be exchanged between these regions.

a planet has a circular orbit of radius a_{in}. We define a dimensionless measure of the orbital separation Δ via

$$a_{\mathrm{out}} = a_{\mathrm{in}}(1 + \Delta), \tag{7.80}$$

and write the mass ratio between the planet and the star as

$$q_{\mathrm{in}} = \frac{M_{\mathrm{p}}}{M_*}. \tag{7.81}$$

Stability is then assured (irrespective of the initial difference in the longitude between test particle and planet) provided that

$$\Delta > 2.4 q_{\mathrm{in}}^{1/3}. \tag{7.82}$$

Since Δ is a measure of the orbital separation in units of (up to a small numerical factor) Hill radii, this is a version of a result used in Section 5.2.2, where we argued that a planet will rapidly perturb small bodies on to crossing trajectories if the separation is less than a few Hill radii. Here, we have demonstrated a more powerful inverse result: if the separation is more than a few Hill radii, close encounters are absolutely forbidden for all time.

Although Eq. (7.82) hints that the stability of a multiple planet system depends upon the separation of the planets measured in units of Hill radii, nothing about the above derivation gives us any right to expect that a *formal proof* of stability can be derived for a general two-planet system. Surprisingly, however, for

q_{in}, $q_{out} \ll 1$ it is possible to derive a sufficient condition for Hill stability[8] that can be applied to planets whose initial orbits have arbitrary eccentricity and inclination (Gladman, 1993; whose analysis draws on work by Marchal & Bozis, 1982). For two planets on initially circular coplanar orbits the system is Hill stable for separations

$$\Delta > 2 \cdot 3^{1/6} \, (q_{in} + q_{out})^{1/3}$$
$$+ \left[2 \cdot 3^{1/3} \, (q_{in} + q_{out})^{2/3} - \frac{11 q_{in} + 7 q_{out}}{3^{1/6} (q_{in} + q_{out})^{1/3}} \right] + \cdots , \tag{7.83}$$

which simplifies to lowest order in the mass ratios to

$$\Delta > 2.4 \, (q_{in} + q_{out})^{1/3} . \tag{7.84}$$

This criterion reduces to that deduced earlier for the equivalent restricted three-body problem (Eq. 7.82). Among the more general results derived and quoted by Gladman (1993) we note only the extension to the case where the planets have equal masses ($q_{in} = q_{out} = q$) and initially small eccentricities,

$$\Delta > \sqrt{\frac{8}{3} \left(e_{in}^2 + e_{out}^2 \right) + 9 q^{2/3}}, \tag{7.85}$$

which, as with Eq. (7.84), is also valid only to lowest order in the masses.

These stability criteria are formally only sufficient conditions for stability – they guarantee that a planetary system with larger Δ is Hill stable but say nothing about whether a more tightly packed system is actually unstable. For initially circular orbits (and only in that case), however, it is found empirically that the Hill limit is also a good general predictor of the onset of instability (with some exceptions such as systems with small Δ that are stable in resonant configurations). Another interesting empirical finding is that two-planet systems that are Hill stable can nevertheless exhibit chaos, which is attributed to the overlap of multiple resonances close to (but beyond) the minimum separation needed for stability.

For systems of three or more planets there are no analytic guarantees of stability. The initial separation between planets in units of their mutual Hill radii,

$$r_{Hill,m} = \left(\frac{M_{p,i} + M_{p,i+1}}{3 M_*} \right)^{1/3} \frac{(a_i + a_{i+1})}{2}, \tag{7.86}$$

is still useful, however, as a guide to the time scale on which instability will result in crossing orbits (Chambers *et al.*, 1996). If we write the separation of the planets in the form

---

[8] As in the circular restricted three-body problem, a system that is "stable" by this criterion is guaranteed to be free of close encounters between the planets for all time. This is not quite the common-sense definition of stability, since it leaves open the loophole that one planet might become unbound from the system via the cumulative action of many distant, weak encounters. A stronger definition of stability, known as *Lagrange stability*, requires that both planets remain bound. Lagrange stability cannot be proven in the same way as Hill stability, though for practical purposes the distinction is immaterial.

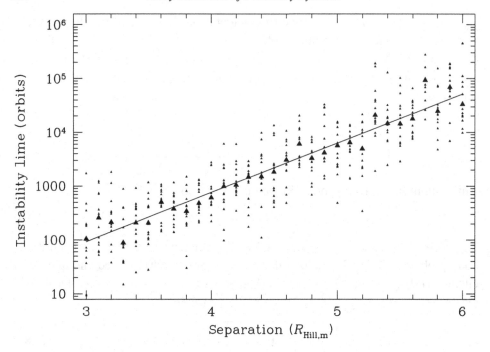

Figure 7.12 The median instability time scale (triangles) for initially circular, coplanar three planet systems, with mass ratio $M_p/M_* = 10^{-6}$, as a function of the separation in units of mutual Hill radii. The dots show the instability time scale obtained from individual numerical realizations. The instability time scale is defined as the time (in units of the initial orbital period of the inner planet) until the first pair of planets approach within one Hill radius.

$$a_{i+1} = a_i + K r_{\text{Hill, m}}, \tag{7.87}$$

then it is possible to derive empirical expressions for the median time scale on which instability will develop. For relatively low mass planetary systems, in particular, the instability time scale is a smooth and essentially monotonic function of separation (Fig. 7.12), which can be represented by an expression of the form,

$$\log_{10}(t_{\text{instability}}/t_0) = a + bK, \tag{7.88}$$

where t_0 is the initial orbital period of the innermost planet and a and b are constants. Numerical values and discussion can be found in Obertas *et al.* (2017). The instability time scale implied by this formula increases rapidly as the separation of the planets increases. As with two-planet systems, this general rule does not apply in the vicinity of strong resonances, but provided that this caveat is borne in mind it provides a good rule of thumb for assessing the stability of a complicated planetary system.

7.4.2 Planet–Planet Scattering

Close encounters between giant planets may have occurred in the early history of the Solar System, but the primary motivation for studying the evolution of unstable planetary systems comes from extrasolar planetary systems whose typically eccentric orbits immediately suggest that the observed planets may be survivors of violent planet–planet scattering events that occurred early on (Rasio & Ford, 1996; Weidenschilling & Marzari, 1996; Lin & Ida, 1997). In general, an initially unstable planetary system can evolve ("relax") via four distinct channels:

- One or more planets are ejected, either as a result of a close encounter between planets or via numerous weaker perturbations. An example of a system that displays this evolution is shown in Fig. 7.13.
- One or more planets have their semi-major axis and eccentricity changed in such a way that the system becomes stable.
- Two planets physically collide and merge.
- One or more planets impact the star.

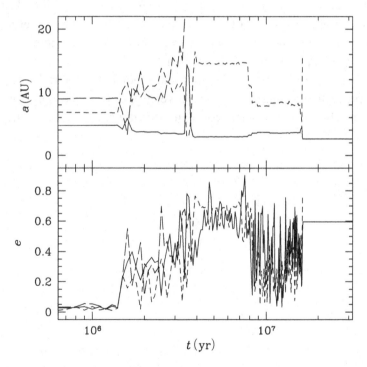

Figure 7.13 Gravitational scattering in an initially marginally unstable system of three massive planets on circular coplanar orbits. The system relaxes via a complex set of interactions whose end result (in this particular realization) is the ejection of two of the planets. The lone survivor is left in a shorter-period orbit with an eccentricity $e \simeq 0.6$. Based on simulations by sean Raymond.

The relative importance of the different channels is a function of the orbital radius at which scattering occurs (physical collisions obviously become more likely at smaller orbital radii) and of the mass distribution of the planets participating in the scattering.

Ideally perhaps, the distribution of planetary masses and orbital radii at some early epoch (when the gas disk has just been dissipated) would be a specified outcome of the theory of giant planet formation. The subsequent N-body evolution of an ensemble of planetary systems would then yield a single prediction for the distribution of final states of planetary systems after scattering. Although such calculations are possible, they require a possibly unwarranted degree of faith in the fidelity of the giant planet formation model, which as we have noted before is quite uncertain. Most authors have therefore followed a less ambitious path and studied the N-body evolution of unstable multiple planet systems starting from well-defined but essentially arbitrary initial conditions (Ford *et al.*, 2001; Chatterjee *et al.*, 2008; Jurić & Tremaine, 2008; Raymond *et al.*, 2008). The most important result from these studies is shown in Fig. 7.14, which shows the comparison between the cumulative eccentricity distribution of known extrasolar planets and that obtained theoretically as the endpoint of relaxation in an ensemble of unstable planetary systems. Good agreement with the observations can be obtained starting from a variety of plausible initial conditions that vary in the assumed number and initial separation of giant planets within the system.

The agreement between the measured and predicted eccentricity distributions does not prove that strong scattering between planets is the only (or even the dominant) mechanism responsible for their typically significant eccentricities, but

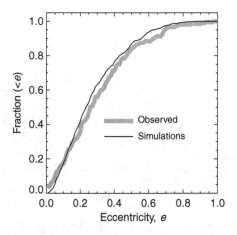

Figure 7.14 The cumulative eccentricity distribution of known extrasolar planets is compared to the predicted distribution that results from scattering in three-planet systems. The simulation results were derived by numerically evolving an ensemble of unstable planetary systems made up of three planets whose masses were drawn from the observed mass function for extrasolar planets in the range $M_{Sat} < M_p < 3\,M_J$. Based on simulations by sean Raymond.

it is consistent with the basic hypothesis that the typical outcome of giant planet formation is an unstable multiple planet system. One should bear in mind, however, that the observational sample is still biased toward massive planets at small orbital radii. At larger radii planetesimal disk scattering is likely to become competitive with strong planet–planet scattering as a mechanism for dynamical evolution, especially for relatively low mass planets more akin to Uranus and Neptune than to Jupiter. It is thus probable that the eccentricity distribution of lower mass extrasolar planets at larger semi-major axis will differ from that currently measured, and it remains possible that the architecture and relatively low eccentricities of the Solar System's giant planets are typical, given the masses and orbital radii of the planets involved.

7.4.3 The Titius–Bode Law

It is not possible to give many talks on planet formation without having to field a question on "Bode's law," an apparent regularity in the orbital spacing of the Solar System's planets. The law is a fitting formula for the semi-major axes of the planets,

$$a_i = 0.4 + 0.15 \times 2^n \text{ AU}, \tag{7.89}$$

where values of $n = [-\infty, 1, \ldots, 7]$ correspond to the planets between Mercury and Uranus, except for $n = 4$ which is identified with the asteroid belt. The law was formulated prior to the discovery of Ceres, the largest asteroid, and Uranus, and correctly predicted their locations (though not, in the case of Ceres, its low mass). Extrapolating to $n = 8$ the formula gives $a = 39$ AU, which is not close to the correct value for Neptune.

There is no plausible physical basis for the detailed form of Bode's law, which is essentially a moderately successful fitting formula. (Attempts to assess its statistical significance more rigorously are inevitably inconclusive, because it is impossible to assess how many similarly simple-looking formulae were, or could have been, considered.) That is not to say that there is *no* physical content. The asymptotic form, in which successive planets are separated by a fixed ratio α in semi-major axis, matches what one expects for a marginally stable system of multiple planets, but one could certainly construct stable systems with $\alpha \neq 2$ or where there were large enough gaps to invalidate laws of Bode's type. In sum, it appears that something like Bode's law is a likely but not guaranteed outcome of planet formation if planets form over a broad range of orbital radii, and have orbits that are constrained by the requirement that they be stable over long periods.

7.5 Solar System Migration Models

One of the key theoretical insights prompted by observations of extrasolar planetary systems is that planetary systems can often evolve dramatically over time. Models for early Solar System evolution predate the discovery of extrasolar planetary systems, but have received renewed attention now that the broader context is understood. Compared to extrasolar planetary systems, the Solar System has well-observed small-body populations and access to in situ measurements that greatly constrain the early migration history. Nonetheless, the extent to which the early dynamical history of the Solar System can be pinned down from current data remains an open question.

7.5.1 Early Theoretical Developments

The idea that scattering of planetesimals could have altered the orbits of planets in the outer Solar System is not new. As with so many other fundamental results in the theory of planet formation, a brief discussion can be found in Safronov (1969). Safronov, however, along with other early investigators, was primarily interested in the ability of planets to *eject* bodies from the Solar System and thereby populate the Oort cloud (Oort, 1950). Ejection requires that the planet transfer energy to the planetesimal and hence results in inward planetary migration. The process is only efficient for Jupiter, whose large mass allows it to eject a substantial mass of material from the Solar System without moving very far inward.

Larger scale orbital migration is possible from scattering (without ejection) initiated by the ice giants. The first important development was the realization by Fernandez and Ip (1984) that the architecture of the outer Solar System lends itself to precisely the type of one-way scattering developed in our model in Section 7.3. If the giant planets formed within an exterior planetesimal disk, Neptune is massive enough to perturb bodies within the disk into orbits that lie between Uranus and Neptune, but is not massive enough to eject them from the Solar System. The same is true of Uranus and Saturn, and the overall result is that objects scattered inward from the disk by Neptune can be moved inward by successive encounters with Uranus and Saturn until eventually they reach Jupiter and are flung out of the Solar System. To conserve energy and angular momentum, the process results in the expansion of the orbits of Neptune, Uranus, and (to a lesser extent) Saturn, while the orbit of Jupiter shrinks slightly. An attractive feature of such a model is that it allows the giant planets to form in a more compact configuration closer to the Sun, ameliorating substantially the difficulty of trying to form Neptune via core accretion at its current location.

These theoretical ideas receive strong support from observations of the distribution of minor bodies in the outer Solar System (Fig. 1.3), which show that Pluto and a host of smaller Kuiper Belt Objects orbit in stable 3:2 mean-motion resonance with Neptune. Malhotra (1993, 1995) showed that the properties of Pluto's orbit are consistent with those expected if it was resonantly captured by Neptune during the latter's slow outward migration, and was able to predict the distribution of the now well-observed population of resonant KBOs.

7.5.2 The Nice Model

The current paradigm for outer Solar System evolution is the Nice model (named after the French city, where some of the original work was done). The first version of the Nice model was introduced by Tsiganis *et al.* (2005), drawing on earlier ideas by Thommes *et al.* (1999) and others. Contemporary versions of the model, although different in significant respects, share common elements. In the Nice model the giant planets of the Solar System, together (possibly) with one or more extra planets that were subsequently lost, formed in a compact configuration that was surrounded by a massive planetesimal belt. A combination of planet–planet scattering and planetesimal-driven migration, schematically shown in Fig. 7.15 led

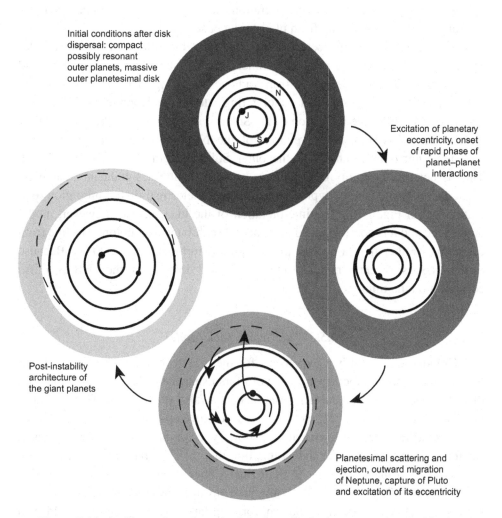

Figure 7.15 An illustration of some of the key physical processes invoked in the Nice Model for the early evolution of the outer Solar System. The timing of the different phases is uncertain, and significantly different models (for example those with an additional giant planet) are possible.

to the evolution of the outer Solar System toward its current configuration, and the dynamical clearing of most of the belt.

Nesvorný (2018) reviews the current status of the Nice model and discusses the main constraints on the model. Many observable properties of the outer Solar System, including the population of Jupiter's Trojan asteroids (Morbidelli *et al.* 2005), the obliquities of the giant planets, and the structure of the Kuiper Belt, can be modeled within the framework of the Nice model. Some of the strongest constraints come from the terrestrial planets, which are coupled dynamically to what happens in the outer Solar System via secular resonances. Requiring that this coupling does not excessively excite the orbits of the terrestrial planets (or their progenitors) favors variants of the Nice model in which the evolution is not solely "slow" planetesimal-driven migration, but rather includes "rapid" changes in orbital properties resulting from planet–planet scattering.

The relationship (if any) between the Nice model evolution of the outer Solar System and the cratering history of inner Solar System bodies is unclear and controversial. Based on evidence from radioactive dating of lunar rock samples, Tera *et al.* (1974) suggested that the cratering rate on the Moon exhibited a spike about 700 Myr after the formation of the Solar System, known as the "Late Heavy Bombardment" (LHB). Early versions of the Nice model associated the LHB with the timing of instability in the outer Solar System (Gomes *et al.*, 2005). This association is highly constraining, because having a significant dynamical event occur so late is not a typical outcome of planetesimal-driven migration (e.g. that shown in Fig. 7.10). The interpretation of the lunar data is, however, model dependent, and it is possible that the impact rate exhibited a monotonic decline from early times rather than a decline plus a spike (Morbidelli *et al.*, 2018). Because of these issues, recent versions of the Nice model do not treat the LHB as a firm constraint on the timing of the instability.

7.5.3 The Grand Tack Model

The Grand Tack model is a hypothesis for the early evolution of the inner Solar System, in which a temporary intrusion of Jupiter shaped the distribution of material that subsequently formed the terrestrial planets (Walsh *et al.* 2011). (This one has a nautical flavor, a sailing boat "tacks" in order to move upwind.[9]) The physical basis of the Grand Tack derives from a quirk in how Type 2 migration would have worked for Jupiter and Saturn, under the assumption that their migration took place when the planets had masses similar to their final values. If Jupiter, the more massive planet, formed first, it could have migrated inward toward the terrestrial planet region at a relatively slow rate. Saturn, forming later but being lighter, could then migrate inward at a faster rate and become trapped into a mean-motion resonance

---

[9] A name needing parenthetical explanation appears to be a prerequisite in this field.

with Jupiter. Once locked in resonance, simulations show that the *direction* of migration can be reversed (Masset & Snellgrove, 2001). The planets, still locked in resonance, then "tack" back out against the inward flow of gas through the disk.

The dynamical effect of having a fully formed Jupiter at 1.5 AU in the Solar System, even briefly, is predictably dramatic. Primordial material in the vicinity of the modern asteroid belt is severely depleted, and planetesimals and planetary embryos in the terrestrial planet-forming region are confined to an annulus whose outer edge is close to 1 AU. These initial conditions are known to give rise to terrestrial planet systems that match our own reasonably well (Hansen, 2009). In particular, they lead to a low mass planet in the location of Mars, in accord with observations, which is hard to reproduce without a depletion of the solid surface density in that region.

7.6 Debris Disks

Infrared emission in excess of that expected from the stellar photosphere is detected from a fraction of main-sequence stars. In cases where the structure producing the emission can be resolved, either in thermal radiation or in scattered light, the morphology is disk-like. These debris disks differ from protoplanetary disks not only in the age of their host stars, but also in having little or no detectable gas. The emission is consistent with a disk of optically thin dust that scatters and absorbs / re-radiates stellar radiation. Stars where this phenomenon was recognized early on include Vega and β Pictoris.

Dust cannot persist for very long in orbit around a star. Small grains will be blown out by radiation pressure, while somewhat larger grains lose angular momentum and spiral inward under the action of Poynting-Robertson drag. The observed dust in debris disks must be continuously replenished from an unseen reservoir of larger bodies. The physical picture is that debris disks contain a population of relatively large bodies, which collide with typical velocities that exceed the threshold for catastrophic disruption. The fragments suffer a succession of further collisions, eventually reducing the debris to dust that is then lost from the system. In some cases a larger planet may be needed to excite random velocities into the catastrophic disruption regime, or to explain the detailed morphology of resolved debris disks.

Wyatt (2008) reviews the physics and observational characteristics of debris disk systems. The key theoretical concept is that of a *collisional cascade*, which given appropriate simplifications leads to a calculable size distribution of fragments. We derive this size distribution below. We then introduce a simple model for the time evolution of debris disks, and discuss the special case of debris disks around white dwarfs.

7.6.1 Collisional Cascades

Gravitational interactions among a population of similar-sized bodies excite their relative velocities up to values comparable to the escape velocity from their surfaces. These values are modest enough that collisions predominantly lead to accretion rather than disruption, allowing larger bodies to form.

The bias toward accretion that results from the self-limiting tendency of gravitational stirring can be reversed at later stages of the planet formation process. An obvious example occurs after planets have formed and the gas disk has been dispersed. The planets can then excite the velocity dispersion of surrounding belts of planetesimals to large values, unimpeded by the damping influence of gas drag. Collisions among the planetesimals will then be destructive, creating smaller fragments with similarly large velocity dispersions that will themselves collide destructively. This situation is described as a *collisional cascade*, and usually results in a size distribution where most of the mass resides in the largest bodies while most of the surface area is in the smallest. Collisional cascades provide an approximation to the dynamics of the Solar System's asteroid belt, where the size distribution is measured directly, and to debris disks, where the largest bodies are not seen and the observable is typically infrared emission from the smallest particles.

The time-dependent evolution of a size distribution under destructive collisions is just the inverse of the coagulation problem discussed in Section 5.5, and can be treated numerically using the same methods. If the cascade has enough time to attain a steady state it is possible to derive (under not too restrictive conditions) an analytic solution for the size distribution (Dohnanyi, 1969). In the case where the strength of the colliding bodies is independent of their size, the size distribution is a power-law with $n(s) \propto s^{-7/2}$. Here, following Wyatt *et al.* (2011), we give a derivation of this classic result, including its generalization to the more realistic case where particle strength is a power-law function of size.

The framework for the calculation is illustrated in Fig. 7.16. We consider an annulus of the disk in which the relative velocities are high enough that essentially

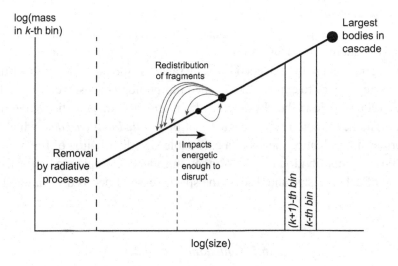

Figure 7.16 Illustration of a collisional cascade.

all collisions between bodies are destructive. A reservoir of large bodies supplies mass to the top of the cascade. For any body in the cascade the first impact with an object that exceeds a critical size leads to catastrophic disruption. (We ignore impacts with smaller objects.) A disrupted body creates a shower of smaller fragments that populate the cascade further down. Mathematically this process could continue indefinitely, though in environments such as debris disks radiation pressure or Poynting–Robertson drag will act to promptly remove particles below some minimum size.

The calculation has two steps. The first step is to show that a steady-state collisional cascade is possible if there is a constant rate of mass loss per logarithmic size bin. We adopt a discrete representation in which the mass in the cascade is divided into bins that are uniformly spaced in $\log(s)$, where s is the particle size. The largest bodies with characteristic size s_1 occupy the $k = 1$ bin, with successively smaller particles in bins that are defined via

$$\frac{s_{k+1}}{s_k} = 1 - \epsilon, \tag{7.90}$$

with $\epsilon \ll 1$. The total mass in bin k is M_k. Mass conservation implies that

$$\dot{M}_k = \dot{M}_k^{\text{gain}} - \dot{M}_k^{\text{loss}}, \tag{7.91}$$

where the dots denote time derivatives, $\dot{M}_k^{\text{gain}}$ is the rate of mass increase in the k-th bin due to destructive collisions among larger bodies, and $\dot{M}_k^{\text{loss}}$ is the rate of mass loss from the bin due to collisions.

When a destructive collision occurs in a bin some fraction of the mass will be redistributed into bins with larger indices that represent particles with smaller sizes. We define the *redistribution function* $f(i,k)$ to be the fraction of mass removed from the i-th bin after a collision that ends up in the k-th bin. The redistribution function must satisfy some general constraints:

$$f(i, k \leq i) = 0 \text{ (no mass flows to larger bins)}, \tag{7.92}$$

$$\sum_{k=i+1}^{\infty} f(i,k) = 1 \text{ (all mass goes somewhere)}. \tag{7.93}$$

These constraints are weak and the resulting freedom in f allows for quite complex collisional physics. For example, if destructive collisions among gravity dominated bodies yielded a different fragment distribution than those among strength dominated bodies, that physics could be captured with a redistribution function that depended separately on i and $(k - i)$. Here though we ignore such possibilities and specialize to redistribution functions that are *scale-free*, which we express mathematically via

$$f(i + n, k + n) = f(i,k), \tag{7.94}$$

for all integers n. In the scale-free limit we can define a redistribution function f' that depends on only one parameter:

$$f'(k - i) = f(i, k). \tag{7.95}$$

With this definition the rate of mass being gained by the k-th bin is equal to the sum over losses in all larger bins:

$$\dot{M}_k = \sum_{i=1}^{k-1} f'(k - i) \dot{M}_i^{\text{loss}} - \dot{M}_k^{\text{loss}}. \tag{7.96}$$

Now assume a steady state, so that $\dot{M}_k = 0$ for all k. We have

$$\dot{M}_k^{\text{loss}} = \sum_{i=1}^{k-1} f'(k - i) \dot{M}_i^{\text{loss}}, \tag{7.97}$$

which we can rewrite equivalently as

$$\dot{M}_k^{\text{loss}} = \sum_{l=1}^{\infty} f'(l) \dot{M}_i^{\text{loss}} - \sum_{l=k}^{\infty} f'(l) \dot{M}_i^{\text{loss}}. \tag{7.98}$$

If the cascade extends over a very broad range of sizes (compared to the range of fragment sizes produced by a collision) we can assume that we are far from the boundaries, take $k \gg 1$, and neglect the second term. Then, noting that

$$\sum_{l=1}^{\infty} f'(l) = 1, \tag{7.99}$$

a solution is always possible with

$$\dot{M}_k^{\text{loss}} = \dot{M}_i^{\text{loss}} = C, \tag{7.100}$$

with C a constant. In words, a solution for a steady-state scale-free collisional cascade of infinite extent can be obtained in which the rate of mass loss from logarithmic bins of particle size is a constant.

Surprisingly, in some circumstances we can use Eq. (7.100) to work out the steady-state size distribution in the cascade even if we don't know the detailed function form of $f'(l)$. To do so we assume that the rate of mass loss from the k-th bin is dominated by catastrophic collisions, with rate $\mathcal{R}_k$,

$$\dot{M}_k^{\text{loss}} = \mathcal{R}_k M_k. \tag{7.101}$$

Using the definition given in Section 5.1.3, a catastrophic collision between bodies of mass M and m, impacting at relative velocity v, occurs when the specific energy of the impact $Q \equiv mv^2/(2M)$ exceeds a threshold value Q_*^D. Given a constant material density, such that $m \propto s^3$, this means that the minimum size of an impactor that can trigger the catastrophic disruption of a body in the bin of size s_k is

$$s_{min} = \left(\frac{2Q_*^D}{v^2}\right)^{1/3} s_k. \tag{7.102}$$

Adopting a power-law form for the catastrophic disruption threshold, $Q_*^D \propto s^b$, the minimum size scales as

$$s_{min} \propto s_k^{1+b/3}. \tag{7.103}$$

Physically $b = 0$ corresponds to the idealized situation where the strength of colliding bodies is independent of their size, while $b > 0$ is appropriate for gravity dominated bodies which become more resistant to disruption as they get bigger.

We can now use statistical arguments to calculate $\mathcal{R}_k$, and then employ Eqs. (7.101) and (7.100) to find the size distribution. For simplicity we consider the case where the relative velocity v is independent of size (this assumption can be easily relaxed if need be). The catastrophic collision rate is then

$$\mathcal{R}_k \propto \int_{s_{min}}^{s_1} n(s)(s_k + s)^2 ds, \tag{7.104}$$

where $n(s)$ is the continuous function describing the number density of bodies of size s. In terms of $n(s)$ the mass in the k-th bin is obtained by integrating the mass-weighted number density across the width of the bin (i.e. between sizes $s_{k+1/2}$ and $s_{k-1/2}$),

$$M_k \propto \int_{s_{k+1/2}}^{s_{k-1/2}} n(s)s^3 ds \propto n(s_k)s_k^4 \epsilon. \tag{7.105}$$

Equation (7.101) then specifies the mass loss rate from the bin as

$$\dot{M}_k^{loss} \propto n(s_k)s_k^4 \int_{s_{min}}^{s_1} n(s)(s_k + s)^2 ds. \tag{7.106}$$

Here and henceforth we drop numerical factors and stick with writing the equations as proportionalities because that suffices to find the size distribution.

The underlying assumption of Eq. (7.100) was that the cascade is scale-free, and consistent with that assumption we have picked a power-law form for Q_*^D. In a scale-free situation there cannot exist any characteristic sizes or masses, and we expect the size distribution to have a power-law form. Writing

$$n(s) \propto s^{-\alpha}, \tag{7.107}$$

with α a constant, the rate of mass loss from the k-th bin becomes

$$\dot{M}_k^{loss} \propto s_k^{4-\alpha} \int_{s_{min}}^{s_1} s^{-\alpha}(s_k + s)^2 ds. \tag{7.108}$$

Let's write out the integral explicitly:

$$\int_{s_{\min}}^{s_1} s^{-\alpha}(s_k + s)^2 ds = \left[\frac{s_k^2 s^{1-\alpha}}{1 - \alpha} + \frac{2 s_k s^{2-\alpha}}{2 - \alpha} + \frac{s^{3-\alpha}}{3 - \alpha} \right]_{s_{\min}}^{s_1}. \tag{7.109}$$

This is fairly complicated, but a solution consistent with the assumptions that we have made previously turns out to be possible for $\alpha > 3$. In this limit the first term of the right-hand-side and the lower limit dominate, and the integral $\propto s_k^2 s_{\min}^{1-\alpha}$. Substituting for $s_{\min}$ from Eq. (7.103) we find,

$$\dot{M}_k^{\text{loss}} \propto s_k^{6-\alpha} s_k^{(1+b/3)(1-\alpha)}. \tag{7.110}$$

Finally we determine the size distribution by requiring that $\dot{M}_k^{\text{loss}}$ be invariant with size, as demanded by Eq. (7.100). We obtain

$$6 - \alpha + (1 + b/3)(1 - \alpha) = 0, \tag{7.111}$$

and solve for α, where $-\alpha$ is the power-law slope of the steady-state size distribution:

$$\alpha = \frac{7 + b/3}{2 + b/3}. \tag{7.112}$$

An idealized collisional cascade among bodies whose strength is independent of size yields a size distribution $n(s) \propto s^{-\alpha} \propto s^{-7/2}$. In the more realistic case where $Q_*^D \propto s^b$ with $b \approx 1$ we obtain a slightly flatter size distribution with α slightly larger than 3.

7.6.2 Debris Disk Evolution

We can develop an idealized model for debris disk evolution based upon the idea of a collisional cascade. To do so, we assume that we have an annulus of debris orbiting at radius r, within which most of the mass resides in large bodies of radius $R_{\max}$. At time t_{stir} the belt is "stirred" – for example due to perturbations from a massive planet in the same system – such that collisions among the large bodies lead to catastrophic disruption and the formation of an approximately steady-state collisional cascade. The observable properties of the debris disk, such as its luminosity, are then tied to the rate at which collisions deplete the source population of large bodies.

If we assume that both the size of the large bodies and the velocity dispersion in the belt, $\sigma \simeq e v_K$, are constant, then the collision time scale,

$$t_c = \frac{1}{n \pi R_{\max}^2 \sigma}, \tag{7.113}$$

varies only as the number density of large bodies n changes. The number density, in turn, is proportional to the instantaneous mass of the belt,

$$n \propto M_{\text{belt}}. \tag{7.114}$$

We therefore write the collision time in the form

$$t_{\text{c}} = \left(\frac{M_0}{M_{\text{belt}}} \right) t_{\text{c},0}, \tag{7.115}$$

where M_0 is the initial mass of the belt at t_{stir} and $t_{\text{c},0}$ is the collision time at that instant. The belt mass evolves according to

$$\frac{\mathrm{d} M_{\text{belt}}}{\mathrm{d} t} = -\frac{M_{\text{belt}}^2}{t_{\text{c},0}}. \tag{7.116}$$

Solving this differential equation gives the belt mass as a function of time,

$$M_{\text{belt}}(t) = \frac{M_0}{1 + (t - t_{\text{stir}})/t_{\text{c},0}}. \tag{7.117}$$

The mass, and quantities that are linked to it such as the luminosity of the dust at the bottom of the collisional cascade, is constant at early times and decays asymptotically as t^{-1} (Dominik & Decin, 2003).

To apply this model to real systems we need only an estimate of the collision time in the belt at the time when the collisional cascade is initiated. This depends on the orbital radius and width of the belt, on the eccentricity, and on the strength and maximum size of the large bodies. The dependences on these parameters are given in Wyatt *et al.* (2007). Holding them fixed, the only remaining dependence is that the initial collision time is inversely proportional to the initial mass, $t_{\text{c},0} \propto M_0^{-1}$. As shown in Fig. 7.17 this leads to convergent evolution, such that the asymptotic behavior of the disk mass is *independent* of its initial mass. A corollary is that at any time there is a maximum disk mass, and maximum debris disk luminosity, that can plausibly be consistent with a steady collisional grinding model. If we observe an old star ($t \gtrsim$ Gyr) with an anomalously large infrared excess we have reasons to suspect that something cataclysmic happened at late times to trigger a transient surge in dust production.

7.6.3 White Dwarf Debris Disks

An interesting subset of debris disks are those that are observed around white dwarfs (for a review, see Farihi, 2016). The presence of metals can be detected spectroscopically from a sizeable fraction (perhaps 25–50%) of isolated white dwarfs' atmospheres. This is surprising. White dwarfs have high surface gravities, and except for the youngest and hottest stars any heavy elements are predicted to sediment below the observable photosphere on a short time scale. A smaller number of white dwarfs show excess emission in the near- and mid-infrared, consistent

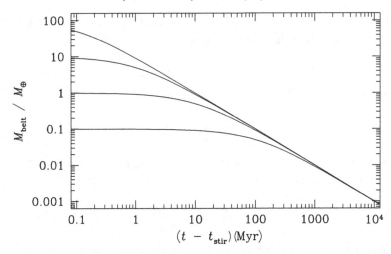

Figure 7.17 The predicted evolution of the mass of a debris disk evolving due to collisional destruction of a population of large bodies. Provided that other parameters (such as the orbital radius of the belt, and the eccentricity of the bodies within it) are fixed, the initial time scale for collisions is inversely proportional to the belt mass. This leads to convergent evolution of the belt mass at late times, and implies that at any time there is a maximum mass of debris that could have survived prior collisional processing.

with the presence of compact disks of debris. Taken together, the interpretation is that debris is relatively common around white dwarfs, and can be detected observationally either relatively directly (as unresolved excess emission) or entirely indirectly (as a polluted atmosphere).

Debris disks around main-sequence stars are typically thought to be co-spatial with the source population of large bodies at the top of the collisional cascade. This model runs into problems when applied to white dwarfs, because in many cases the inferred spatial extent of the emission is more compact than the radius of the giant star that evolved into the white dwarf. Instead, the standard scenario for white dwarf debris invokes the tidal disruption of asteroid-like bodies that have been scattered onto highly eccentric orbits. The debris disk, and subsequent accretion onto the star, is then formed from the combined effect of tidal disruption and a collisional cascade (Debes & Sigurdsson, 2002; Jura, 2003).

7.7 Further Reading

"Planet-Disk Interaction and Orbital Evolution," W. Kley & R. Nelson (2012), *Annual Review of Astronomy and Astrophysics*, **50**, 211.

"The Eccentric Kozai–Lidov Effect and Its Applications," S. Naoz (2016), *Annual Review of Astronomy and Astrophysics*, **54**, 441.

"Dynamical Evolution of the Early Solar System," D. Nesvorný (2018), *Annual Review of Astronomy and Astrophysics*, **56**, 137.

"Origins of Hot Jupiters," R. Dawson & J. Johnson (2018), *Annual Review of Astronomy and Astrophysics*, **56**, 175.

Appendix A

Physical and Astronomical Constants

Physical Constants

Speed of light	$c = 2.998 \times 10^{10} \, \text{cm s}^{-1}$
Newtonian gravitational constant	$G = 6.67 \times 10^{-8} \, \text{cm}^3 \, \text{g}^{-1} \, \text{s}^{-2}$
Planck constant	$h = 6.626 \times 10^{-27} \, \text{erg s}$
Proton mass	$m_\text{p} = 1.673 \times 10^{-24} \, \text{g}$
Boltzmann constant	$k_\text{B} = 1.381 \times 10^{-16} \, \text{erg K}^{-1}$
Stefan–Boltzmann constant	$\sigma = 5.670 \times 10^{-5} \, \text{erg cm}^{-2} \, \text{K}^{-4} \, \text{s}^{-1}$
Molar gas constant	$\mathcal{R} = 8.314 \times 10^7 \, \text{erg mol}^{-1} \, \text{K}^{-1}$

Astronomical Constants

Solar mass	$M_\odot = 1.989 \times 10^{33} \, \text{g}$
Jupiter mass	$M_\text{J} = 1.899 \times 10^{30} \, \text{g}$
Earth mass	$M_\oplus = 5.974 \times 10^{27} \, \text{g}$
Solar radius	$R_\odot = 6.96 \times 10^{10} \, \text{cm}$
Jupiter radius	$R_\text{J} = 7.15 \times 10^9 \, \text{cm}$
Earth radius	$R_\oplus = 6.38 \times 10^8 \, \text{cm}$
Solar luminosity	$L_\odot = 3.83 \times 10^{33} \, \text{erg s}^{-1}$
Astronomical unit	$1 \, \text{AU} = 1.496 \times 10^{13} \, \text{cm}$
Parsec	$1 \, \text{pc} = 3.086 \times 10^{18} \, \text{cm}$

Appendix B

The Two-Body Problem

The problem of how two point masses interact under Newtonian gravity forms the foundation both for planet formation (where physical collisions and nongravitational forces are important) and for more complex celestial mechanics problems with larger numbers of bodies. It is the most complex problem in Newtonian gravity that is amenable to a fully analytic solution. This appendix steps through the elementary derivation of the solution, along the way defining formally the quantities that describe Keplerian orbits.

B.1 Solution to the Two-Body Problem

Consider two point masses interacting via Newtonian gravity, as shown in Fig. B.1. The forces acting between them are given by

$$\mathbf{F}_1 = m_1\ddot{\mathbf{r}}_1 = \frac{Gm_1m_2}{r^3}\mathbf{r},$$
$$\mathbf{F}_2 = m_2\ddot{\mathbf{r}}_2 = -\frac{Gm_1m_2}{r^3}\mathbf{r},$$

(B.1)

where $r = |\mathbf{r}|$. We reduce the problem to a single equation for the relative motion, $\mathbf{r} = \mathbf{r}_2 - \mathbf{r}_1$,

$$\ddot{\mathbf{r}} = \ddot{\mathbf{r}}_2 - \ddot{\mathbf{r}}_1 = -\frac{G(m_1 + m_2)}{r^3}\mathbf{r}.$$

(B.2)

Defining $\mu \equiv G(m_1 + m_2)$ the equation becomes

$$\ddot{\mathbf{r}} = -\frac{\mu}{r^3}\mathbf{r}.$$

(B.3)

To solve this equation we work in the plane of the orbit and define a polar coordinate system (r, θ) that is centered on m_1. We will need the time derivatives of the radius vector $\mathbf{r}$, which are,

$$\mathbf{r} = r\hat{\mathbf{r}},$$
$$\dot{\mathbf{r}} = \dot{r}\hat{\mathbf{r}} + r\dot{\theta}\hat{\boldsymbol{\theta}},$$
$$\ddot{\mathbf{r}} = \left(\ddot{r} - r\dot{\theta}^2\right)\hat{\mathbf{r}} + \frac{1}{r}\frac{\mathrm{d}}{\mathrm{d}t}\left(r^2\dot{\theta}\right)\hat{\boldsymbol{\theta}}.$$

(B.4)

Here $\hat{\mathbf{r}}$ is a unit vector in the radial direction and $\hat{\boldsymbol{\theta}}$ is an azimuthal unit vector perpendicular to $\hat{\mathbf{r}}$.

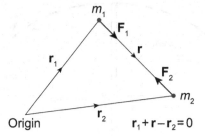

Figure B.1 Geometry for the Newtonian two-body problem.

The azimuthal component of the equation of motion (Eq. B.3) can be solved immediately,

$$\frac{1}{r}\frac{d}{dt}\left(r^2\dot\theta\right) = 0,$$

$$r^2\dot\theta = h. \tag{B.5}$$

We will return to h, the constant of integration, later, but it is clearly related to the angular momentum of the binary system.

Turning to the radial component,

$$\ddot r - \frac{h^2}{r^3} = -\frac{\mu}{r^2}, \tag{B.6}$$

it is helpful to make the substitution $u \equiv r^{-1}$. We are after the *shape* of the orbit, $r(\theta)$, so we use the chain rule to write

$$\frac{dr}{dt} = \frac{dr}{du}\frac{du}{d\theta}\frac{d\theta}{dt} = -h\frac{du}{d\theta}, \tag{B.7}$$

and similarly for $\ddot r$. We obtain a linear second-order differential equation,

$$\frac{d^2u}{d\theta^2} + u = \frac{\mu}{h^2}, \tag{B.8}$$

which has a solution

$$u = \left(\mu/h^2\right)[1 + e\cos(\theta - \varpi)],$$

$$r = \frac{h^2/\mu}{1 + e\cos(\theta - \varpi)}. \tag{B.9}$$

The solution defines a conic section (either an ellipse, a parabola, or a hyperbola), and we identify the constant of integration e as the eccentricity. The second constant of integration ϖ is called the longitude of pericenter.[1]

Bound Orbits

In the case where the two bodies are bound, Eq. (B.9) describes an ellipse with m_1 occupying one of the foci. We can use various geometric properties of the ellipse,

---

[1] The symbol ϖ is a cursive form of π, called variously "varpi", "curly pi," or "pomega".

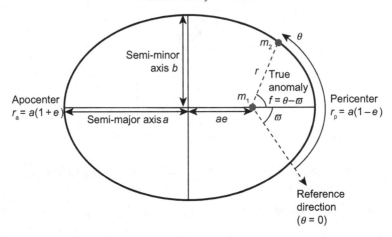

Figure B.2 Definition of quantities describing the two-body elliptical orbit of m_2 in a noninertial frame centered on m_1.

shown in Fig. B.2, to rewrite the orbit in terms of the ellipse's semi-major axis a and eccentricity e. The point of closest approach between the bodies is called pericenter, at a separation $r_p = a(1 - e)$ which we identify with the time when $\cos(\theta - \varpi) = 1$. Equation (B.9) then gives

$$h = \sqrt{\mu a(1 - e^2)}, \tag{B.10}$$

and the shape of the orbit has the standard form describing an ellipse in polar coordinates,

$$r = \frac{a(1 - e)^2}{1 + e\cos(\theta - \varpi)}. \tag{B.11}$$

It can be useful to measure the progress of m_2 around its orbit relative to pericenter rather than to an arbitrary reference direction, and so we define an angle called the true anomaly,

$$f \equiv \theta - \varpi. \tag{B.12}$$

The other important geometric properties of the orbit are the apocenter distance $r_a = a(1 + e)$ (the maximum separation between the bodies), the semi-minor axis $b = a\sqrt{1 - e^2}$, and the area within the ellipse $A = \pi ab = \pi a^2\sqrt{1 - e^2}$.

To compute the orbital period we note that the area swept out by m_2 as it orbits m_1 can be written as

$$\frac{dA}{dt} = \frac{1}{2}r^2\dot\theta = \frac{1}{2}h, \tag{B.13}$$

which is a constant of the motion. After time t the area $A = (1/2)h(t - t_p)$, where t_p is the time of a pericenter passage. A complete orbit occurs once $A = \pi ab$, at time $t = P$. Substituting for h gives the familiar relation for the orbital period, which does not depend on the eccentricity,

$$P = 2\pi\sqrt{\frac{a^3}{\mu}}. \tag{B.14}$$

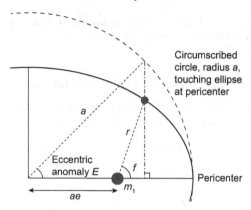

Figure B.3 Geometric construction defining the *eccentric anomaly E* for an elliptical orbit.

Finally we define the mean motion n as an average angular velocity of m_2 around its orbit,

$$n \equiv \frac{2\pi}{P}. \tag{B.15}$$

The mean motion is *the* angular velocity only in the special case of a circular orbit. In general $\dot{\theta}$ varies around the orbit, but even in this case n is physically significant, for example in defining the conditions for mean-motion resonances in more complex systems with two or more orbiting bodies.

By eliminating the time dependence in the original equation of motion via Eq. (B.7) we have side-stepped the question of *where m_2 is along its orbit at a given time*. This question is conventionally approached via the *eccentric anomaly*, which can be defined via the rather nonintuitive geometric construction shown in Fig. B.3. In an $x-y$ coordinate system with the x-axis aligned with the semi-major axis, the location of m_2 around its orbit is projected along the y-direction until it meets a circle of radius a that touches the orbit at pericenter and apocenter. The eccentric anomaly E is then the angle between that point and the x-axis, as seen from the center of the ellipse.

Given this construction, the relation between the true anomaly f and the eccentric anomaly E is

$$\tan\left(\frac{f}{2}\right) = \sqrt{\frac{1+e}{1-e}} \tan\left(\frac{E}{2}\right), \tag{B.16}$$

while the time dependence of E (which we won't prove, though the proof is not hard) is given by

$$E - e \sin E = n(t - t_p). \tag{B.17}$$

This is *Kepler's equation* and the right-hand-side is called the *mean anomaly*, $M \equiv n(t - t_p)$. The mean longitude is defined in terms of the mean anomaly as $\lambda = M + \varpi$. Conceptually then, given some time t we solve Kepler's equation (Eq. B.17) to find E, and then use Eq. (B.16) to get f and the instantaneous position of m_2.

Kepler's equation does not have an analytic solution in terms of simple functions, but it is not difficult to solve numerically. Standard root-finding schemes suffice, and for many purposes the extensive literature on approximate solutions is of historical interest only. An important exception is the class of orbit integration schemes that require repeated transformations between orbital elements and Cartesian coordinates. These schemes require the most efficient possible solution of Kepler's equation. Rein and Spiegel (2015) describe how this is done in a modern N-body integration code.

Orbital Energy and Angular Momentum

The center of mass (or barycenter) of the two bodies is defined as in Fig. B.4. If $\mathbf{R}_1$ is the vector pointing from the center of mass to m_1, and $\mathbf{R}_2$ the vector to m_2, then

$$m_1\mathbf{R}_1 + m_2\mathbf{R}_2 = 0, \tag{B.18}$$

and $\mathbf{r} = \mathbf{R}_2 - \mathbf{R}_1$. It then follows immediately that

$$\mathbf{R}_1 = -\left(\frac{m_2}{m_1 + m_2}\right)\mathbf{r},$$
$$\mathbf{R}_2 = \left(\frac{m_1}{m_1 + m_2}\right)\mathbf{r}. \tag{B.19}$$

The orbits of each of the masses in an inertial coordinate system thus have the same shape as the relative motion orbits previously derived, but with rescaled (and reduced) semi-major axes and pericenters that are offset by 180°.

The gravitational force between the bodies acts along the line joining their centers, so the angular momenta are conserved individually. The total angular momentum is

$$
\begin{aligned}
L_{\text{total}} &= m_1 R_1^2 \dot{\theta} + m_2 R_2^2 \dot{\theta} \\
&= \left[m_1 \left(\frac{m_2}{m_1 + m_2}\right)^2 + m_2 \left(\frac{m_1}{m_1 + m_2}\right)^2 \right] r^2 \dot{\theta} \\
&= \left(\frac{m_1 m_2}{m_1 + m_2}\right) h, \tag{B.20}
\end{aligned}
$$

with h as given in Eq. (B.10). To compute the energy, we first calculate the *relative* velocity v of m_2 in the noninertial frame centered on m_1. Starting from Eq. (B.11),

$$r = \frac{a(1 - e^2)}{1 + e \cos f}, \tag{B.21}$$

the time derivative of the separation is

$$
\begin{aligned}
\dot{r} &= \frac{dr}{df}\dot{f} \\
&= \frac{r^2 \dot{f}}{a(1 - e^2)} e \sin f. \tag{B.22}
\end{aligned}
$$

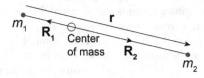

Figure B.4 Setup for converting the orbit to an inertial coordinate system with the center of mass at the origin.

Using the fact that $\dot{f} = \dot{\theta}$ we can replace $r^2 \dot{f}$ with $h = \sqrt{\mu a(1 - e^2)}$. The components of the velocity in the radial and azimuthal direction are then

$$\dot{r} = \sqrt{\frac{\mu}{a(1 - e^2)}} e \sin f, \tag{B.23}$$

$$r\dot{f} = \sqrt{\frac{\mu}{a(1 - e^2)}} (1 + e \cos f), \tag{B.24}$$

and the square of the relative velocity is

$$v^2 = \dot{r}^2 + (r\dot{f})^2,$$

$$= \mu \left(\frac{2}{r} - \frac{1}{a} \right). \tag{B.25}$$

Returning to the center of mass frame the total energy of the system is

$$E_{total} = \frac{1}{2} m_1 v_1^2 + \frac{1}{2} m_2 v_2^2 - \frac{Gm_1 m_2}{r}, \tag{B.26}$$

where v_1 and v_2 are the velocities of the bodies in this frame. We can write v_1 in terms of v via

$$v_1^2 = \dot{R}_1^2 + R_1^2 \dot{\theta}^2$$

$$= \left(\frac{m_2}{m_1 + m_2} \right)^2 \left(\dot{r}^2 + r^2 \dot{\theta}^2 \right)$$

$$= \left(\frac{m_2}{m_1 + m_2} \right)^2 v^2, \tag{B.27}$$

and similarly for v_2. Substituting for v^2 (Eq. B.25) we obtain a familiar but important result,

$$E_{total} = -\frac{Gm_1 m_2}{2a}. \tag{B.28}$$

The energy of a bound elliptical orbit depends upon the semi-major axis but is independent of the eccentricity.

Inclined Orbits

In its own plane the shape of a bound orbit and the position of m_2 along that orbit are fully defined by (a, e, ϖ, f). Often, however, we wish to describe such an orbit in the context of a larger system in which two or more bodies orbiting the central mass do not share a common orbital plane. This requires thinking about two additional complications. The first is simply how to describe an orbit with respect to an arbitrary reference plane in three-dimensional space. Two extra quantities are needed, one to measure the tilt of the orbit with respect to the plane and one to describe the orientation of the tilt with respect to a reference direction. Conventionally, as shown in Fig. B.5, we define a reference plane $(\mathbf{X}, \mathbf{Y})$ containing a reference direction along the $\hat{\mathbf{X}}$ axis. The inclination of the orbit i with respect to this plane defines the tilt, with $i < 90°$ corresponding to prograde orbits and $i \geq 90°$ corresponding to retrograde orbits. The angle in the reference plane between the reference direction and the point where the orbit crosses the plane in the positive $\hat{\mathbf{Z}}$ direction is called the *longitude of the ascending node*, and denoted by the symbol Ω. The angle in the orbital plane between the ascending node and presenter is called the *argument*

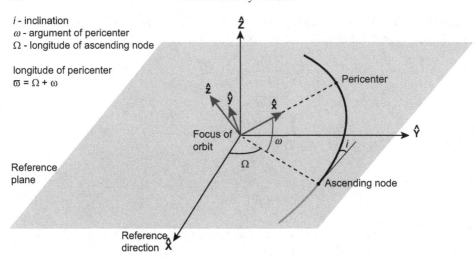

i - inclination
ω - argument of pericenter
Ω - longitude of ascending node

longitude of pericenter
ϖ = Ω + *ω*

Figure B.5 Definition of the orbital angles for a three-dimensional orbit inclined with respect to a reference plane.

of pericenter, and denoted by ω. The longitude of pericenter, which in the two-dimensional description was just the in-plane angle between the reference direction and pericenter, is more generally

$$\varpi = \Omega + \omega. \tag{B.29}$$

For an orbit that has nonzero i this is a nonobvious definition, as the longitude of the ascending node and the argument of pericenter are two angles that do not lie in the same plane. The three-dimensional orbit is then fully described by the set $(a, e, \varpi, i, \Omega, f)$.

The second complication that arises when thinking about three-dimensional orbits is the meaning of the orbital elements themselves. It is evidently most economical to describe a system of only two bodies in its own orbital plane, so if we are thinking about inclined orbits we are normally doing so in the context of systems of three or more bodies (or other situations, such as orbits around an oblate star, where the simple two-body problem is only an approximation to the true dynamics). In a three-body system, the true orbits are not closed ellipses/parabolae/hyperbolae, but rather something more complex which in general can only be determined numerically. Nonetheless, given the instantaneous positions and velocities of any pair of bodies, together with a reference plane and reference direction, we can go ahead and calculate orbital elements $(a, e, \varpi, i, \Omega, f)$ for that pair *as if the rest of the system did not exist*. Elements defined in this way are known as osculating elements, and represent an instantaneous two-body Keplerian approximation to the true orbit. In the common situation where m_1 is a star and m_i with $i = 2, 3, \ldots$ are planets on stable or quasi-stable orbits, the planet–planet perturbations are weak and the osculating elements define planetary orbits about the star that are a good approximation to the actual trajectories. One can then go further and derive equations (Lagrange's planetary equations) for the time evolution of the osculating elements under the action of the planet–planet perturbations, though we will not follow that line of rather technical development here.

Relativistic Precession

General relativity predicts that two-body orbits are qualitatively different from expectations based on Newtonian gravity. The differences include precession of the pericenter of elliptical orbits, precession of the longitude of the ascending node of orbits inclined with respect to the equatorial plane of spinning masses (Lense–Thirring precession), and loss of energy and angular momentum due to the radiation of gravitational waves. In an orbiting system in which the characteristic velocity is v these effects become significant at different orders in $(v/c)^2$, and for planetary systems – where $v \ll c$ – only the lowest order effect of pericenter precession normally needs to be considered. For a binary with masses m_1 and m_2 and eccentricity e the longitude of pericenter of the orbit, which is fixed in space in Newtonian gravity, advances at a rate

$$\frac{d\varpi}{dt} = \frac{3G^{3/2}(m_1 + m_2)^{3/2}}{c^2 a^{5/2}(1 - e^2)}. \tag{B.30}$$

The pericenter advances by 2π radians in a period of

$$P_{\mathrm{GR}} = \frac{2\pi c^2 a^{5/2}(1 - e^2)}{3G^{3/2}(m_1 + m_2)^{3/2}}. \tag{B.31}$$

For Mercury ($a = 0.387$ AU, $e = 0.206$) $P_{\mathrm{GR}} \simeq 3$ Myr, corresponding to the "anomalous" (i.e. non-Newtonian) but measured pericenter advance of 43 arcseconds per century that Einstein sought to reproduce as he developed general relativity. Hot Jupiters display a much stronger effect. A low eccentricity planet at $a = 0.05$ AU around a solar mass star has $P_{\mathrm{GR}} \simeq 1.9 \times 10^4$ yr. Although long, these time scales can be of the same order of magnitude as those that are associated with other secular effects, and when that secular dynamics is of interest relativity cannot be ignored. Examples include the long-term stability of Mercury, which is substantially altered by relativistic precession (Laskar & Gastineau, 2009), and Kozai–Lidov dynamics of extrasolar planets, which is suppressed when P_{GR} is less than the time scale of eccentricity oscillations (Naoz, 2016).

Relativistic dynamical effects can be incorporated into essentially Newtonian calculations in a number of ways. The simplest is to add Eq. (B.30) as an explicit evolution equation for the longitude of pericenter. However, it is also possible to modify the gravitational potential or the Hamiltonian in such a way as to reproduce the right rate of perihelion advance, or to integrate equations of motion that include additional accelerations representing post-Newtonian effects up to a specified order in $(v/c)^2$. These approaches will not generally yield identical (or correct) trajectories, but are almost functionally equivalent if the sole focus is on secular planetary dynamics.

Parabolic and Hyperbolic Orbits

The general solution to the two-body problem (Eq. B.9) describes hyperbolic and parabolic orbits as well as elliptical ones, though the geometric meaning of a and e in these cases is different. A hyperbolic orbit, with $e > 1$, is defined by the equation

$$r = \frac{a(e^2 - 1)}{1 + e\cos(\theta - \varpi)}. \tag{B.32}$$

Figure B.6 shows the geometry of such an orbit. In this case a is the distance, measured along the symmetry axis passing through the point of closest approach, between m_1 and the *unperturbed* trajectory that m_2 would have taken in the absence of gravity. More frequently, we characterize hyperbolic orbits via a combination of the impact parameter b, the distance of closest approach $r_{\min}$, and the velocity at infinity v_∞. Some simple geometry gives the impact parameter, defined as the closest approach along unperturbed trajectories, as

$$b = a\sqrt{e^2 - 1}. \tag{B.33}$$

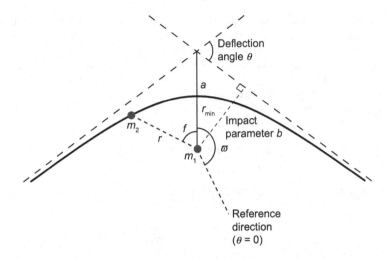

Figure B.6 Geometry of a hyperbolic orbit.

The closest approach along the actual orbit is

$$r_{\min} = a(e - 1),$$ (B.34)

occurring when $f \equiv \theta - \varpi = 0$. To obtain the velocity, the derivation leading to Eq. (B.25) for an elliptical orbit carries over identically apart from the definition of h, which in the hyperbolic case is

$$h = \sqrt{\mu a(e^2 - 1)}.$$ (B.35)

The result is that

$$v^2 = \mu \left(\frac{2}{r} + \frac{1}{a} \right),$$ (B.36)

and the velocity as $r \to \infty$ is $v_\infty = \sqrt{\mu/a}$.

Another quantity of interest is the deflection angle θ between the inbound and outbound trajectories of m_2 at large distances from m_1. To compute this, we note that $r \to \infty$ as $(1 + e \cos f) \to 0$. The limiting values of the true anomaly at large r are thus

$$f_\infty = \cos^{-1}(-e^{-1}) = \pi \pm \cos^{-1}(e^{-1}).$$ (B.37)

Inspection of Fig. B.6 then gives

$$\theta = \pi - 2\cos^{-1}(e^{-1}).$$ (B.38)

The deflection angle decreases with increasing values of e, tending to zero as $e \to \infty$.

For the special case of a parabolic orbit ($e = 1$) replacing a with $r_{\min}$ in Eq. (B.32) gives a well-defined expression,

$$r = \frac{2r_{\min}}{1 + \cos(\theta - \varpi)}.$$ (B.39)

A parabolic orbit has zero total energy.

Appendix C

N-Body Methods

Solving for the evolution of a system composed of N bodies interacting via their mutual gravitational force is a common problem in astrophysics. Although analytic or approximate numerical treatments (for example those based on the Fokker–Planck equation) can be useful, in many cases solution of N-body problems requires the explicit numerical integration of the trajectories traced by the bodies. A large body of technical literature (and at least one textbook, Sverre Aarseth's *Gravitational N-Body Simulations*[1]) is devoted to N-body methods, and codes that implement many of the most useful methods are readily available. This brief summary is intended to help guide aspiring N-body practitioners to the methods and literature that may be most suitable for their problem.

Specification of the N-body Problem

The state of a system of N point masses interacting only via Newtonian gravitational forces is fully specified once the masses m_i $(i = 1, N)$, positions $\mathbf{r}_i$, and velocities $\mathbf{v}_i$ are given at some reference time. The evolution of the system is then described by Newton's laws:

$$\dot{\mathbf{r}}_i = \mathbf{v}_i, \tag{C.1}$$

$$\dot{\mathbf{v}}_i = \mathbf{F}_i = -G \sum_{j=1;\, j \neq i}^{N} \frac{m_j(\mathbf{r}_i - \mathbf{r}_j)}{|\mathbf{r}_i - \mathbf{r}_j|^3}, \tag{C.2}$$

where the dots denote time derivatives and we have defined $\mathbf{F}_i$ as the force per unit mass felt by body i due to the gravitational interaction with all the remaining bodies. The problem of numerically integrating an N-body system can then be split into two parts. First, how do we calculate the $\mathbf{F}_i$? Although it is trivial to perform the explicit sum indicated in Eq. (C.2) this type of *direct* summation is computationally prohibitive for very large N. Approximate schemes which are much faster must then be used, but care is needed to ensure that the errors involved do not invalidate the results. Second, given the positions and velocities at some time t, together with an algorithm to compute the forces, how do we update the state of the system to some later time $(t + \delta t)$? There exist an arbitrary number of finite difference representations that reduce to the ordinary differential equations (ODEs) as $\delta t \to 0$, and it is far from a trivial matter to decide which is best. Long-term planetary integrations pose particular challenges since long-term error control is paramount and standard schemes that are accurate for the integration of other ODEs can perform very poorly.

---

[1] Unfortunately Aarseth (2003), one of the pioneers of the field, does not discuss the methods of greatest interest for planetary integrations.

Exact and Approximate Force Evaluation

Evaluation of the force by direct summation over all pairs of particles is straightforward and yields a force calculation that is exact up to the inevitable round-off errors that result from representing real numbers with a finite number of bits. The number of force evaluations per time step scales with the particle number as $\mathcal{O}(N^2)$. This scaling limits the use of direct summation schemes to relatively modest particle numbers.

Exact force evaluation is necessary for the long-term evolution of a small N system of planets, and is preferable in all circumstances. There can, however, be situations where tolerating some inaccuracy in the evaluation of $\mathbf{F}_i$ is beneficial for the overall fidelity of a calculation. If, for example, the physical number of bodies in a system exceeds the maximum number that can be integrated using direct summation, one is left with two unpalatable choices. One can either try to represent the system with a numerical proxy that has a smaller number of bodies, or integrate the true number of bodies approximately. Either choice may be appropriate depending on the problem at hand.

There are many different ways to efficiently compute an approximation to $\mathbf{F}_i$. In a system with a large number of particles it is clear that the force felt by particle i due to its immediate neighbors will vary on a much shorter time scale than the force due to distant particles. We could therefore split the force into two pieces,

$$\mathbf{F}_i = \mathbf{F}_{i,\text{close}} + \mathbf{F}_{i,\text{far}}, \tag{C.3}$$

and save effort by recalculating the force from the distant particles less frequently than for the nearby bodies. This venerable approach, known as the Ahmad–Cohen (1973) scheme, results in a modest improvement in the scaling to $\mathcal{O}(N^{7/4})$ at the expense of a small decrease in accuracy.

More aggressive trade-offs of accuracy for speed are not only possible but also necessary if the system of interest has very large N (at the time of writing, in 2019, the largest planet formation calculations using direct summation have $N \sim 10^6$). A powerful approach is to approximate the *spatial* distribution of the distant particles by grouping them together into ever larger clumps, and evaluating the force from the center of mass of each clump rather than from the particles directly. There are many ways in which this could be done, of which the most elegant is the *tree code*, introduced to astrophysics by Josh Barnes and Piet Hut (1986). The basic principle is illustrated (in two dimensions) in Fig. C.1. Starting with a cube that is large enough to enclose all of the particles within the simulation, space is first divided into eight sub-cubes by splitting the original cube in half along each of the three coordinate axes. This division of space is continued recursively until every cell contains either one or zero particles. The lowest level cells that contain just one particle are known as the *leaves* on the tree. We then work backward, calculating the total mass and center of mass of the cells at all higher levels of the tree, until eventually we return to the original or *root* cube whose mass and center of mass are just the mass and center of mass of the whole N-body system. When this procedure has been completed the resulting data structure, known as an oct-tree, encodes a unique prescription for how particles can be clumped into successively larger groups by moving up from the leaves toward the root level of the tree. Unless the particle distribution is pathological, the recursive nature of the division scheme ensures that the depth of the tree (the number of subdivisions needed until each cell contains at most one particle) will scale with particle number as $\mathcal{O}(\log N)$.

With the tree in hand the force $\mathbf{F}_i$ on any individual particle can be calculated by traversing the tree from the top down. At each level in the hierarchy of nested cells we evaluate the opening angles θ subtended by the cells as seen from the location of particle i. If for a particular cell

$$\theta > \theta_{\text{open}}, \tag{C.4}$$

Tree construction   Force evaluation

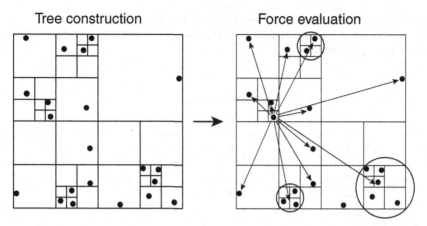

Figure C.1 Illustration of how a tree structure can be imposed upon a particle distribution by recursively partitioning space into cells (squares in this two-dimensional quad-tree example). Once the tree has been built, the force on any given particle is determined by summing contributions from nearby particles directly, while that from distant particles is evaluated approximately by grouping particles into larger clumps and calculating the force from the center of mass of each clump.

then the cell in question is too close for its contents to be treated as a single object and we proceed to the next level down in the tree. If, conversely, $\theta < \theta_{open}$ then it is acceptable to evaluate the force as if the contents of the cell were a single particle located at the center of mass of the true particle distribution within the cell.[2] Using this scheme it is possible to compute an approximate set of forces for the particles in $\mathcal{O}(N \log N)$ time, which for large N is a vast improvement over the N^2 scaling of direct summation methods. Moreover, by varying the cell opening criterion θ_{open}, the accuracy of the approximation can be adjusted.

Tree codes have many advantages for large N simulations. The algorithm is very nearly coordinate independent, works well for arbitrarily clustered mass distributions which may have an unusual geometry, and can be parallelized with good efficiency (Salmon, 1991; Dave *et al.*, 1997; Springel, 2005). There is, however, a significant computational overhead inherent in the construction of the tree. As a result there are situations – particularly those where the particle distribution is not too strongly clustered – when it can be substantially more efficient to compute the forces from a *potential* which has been pre-computed on a regular lattice of points that encloses the mass distribution. This method (Hockney & Eastwood, 1981) involves three steps:

- Starting from the locations $\mathbf{r}_i$ and masses m_i of the particles, a density field $\rho(x_i, y_j, z_k)$ is defined on a uniformly spaced lattice by assigning the masses of the particles to the vertices of nearby lattice cells.
- The gravitational potential in Fourier space is calculated by taking the discrete Fourier transform of the density on the lattice, multiplied by the Green's function for the Poisson equation. The real space potential on the lattice is then obtained by an inverse transform.
- The force at the location of each particle is found by finite differencing the potential on the lattice.

[2] It is also possible to pre-compute and store higher order multipoles of the mass distribution within cells, and to use this information to improve the accuracy of the force calculation.

A discrete Fourier transform can be computed efficiently using readily available and highly optimized libraries. As a consequence, grid methods of this kind are both fast and comparatively easy to implement.

Softening

Two classes of problem are commonly attacked using N-body methods. In the first class, the number of bodies in the physical system – which may be planets in a planetary system or stars in a small star cluster – is small enough that the individual particles in the N-body simulation can be considered to represent real bodies. Unless the bodies experience such close encounters that physical collisions or tidal effects become important, the only force that needs to be considered is Newtonian gravity with its $1/r$ potential, and we can aspire to integrate the orbits given this potential as accurately as possible. A common problem is that binaries may form during the simulation even if none is present initially. This will often require the use of special methods (known as *regularization*) that can handle the large disparity in time scales between the internal motion of a single tight binary and the evolution time scale for the overall system (see e.g. Aarseth, 2003).

A second class of problems is composed of those where the number of bodies in the physical system is far too large to represent on a one to one basis in an N-body simulation. We cannot presently hope to simulate *every* planetesimal in the terrestrial planet forming region of the disk, and we will never be able to simulate galaxy formation using N-body particles that have the same mass as dark matter particles. The particles in a simulation of one of these problems should not then be thought of as real bodies within the physical system, but rather as tracers whose trajectories are intended to be a fair statistical sample of the dynamics of the real system. In this regime we may need to suppress effects – such as large-angle gravitational scattering or binary formation – that would occur unphysically in the N-body system due to the fact that the particle mass is much larger than the mass of the physical bodies. This can be done by replacing the Newtonian potential $\Phi \propto 1/r$ with a *softened* potential, of which a simple example is

$$\Phi \propto \frac{1}{(r^2 + \epsilon^2)^{1/2}}. \tag{C.5}$$

The parameter ϵ is a user-variable softening length, which needs to be chosen so as to suppress two-body interactions while leaving the large-scale dynamics under the action of the smooth potential unaffected.

Time Stepping

Once the algorithm for calculating the forces has been specified, it remains for us to decide how to advance the system in time. This requires settling on a discrete representation of the ordinary differential equations (Eq. C.2). Advancing the discrete form of the equation will introduce errors, even if (as we assume henceforth) the forces have been computed exactly via direct summation. There are an infinite number of possible schemes, starting with the most naive:

$$\mathbf{r}_i^{t+\delta t} = \mathbf{r}_i^t + \mathbf{v}_i^t \delta t, \tag{C.6}$$

$$\mathbf{v}_i^{t+\delta t} = \mathbf{v}_i^t + \mathbf{F}_i^t \delta t. \tag{C.7}$$

This scheme, known as Euler's method, updates the position and the velocity simultaneously using only information available at the initial time t. Euler's method is inaccurate and should never be used. By making just a small modification, however, and computing the new velocity at a time that is offset from the calculation of the new position, we arrive at the very useful leapfrog method:

$$\mathbf{r}_i^{t+\delta t/2} = \mathbf{r}_i^t + \frac{1}{2}\mathbf{v}_i^t \delta t, \tag{C.8}$$

$$\mathbf{v}_i^{t+\delta t} = \mathbf{v}_i^t + \mathbf{F}_i(\mathbf{r}_i^{t+\delta t/2})\delta t, \tag{C.9}$$

$$\mathbf{r}_i^{t+\delta t} = \mathbf{r}_i^{t+\delta t/2} + \frac{1}{2}\mathbf{v}_i^{t+\delta t}\delta t. \tag{C.10}$$

How can we compare different methods quantitatively? One way is to study how the error (the difference between the numerical solution to the discrete equations and the true solution) scales as the time step is reduced. This scaling is known as the *order* of the method. Higher-order methods converge more rapidly toward the true solution with decreasing time step than low-order schemes.

Long-term Planetary Integrations

The trade-offs involved in choosing an integration scheme are most difficult when the problem involves the extremely long-term integration of planetary orbits. These are small-N systems, where the force integration can be completed using direct summation. Several considerations enter into the choice of time integration and numerical scheme.

- **Long-term error accumulation**. Evolution of a system over a billion orbits requires typically 10^{10}–10^{11} time steps. Error in the solution accumulates due to inaccuracy in the numerical scheme, and (unavoidably) due to the use of finite-precision arithmetic. In principle, we ought to choose the most efficient scheme that keeps errors below the level where they affect the fidelity of the results. This is a difficult criterion to assess in practice.
- **Eccentricity**. Bodies in systems where the Kozai–Lidov effect is operating, among others, can develop very high eccentricities. Some schemes that work well for near-circular orbits perform poorly as $e \to 1$.
- **Collisions and nonconservative forces**. Systems in which N point particles interact solely through Newtonian gravity have a special dynamical structure, which it can be advantageous to preserve numerically through the use of *symplectic* integrators. Real systems, however, can evolve such that physical collisions occur, and can be affected by other forces including tides and (for small particles) radiation forces. Consideration needs to be given to how to best represent these additional effects.

Chaos introduces additional considerations. Planetary systems are frequently chaotic, such that neighboring orbits diverge on a time scale that is short compared to the system lifetime. Physically, this means that the very long-term evolution of a specific system is inherently unpredictable, as we can neither know the initial conditions to arbitrary precision nor account for small external forces that would eventually affect the outcome. Numerically, the situation is a little different. A model N-body system (with no other forces) evolves deterministically from specified initial conditions, and even in the presence of chaos that deterministic evolution can in principle be computed over any finite time interval. Doing so, however, requires the use of arbitrary precision arithmetic, whose cost is computationally prohibitive. Invariably the numerical solution is instead computed with fixed precision arithmetic, in which case the solution differs from the deterministic one in the presence of chaos. The numerical solution must then be interpreted in a statistical sense, as one member of an ensemble of solutions that would result from slightly perturbed initial conditions.

High-Order Methods

Integrators that advance the N-body equations using high-order numerical schemes can be used profitably for long-term planetary integrations. IAS15, for example, is a modified version of a 15th-order integrator introduced by Everhart (1985), which is implemented within the REBOUND code (Rein & Liu, 2012; Rein & Spiegel, 2015). This type of integrator can evolve systems, over long periods, without introducing numerical errors that exceed the unavoidable lower bound imposed by the use of finite-precision arithmetic. It is therefore a good option in situations where the accuracy of an individual calculation is of paramount importance. High-order numerical schemes can also be extended relatively easily to include other forces, without damage to their favorable numerical properties.

Symplectic Methods

Geometric integration schemes represent an alternative and widely used set of methods for planetary integrations (Kinoshita *et al.*, 1991; Wisdom & Holman, 1991). The central realization behind these methods is that for long-term integrations it is desirable that the discrete representation of the equations preserves as much as possible of the structure of the true equations. We know, for example, that the true dynamical system is both time reversible and energy conserving, and we might seek out integration schemes that share such properties. Schemes that exactly conserve energy can be designed, but global energy conservation is a relatively weak constraint on the dynamics and this avenue is not very profitable. Time reversibility, on the other hand, turns out to be an extremely desirable property of an integration scheme, since the dissipation that leads to drift in the total energy is a numerical analog for a physically irreversible phenomenon. The simple leapfrog scheme is an example of a time reversible algorithm.

In addition to time reversibility, more subtle geometric properties of the true dynamics can serve as a further guide to constructing good integration schemes. In particular, the N-body problem is an example of a dynamical system that can be described by a Hamiltonian $H(\mathbf{p}, \mathbf{q}, t)$ that is a function of the generalized coordinates $\mathbf{q}$ and generalized momenta $\mathbf{p}$. The evolution of the state of a Hamiltonian system in the phase space described by these coordinates is given by

$$\dot{\mathbf{p}} = -\frac{\partial H}{\partial \mathbf{q}}, \tag{C.11}$$

$$\dot{\mathbf{q}} = \frac{\partial H}{\partial \mathbf{p}}. \tag{C.12}$$

Hamilton's equations describe how the state of the system at one time can be mapped into the state at some later time, and we can look for properties of this map that can be carried over to an integration scheme. Let us define the state of the system at time t to be $z = (\mathbf{q}, \mathbf{p})$. At some later time $(t+\delta t)$ the state is $z' = (\mathbf{q}', \mathbf{p}')$. One can then show that the map between these states $z \to z'$ has the property that it is a *symplectic transformation*, which means that it obeys the relation

$$\mathbf{MJM}^T = \mathbf{J}. \tag{C.13}$$

Here the matrices $\mathbf{M}$ and $\mathbf{J}$ are defined as

$$\mathbf{M}_{ij} = \frac{\partial z_i'}{\partial z_j}, \tag{C.14}$$

$$\mathbf{J} = \begin{bmatrix} 0 & -\mathbf{I} \\ \mathbf{I} & 0 \end{bmatrix}, \tag{C.15}$$

with **0** and **I** being respectively the $N \times N$ zero and unit matrix. The important point is that we can apply this definition of a symplectic transformation not just to the real equations describing the evolution of an N-body system (which, since the system is Hamiltonian, are guaranteed to yield symplectic maps), but also to any discrete representation of those equations. Demanding that the numerical scheme generate a symplectic transformation of the state of the system from one time step to the next enforces a very powerful constraint on the trajectories that the system can describe in phase space, and ensures that at least qualitatively they must behave in the same way as do trajectories of the true system. For this reason symplectic integration schemes are often preferable for very long-term integrations of planetary dynamics, where preserving the qualitative dynamics of the system is more important than obtaining the smallest error in the trajectory over the course of a few orbits. Somewhat amazingly the leapfrog scheme is not merely time reversible but also symplectic, and it is the simplest scheme that shares these two desirable traits.[3]

It is possible and often recommended to use symplectic methods for arbitrary N-body problems. Nonetheless, the problem for which symplectic methods are best suited remains the small N direct integration of the Solar System or other planetary systems over very long time scales. In this situation, it is possible to make further optimizations by noting that the Sun is, by far, the dominant body in the system, and that as a result the system is *almost* integrable. Formally we could write

$$H = H_{\text{Kepler}} + \epsilon H_{\text{planets}}, \tag{C.16}$$

where H_{Kepler} is the Hamiltonian describing two-body motion of a planet around the Sun, and the perturbing term due to the mutual interactions between the planets is smaller by a factor that is of the order of the mass ratio between the planets and the star. To construct an integration scheme, we now approximate this Hamiltonian in the form (Wisdom & Holman, 1991; Saha & Tremaine, 1992)

$$H = H_{\text{Kepler}} + \epsilon H_{\text{planets}} \tau \sum_{n=-\infty}^{\infty} \delta(t - t_0 - n\tau), \tag{C.17}$$

where t_0 is some reference time and τ is some small fraction of an orbital period. Written this way, the motion of the body is *exactly* Keplerian except at a set of closely spaced discrete intervals when the planetary perturbations are applied. The resulting integration scheme has three steps:

- The bodies are evolved under the action of the Keplerian Hamiltonian for half a time step $\tau/2$. This amounts to translating the bodies along ellipses (if the orbits are bound) and can be done analytically, either in terms of the normal Keplerian elements or in some other advantageous (non-Cartesian) coordinate system.
- The perturbing forces are evaluated in Cartesian coordinates, and each body receives an impulse that corresponds to the velocity change that would accumulate across a full time step.
- The step is completed with another half step under the Keplerian Hamiltonian.

Although this scheme requires continual conversion between coordinate systems, the extra work that this involves is more than compensated by the fact that it is possible to take a much longer time step than would be acceptable when using an unsplit scheme.

---

[3] We have noted that it is possible to construct symplectic integration schemes – leapfrog being an example – and that it is also possible to construct schemes that exactly preserve the total energy (though we have given no explicit example of the latter). It may occur to the reader that it would be best of all to have a scheme that was *both* symplectic and exactly energy conserving. Alas such schemes do not exist, since it has been proved (Zhong & Marsden, 1988) that an energy conserving symplectic scheme would amount to a solution of the *exact* dynamical system, which is not possible if the system is nonintegrable.

Integrators of this class are known as *mixed variable symplectic* (MVS) integrators (Wisdom & Holman, 1991; Levison & Duncan, 1994; Saha & Tremaine, 1994; Chambers, 1999), and efficient implementations are available in several publicly released packages. One should note that the simplest MVS schemes are not well suited to handling close approaches between planets (since during flybys the "perturbation" due to the planets' mutual gravity can dominate over the force from the star), and a more sophisticated version must be employed if the problem might involve close encounters or collisions.

References

Aarseth, S. J. 2003, *Gravitational N-Body Simulations*, Cambridge: Cambridge University Press.

Abod, C. P., Simon, J. B., Li, R., *et al.* 2019, *Astrophysical Journal*, submitted.

Adachi, I., Hayashi, C., & Nakazawa, K. 1976, *Progress of Theoretical Physics*, **56**, 1756.

Adams, F. C., Lada, C. J., & Shu, F. H. 1987, *Astrophysical Journal*, **312**, 788.

Adams, F. C., Laughlin, G., & Bloch, A. M. 2008, *Astrophysical Journal*, **683**, 1117.

Agol, E., Steffen, J., Sari, R., & Clarkson, W. 2005, *Monthly Notices of the Royal Astronomical Society*, **359**, 567.

Ahmad, A. & Cohen, L. 1973, *Journal of Computational Physics*, **12**, 389.

Alexander, R. D., Clarke, C. J., & Pringle, J. E. 2006, *Monthly Notices of the Royal Astronomical Society*, **369**, 229.

Alexander, R., Pascucci, I., Andrews, S., Armitage, P., & Cieza, L. 2014, in *Protostars and Planets VI*, H. Beuther, R. S. Klessen, C. P. Dullemond, & T. Henning (eds.), Tucson: University of Arizona Press, p. 475.

ALMA Partnership, *et al.* 2015, *Astrophysical Journal*, **808**, article id. L3.

Altwegg, K., Balsiger, H., Bar-Nun, A., *et al.* 2015, *Science*, **347**, article id. 1261952.

Andrews, S. M. & Williams, J. P. 2005, *Astrophysical Journal*, **631**, 1134.

Andrews, S. M., Huang, J., Pérez, L. M., *et al.* 2018, *Astrophysical Journal*, **869**, article id. L41.

Arakawa, M., Leliwa-Kopystynski, J., & Maeno, N. 2002, *Icarus*, **158**, 516.

Armitage, P. J., Livio, M., Lubow, S. H., & Pringle, J. E. 2002, *Monthly Notices of the Royal Astronomical Society*, **334**, 248.

Artymowicz, P., Clarke, C. J., Lubow, S. H., & Pringle, J. E. 1991, *Astrophysical Journal*, **370**, L35.

Baehr, H., Klahr, H., & Kratter, K. M. 2017, *Astrophysical Journal*, **848**, article id. 40.

Bai, X.-N. & Stone, J. M. 2011, *Astrophysical Journal*, **736**, article id. 144.

Balbus, S. A. 2011, in *Physical Processes in Circumstellar Disks around Young Stars*, P. J. V. Garcia (ed.), Chicago: University of Chicago Press, p. 237.

Balbus, S. A. & Hawley, J. F. 1991, *Astrophysical Journal*, **376**, 214.

Balbus, S. A. & Hawley, J. F. 1998, *Reviews of Modern Physics*, **70**, 1.

Baraffe, I., Chabrier, G., Allard, F., & Hauschildt, P. H. 2002, *Astronomy & Astrophysics*, **382**, 563.

Barge, P. & Sommeria, J. 1995, *Astronomy & Astrophysics*, **295**, L1.

Barnes, J. & Hut, P. 1986, *Nature*, **324**, 446.

Bate, M. R., Lodato, G., & Pringle, J. E. 2010, *Monthly Notices of the Royal Astronomical Society*, **401**, 1505.

Batygin, K. 2012, *Nature*, **491**, 418.

Beckwith, S. V. W. & Sargent, A. I. 1991, *Astrophysical Journal*, **381**, 250.

Begelman, M. C., McKee, C. F., & Shields, G. A. 1983, *Astrophysical Journal*, **271**, 70.

Bell, K. R. & Lin, D. N. C. 1994, *Astrophysical Journal*, **427**, 987.

Bell, K. R., Cassen, P. M., Klahr, H. H., & Henning, Th. 1997, *Astrophysical Journal*, **486**, 372.

Benz, W. & Asphaug, E. 1999, *Icarus*, **142**, 5.

Berta-Thompson, Z. K., Irwin, J., Charbonneau, D., *et al.* 2015, *Nature*, **527**, 204.

Binney, J. & Tremaine, S. 1987, *Galactic Dynamics*, Princeton, NJ: Princeton University Press.

Birnstiel, T., Klahr, H., & Ercolano, B. 2012, *Astronomy & Astrophysics*, **539**, id. A148.

Birnstiel, T., Ormel, C. W., & Dullemond, C. P. 2011, *Astronomy & Astrophysics*, **525**, id. A11.

Bitsch, B., Johansen, A., Lambrechts, M., & Morbidelli, A. 2015, *Astronomy & Astrophysics*, **575**, id. A28.

Blandford, R. D. & Payne, D. G. 1982, *Monthly Notices of the Royal Astronomical Society*, **199**, 883.

Blum, J. & Wurm, G. 2008, *Annual Review of Astronomy & Astrophysics*, **46**, 21.

Borderies, N. & Goldreich, P. 1984, *Celestial Mechanics*, **32**, 127.

Boss, A. P. 1997, *Science*, **276**, 1836.

Brownlee, D., Tsou, P., Aléon, J., *et al.* 2006, *Science*, **314**, 1711.

Burton, M. R., Sawyer, G. M., & Granieri, D. 2013, *Reviews in Mineralogy & Geochemistry*, **75**, 323.

Butler, R. P., Marcy, G. W., Williams, E., *et al.* 1996, *Publications of the Astronomical Society of the Pacific*, **108**, 500.

Cameron, A. G. W. 1978, *The Moon and the Planets*, **18**, 5.

Cameron, A. G. W. & Ward, W. R. 1976, *Abstracts of the Lunar and Planetary Science Conference*, **7**, 120.

Campbell, B., Walker, G. A. H., & Yang, S. 1988, *Astrophysical Journal*, **331**, 902.

Chambers, J. 2006, *Icarus*, **180**, 496.

Chambers, J. E. 1999, *Monthly Notices of the Royal Astronomical Society*, **304**, 793.

Chambers, J. E., Wetherill, G. W., & Boss, A. P. 1996, *Icarus*, **119**, 261.

Chandrasekhar, S. 1961, *Hydrodynamic and Hydromagnetic Stability*, International Series of Monographs on Physics, Oxford: Clarendon.

Chatterjee, S., Ford, E. B., Matsumura, S., & Rasio, F. A. 2008, *Astrophysical Journal*, **686**, 580.

Chavanis, P. H. 2000, *Astronomy & Astrophysics*, **356**, 1089.

Chenciner, A. & Montgomery, R. 2000, *Annals of Mathematics*, **152**, 881.

Chiang, E. 2008, *Astrophysical Journal*, **675**, 1549.

Chiang, E., Lithwick, Y., Murray-Clay, R., *et al.* 2007, in *Protostars and Planets V*, B. Reipurth, D. Jewitt, and K. Keil (eds.), Tucson: University of Arizona Press, p. 895.

Chiang, E. I. & Goldreich, P. 1997, *Astrophysical Journal*, **490**, 368.

Clarke, C. J. 2009, *Monthly Notices of the Royal Astronomical Society*, **396**, 1066.

Clarke, C. J. & Pringle, J. E. 1988, *Monthly Notices of the Royal Astronomical Society*, **235**, 635.

Clarke, C. J., Gendrin, A., & Sotomayor, M. 2001, *Monthly Notices of the Royal Astronomical Society*, **328**, 485.

Connelly, J. N., Bollard, J., & Bizzarro, M. 2017, *Geochimica et Cosmochimica Acta*, **201**, 345.

Cumming, A., Butler, R. P., Marcy, G. W., *et al.* 2008, *Publications of the Astronomical Society of the Pacific*, **120**, 531.

Cuzzi, J. N., Dobrovolskis, A. R., & Champney, J. M. 1993, *Icarus*, **106**, 102.

Damjanov, I., Jayawardhana, R., Scholz, A., *et al.* 2007, *Astrophysical Journal*, **670**, 1337.

D'Angelo, G., Lubow, S. H., & Bate, M. R. 2006, *Astrophysical Journal*, **652**, 1698.

Dave, R., Dubinski, J., & Hernquist, L. 1997, *New Astronomy*, **2**, 277.

Dawson, R. I. & Johnson, J. A. 2018, *Annual Review of Astronomy & Astrophysics*, **56**, 175.

Dawson, R. I. & Murray-Clay, R. 2012, *Astrophysical Journal*, **750**, article id. 43.

Debes, J. H. & Sigurdsson, S. 2002, *Astrophysical Journal*, **572**, 556.

Deng, H., Mayer, L., & Meru, F. 2017, *Astrophysical Journal*, **847**, article id. 43.

Desch, S. J., Morris, M. A., Connolly, H. C., & Boss, A. P. 2012, *Meteoritics & Planetary Science*, **47**, 1139.

Dohnanyi, J. S. 1969, *Journal of Geophysical Research*, **74**, 2531.

Dominik, C. & Decin, G. 2003, *Astrophysical Journal*, **598**, 626.

Doyle, L. R., Carter, J. A., Fabrycky, D. C., *et al.* 2011, *Science*, **333**, 1602.

Draine, B. T., Roberge, W. G., & Dalgarno, A. 1983, *Astrophysical Journal*, **264**, 485.

Dubrulle, B., Morfiull, G., & Sterzik, M. 1995, *Icarus*, **114**, 237.

Duffell, P. C. & Chiang, E. 2015, *Astrophysical Journal*, **812**, article id. 94.

Duffell, P. C., Haiman, Z., MacFadyen, A. I., D'Orazio, D. J., & Farris, B. D. 2014, *Astrophysical Journal*, **792**, article id. L10.

Dullemond, C. P. & Dominik, C. 2005, *Astronomy & Astrophysics*, **434**, 971.

Dumusque, X., Udry, S., Lovis, C., Santos, N. C., & Monteiro, M. J. P. F. G. 2011, *Astronomy & Astrophysics*, **525**, article id. A140.

Duquennoy, A. & Mayor, M. 1991, *Astronomy & Astrophysics*, **248**, 485.

Dürmann, C. & Kley, W. 2015, *Astronomy & Astrophysics*, **574**, article id. A52.

Eggleton, P. P. & Kiseleva-Eggleton, L. 2001, *Astrophysical Journal*, **562**, 1012.

Eisner, J. A., Hillenbrand, L. A., Carpenter, J. M., & Wolf, S. 2005, *Astrophysical Journal*, **635**, 396.

Ercolano, B. & Pascucci, I. 2017, *Royal Society Open Science*, **4**, article id. 170114.

Espaillat, C., Muzerolle, J., Najita, J., *et al.* 2014, in *Protostars and Planets VI*, H. Beuther, R. S. Klessen, C. P. Dullemond, & T. Henning (eds.), Tucson: University of Arizona Press, p. 497.

Evans, N. J., Dunham, M. M., Jorgensen, J. K., *et al.* 2009, *Astrophysical Journal Supplement*, **181**, 321.

Everhart, E. 1985, in *Dynamics of Comets: Their Origin and Evolution*. A. Carusi & G. B. Valsecchi (eds.), Astrophysics and Space Science Library, **115**, Dordrecht: Reidel, p. 185.

Fabrycky, D. & Tremaine, S. 1997, *Astrophysical Journal*, **669**, 1298.

Fabrycky, D. C., Lissauer, J. J., Ragozzine, D., *et al.* 2014, *Astrophysical Journal*, 790, article id. 146.

Farihi, J. 2016, *New Astronomy Reviews*, **71**, 9.

Feigelson, E. D. & Montmerle, T. 1999, *Annual Review of Astronomy & Astrophysics*, **37**, 363.

Feigelson, E., Townsley, L., Güdel, M., & Stassun, K. 2007, in *Protostars and Planets V*, B. Reipurth, D. Jewitt, & K. Keil (eds.), Tucson: University of Arizona Press, p. 313.

Fernandez, J. A. & Ip, W.-H. 1984, *Icarus*, **58**, 109.

Fischer, D. A. & Valenti, J. 2005, *Astrophysical Journal*, **622**, 1102.

Flaherty, K. M., Hughes, A. M., Rose, S. C., *et al.* 2017, *Astrophysical Journal*, **843**, article id. 150.

Font, A. S., McCarthy, I. G., Johnstone, D., & Ballantyne, D. R. 2004, *Astrophysical Journal*, **607**, 890.

Ford, E. B., Havlickova, M., & Rasio, F. A. 2001, *Icarus*, **150**, 303.

Fricke, K. 1968, *Zeitschrift für Astrophysik*, **68**, 317.

Fromang, S. & Papaloizou, J. 2006, *Astronomy & Astrophysics*, **452**, 751.

Fromang, S., Terquem, C., & Balbus, S. A. 2002, *Monthly Notices of the Royal Astronomical Society*, **329**, 18.

Gammie, C. F. 1996, *Astrophysical Journal*, **457**, 355.

Gammie, C. F. 2001, *Astrophysical Journal*, **553**, 174.

Garaud, P. & Lin, D. N. C. 2004, *Astrophysical Journal*, **608**, 1050.
Garaud, P. & Lin, D. N. C. 2007, *Astrophysical Journal*, **654**, 606.
Ghosh, P. & Lamb, F. K. 1979, *Astrophysical Journal*, **232**, 259.
Gillon, M., Triaud, A. H. M. J., Demory, B.-O., *et al.* 2017, *Nature*, **542**, 456.
Gladman, B. 1993, *Icarus*, **106**, 247.
Godon, P. & Livio, M. 1999, *Astrophysical Journal*, **523**, 350.
Goldreich, P. & Peale, S. 1966, *Astronomical Journal*, **71**, 425.
Goldreich, P. & Sari, R. 2003, *Astrophysical Journal*, **585**, 1024.
Goldreich, P. & Schubert, G. 1967, *Astrophysical Journal*, **150**, 571.
Goldreich, P. & Tremaine, S. 1979, *Astrophysical Journal*, **233**, 857.
Goldreich, P. & Tremaine, S. 1980, *Astrophysical Journal*, **241**, 425.
Goldreich, P. & Ward, W. R. 1973, *Astrophysical Journal*, **183**, 1051.
Gómez, G. C. & Ostriker, E. C. 2005, *Astrophysical Journal*, **630**, 1093.
Gómes, R., Levison, H. F., Tsiganis, K., & Morbidelli, A. 2005, *Nature*, **435**, 466.
Goodman, A. A., Benson, P. J., Fuller, G. A., & Myers, P. C. 1993, *Astrophysical Journal*, **406**, 528.
Greenberg, R., Bottke, W. F., Carusi, A., & Valsecchi, G. B. 1991, *Icarus*, **94**, 98.
Greenberg, R., Wacker, J. F., Hartmann, W. K., & Chapman, C. R. 1978, *Icarus*, **35**, 1.
Guilet, J. & Ogilvie, G. I. 2014, *Monthly Notices of the Royal Astronomical Society*, **441**, 852.
Guillot, T., Stevenson, D. T., Hubbard, W. B., & Saumon, D. 2004, in *Jupiter: The Planet, Satellites and Magnetosphere*, F. Bagenal, T. Dowling, & W. McKinnon (eds.), Cambridge: Cambridge University Press.
Gullbring, E., Hartmann, L., Briceno, C., & Calvet, N. 1998, *Astrophysical Journal*, **492**, 323.
Gundlach, B. & Blum, J. 2015, *Astrophysical Journal*, **798**, article id. 34.
Gutermuth, R. A., Myers, P. C., Megeath, S. T., *et al.* 2008, *Astrophysical Journal*, **674**, 336.
Güttler, C., Blum, J., Zsom, A., Ormel, C. W., & Dullemond, C. P. 2010, *Astronomy & Astrophysics*, **513**, article id. A56.
Haisch, K. E., Lada, E. A., & Lada, C. J. 2001, *Astrophysical Journal*, **553**, 153.
Hansen, B. M. S. 2009, *Astrophysical Journal*, **703**, 1131.
Hartmann, L., Calvet, N., Gullbring, E., & D'Alessio, P. 1998, *Astrophysical Journal*, **495**, 385.
Hartmann, L., Hewett, R., & Calvet, N. 1994, *Astrophysical Journal*, **426**, 669.
Hartmann, W. K. & Davis, D. R. 1975, *Icarus*, **24**, 504.
Hartogh, P., Lis, D. C., Bockelée-Morvan, D., *et al.* 2011, *Nature*, **478**, 218.
Hawley, J. F., Balbus, S. A., & Winters, W. F. 1999, *Astrophysical Journal*, **518**, 394.
Hayashi, C. 1981, *Progress of Theoretical Physics Supplement*, **70**, 35.
Hernández, J., Hartmann, L., Megeath, T., *et al.* 2007, *Astrophysical Journal*, **662**, 1067.
Hirayama, K. 1918, *Astronomical Journal*, **31**, 185.
Hirose, S. & Shi, J.-M. 2019, *Monthly Notices of the Royal Astronomical Society*, **485**, 266.
Hockney, R. W. & Eastwood, J. W. 1981, *Computer Simulation Using Particles*, New York: McGraw-Hill.
Holczer, T., Mazeh, T., Nachmani, G., *et al.* 2016, *Astrophysical Journal Supplement Series*, **225**, article id. 9.
Hollenbach, D., Johnstone, D., Lizano, S., & Shu, F. 1994, *Astrophysical Journal*, **428**, 654.
Holman, M. J. & Murray, N. W. 1996, *Astronomical Journal*, **112**, 1278.
Holman, M. J. & Murray, N. W. 2005, *Science*, **307**, 1288.
Holman, M., Touma, J., & Tremaine, S. 1997, *Nature*, **386**, 254.
Hornung, P., Pellat, R., & Barge, P. 1985, *Icarus*, **64**, 295.
Ida, S. 1990, *Icarus*, **88**, 129.
Ida, S. & Makino, J. 1993, *Icarus*, **106**, 210.
Ida, S., Bryden, G., Lin, D. N. C., & Tanaka, H. 2000, *Astrophysical Journal*, **534**, 428.

Iess, L., Folkner, W. M., Durante, D., *et al.* 2018, *Nature*, **555**, 220.

Igea, J. & Glassgold, A. E. 1999, *Astrophysical Journal*, **518**, 848.

Ikoma, M., Nakazawa, K., & Emori, H. 2000, *Astrophysical Journal*, **537**, 1013.

Ilgner, M. & Nelson, R. P. 2006, *Astronomy & Astrophysics*, **445**, 205.

Ilgner, M. & Nelson, R. P. 2008, *Astronomy & Astrophysics*, **483**, 815.

Inaba, S., Tanaka, H., Nakazawa, K., Wetherill, G. W., & Kokubo, E. 2001, *Icarus*, **149**, 235.

Ji, H., Burin, M., Schartman, E., & Goodman, J. 2006, *Nature*, **444**, 343.

Jiménez, M. A. & Masset, F. S. 2017, *Monthly Notices of the Royal Astronomical Society*, **471**, 4917.

Johansen, A. & Lacerda, P. 2010, *Monthly Notices of the Royal Astronomical Society*, **404**, 475.

Johansen, A., Oishi, J. S., Low, M.-M. M., *et al.* 2007, *Nature*, **448**, 1022.

Johansen, A., Youdin, A., & Klahr, H. 2009a, *Astrophysical Journal*, **697**, 1269.

Johansen, A., Youdin, A., & Mac Low, M.-M. 2009b, *Astrophysical Journal*, **704**, L75.

Johns-Krull, C. M. 2007, *Astrophysical Journal*, **664**, 975.

Johnson, B. M. & Gammie, C. F. 2005, *Astrophysical Journal*, **635**, 149.

Johnstone, D., Hollenbach, D., & Bally, J. 1998, *Astrophysical Journal*, **499**, 758.

Joy, A. H. 1945, *Astrophysical Journal*, **102**, 168.

Jura, M. 2003, *Astrophysical Journal*, **584**, L91.

Jurić, M. & Tremaine, S. 2008, *Astrophysical Journal*, **686**, 603.

Kanagawa, K. D., Muto, T., Tanaka, H., *et al.* 2015, *Astrophysical Journal*, **806**, article id. L15.

Kanagawa, K. D., Tanaka, H., & Szuszkiewicz, E. 2018, *Astrophysical Journal*, **861**, article id. 140.

Kasting, J. F., Whitmire, D. P., & Reynolds, R. T. 1993, *Icarus*, **101**, 108.

Kenyon, S. J. & Hartmann, L. 1987, *Astrophysical Journal*, **323**, 714.

Kenyon, S. J. & Luu, J. X. 1998, *Astronomical Journal*, **115**, 2136.

Kerr, M., Johnston, S., Hobbs, G., & Shannon, R. M. 2015, *Astrophysical Journal*, **809**, article id. L11.

Kida, S. 1981, *Physical Society of Japan*, **50**, 3517.

Kinoshita, H., Yoshida, H., & Nakai, H. 1991, *Celestial Mechanics and Dynamical Astronomy*, **50**, 59.

Kippenhahn, R. & Weigert, A. 1990, *Stellar Structure and Evolution*, Berlin: Springer-Verlag.

Kirkwood, D., 1867, *Meteoric Astronomy: A Treatise on Shooting-stars, Fireballs, and Aerolites*, Philadelphia: J. B. Lippincott & Co.

Kirsh, D. R., Duncan, M., Brasser, R., & Levison, H. F. 2009, *Icarus*, **199**, 197.

Kley, W. & Nelson, R. P. 2012, *Annual Review of Astronomy & Astrophysics*, **50**, 211.

Kokubo, E., Kominami, J., & Ida, S. 2006, *Astrophysical Journal*, **642**, 1131.

Königl, A. 1991, *Astrophysical Journal*, **370**, L39.

Kopparapu, R. K., Ramirez, R., Kasting, J. F, *et al.* 2013, *Astrophysical Journal*, **765**, 131.

Korycansky, D. G. & Asphaug, E. 2006, *Icarus*, **181**, 605.

Kozai, Y. 1962, *Astronomical Journal*, **67**, 591.

Kratter, K. & Lodato, G. 2016, *Annual Review of Astronomy & Astrophysics*, **54**, 271.

Kratter, K. M., Matzner, C. D., Krumholz, M. R., & Klein, R. I. 2010, *Astrophysical Journal*, **708**, 1585.

Krolik, J. H. & Kallman, T. R. 1983, *Astrophysical Journal*, **267**, 610.

Kuiper, G. P. 1951, *Proceedings of the National Academy of Sciences*, **37**, 1.

Kunz, M. W. 2008, *Monthly Notices of the Royal Astronomical Society*, **385**, 1494.

Kwok, S. 1975, *Astrophysical Journal*, **198**, 583.

Lada, C. J. & Lada, E. A. 2003, *Annual Review of Astronomy & Astrophysics*, **41**, 57.

Lada, C. J. & Wilking, B. A. 1984, *Astrophysical Journal*, **287**, 610.

Lada, C. J., Muench, A. A., Luhman, K. L., *et al.* 2006, *Astronomical Journal*, **131**, 1574.
Lambrechts, M. & Johansen, A. 2012, *Astronomy & Astrophysics*, **544**, article id. A32.
Larson, R. B. 1981, *Monthly Notices of the Royal Astronomical Society*, **194**, 809.
Laskar, J. 1997, *Astronomy & Astrophysics*, **317**, L75.
Laskar, J. & Gastineau, M. 2009, *Nature*, **459**, 817.
Lecar, M., Franklin, F. A., Holman, M. J., & Murray, N. J. 2001, *Annual Review of Astronomy & Astrophysics*, **39**, 581.
Lee, M. H. & Peale, S. J. 2001, *Astrophysical Journal*, **567**, 596.
Leinhardt, Z. M. & Richardson, D. C. 2002, *Icarus*, **159**, 306.
Leinhardt, Z. M. & Stewart, S. T. 2009, *Icarus*, **199**, 542.
Leinhardt, Z. M. & Stewart, S. T. 2012, *Astrophysical Journal*, **745**, article id. 79.
Lesur, G., Kunz, M. W., & Fromang, S. 2014, *Astronomy & Astrophysics*, **566**, article id. A56.
Lesur, G. & Papaloizou, J. C. B. 2009, *Astronomy & Astrophysics*, **498**, 1.
Lesur, G., & Papaloizou, J. C. B. 2010, *Astronomy & Astrophysics*, **513**, article id. A60.
Levin, Y. 2007, *Monthly Notices of the Royal Astronomical Society*, **374**, 515.
Levison, H. F. & Duncan, M. J. 1994, *Icarus*, **108**, 18.
Levison, H. F., Kretke, K. A., & Duncan, M. J. 2015, *Nature*, **524**, 322.
Lidov, M. L. 1962, *Planetary and Space Science*, **9**, 719.
Lin, D. N. C. & Ida, S. 1997, *Astrophysical Journal*, **477**, 781.
Lin, D. N. C. & Papaloizou, J. 1979, *Monthly Notices of the Royal Astronomical Society*, **186**, 799.
Lin, D. N. C. & Papaloizou, J. 1986, *Astrophysical Journal*, **309**, 846.
Lin, M.-K. & Youdin, A. N. 2015, *Astrophysical Journal*, **811**, article id. 17.
Lissauer, J. J. 1993, *Annual Review of Astronomy & Astrophysics*, **31**, 129.
Lissauer, J. J. & Stevenson, D. J. 2007, in *Protostars and Planets V*, B. Reipurth, D. Jewitt, & K. Keil (eds.), Tucson: University of Arizona Press, p. 591.
Lissauer, J. J. & Stewart, G. R. 1993, in *Protostars and Planets III*, E. H. Levy & J. I. Lunine (eds.), Tucson: University of Arizona Press, p. 1061.
Lodders, K. 2003, *Astrophysical Journal*, **591**, 1220.
Lovelace, R. V. E., Li, H., Colgate, S. A., & Nelson, A. F. 1999, *Astrophysical Journal*, **513**, 805.
Lubow, S. H., Papaloizou, J. C. B., & Pringle, J. E. 1994, *Monthly Notices of the Royal Astronomical Society*, **267**, 235.
Lynden-Bell, D. & Boily, C. 1994, *Monthly Notices of the Royal Astronomical Society*, **267**, 146.
Lynden-Bell, D. & Pringle, J. E. 1974, *Monthly Notices of the Royal Astronomical Society*, **168**, 603.
Malhotra, R. 1993, *Nature*, **365**, 819.
Malhotra, R. 1995, *Astronomical Journal*, **110**, 420.
Manara, C. F., Fedele, D., Herczeg, G. J., & Teixeira, P. S. 2016, *Astronomy & Astrophysics*, **585**, article id. A136.
Mandel, K. & Agol, E. 2002, *Astrophysical Journal*, **580**, L171.
Marchal, C. & Bozis, G. 1982, *Celestial Mechanics*, **26**, 311.
Marcy, G. W., Butler, R. P., Fischer, D., *et al.* 2001, *Astrophysical Journal*, **556**, 296.
Marois, C., Lafreniére, D., Doyon, R., Macintosh, B., & Nadeau, D. 2005, *Astrophysical Journal*, **641**, 556.
Marois, C., Macintosh, B., Barman, T., *et al.* 2008, *Science*, **322**, 1348.
Masset, F. S. & Ogilvie, G. I. 2004, *Astrophysical Journal*, **615**, 1000.
Masset, F. & Snellgrove, M. 2001, *Monthly Notices of the Royal Astronomical Society*, **320**, L55.
Mathis, J. S., Rumpl, W., & Nordsieck, K. H. 1977, *Astrophysical Journal*, **217**, 425.
Mayor, M. & Queloz, D. 1995, *Nature*, **378**, 355.

McClure, M. K., Bergin, E. A., Cleeves, L. I., *et al.* 2016, *Astrophysical Journal*, **831**, article id. 167.

McLaughlin, D. B. 1924, *Astrophysical Journal*, **60**, 22.

Meyer-Vernet, N. & Sicardy, B. 1987, *Icarus*, **69**, 157.

Millholland, S., Laughlin, G., Teske, J., *et al.* 2018, *Astrophysical Journal*, 155, article id. 106.

Mizuno, H. 1980, *Progress of Theoretical Physics*, **64**, 544.

Morbidelli, A. & Nesvorny, D. 2012, *Astronomy & Astrophysics*, **546**, article id. A18.

Morbidelli, A., Chambers, J., Lunine, J. I., *et al.* 2000, *Meteoritics & Planetary Science*, **35**, 1309.

Morbidelli, A., Nesvorny, D., Laurenz, V., *et al.* 2018, *Icarus*, **305**, 262.

Morbidelli, A., Levison, H. F., Tsiganis, K., & Gómes, R. 2005, *Nature*, **435**, 462.

Movshovitz, N., Bodenheimer, P., Podolak, M., & Lissauer, J. J. 2010, *Icarus*, **209**, 616.

Murray, C. D. & Dermott, S. F. 1999, *Solar System Dynamics*, Cambridge: Cambridge University Press.

Murray-Clay, R. A. & Chiang, E. I. 2006, *Astrophysical Journal*, **651**, 1194.

Nakagawa, Y., Sekiya, M., & Hayashi, C. 1986, *Icarus*, **67**, 375.

Naoz, S. 2016, *Annual Review of Astronomy & Astrophysics*, **54**, 441.

Naoz, S., Farr, W. M., & Rasio, F. A. 2012, *Astrophysical Journal*, **754**, article id. L36.

Nayakshin, S. 2010, *Monthly Notices of the Royal Astronomical Society*, **408**, L36.

Nelson, R. P., Gressel, O., & Umurhan, O. M. 2013, *Monthly Notices of the Royal Astronomical Society*, **435**, 2610.

Nesvorný, D. 2018, *Annual Review of Astronomy & Astrophysics*, **56**, 137.

Nesvorný, D., Bottke, W. F., Jr., Dones, L., & Levison, H. F. 2002, *Nature*, **417**, 720.

Nittler, L. R. & Ciesla, F. 2016, *Annual Review of Astronomy & Astrophysics*, **54**, 53.

Obertas, A., Van Laerhoven, C., & Tamayo, D. 2017, *Icarus*, **293**, 52.

O'Brien, D. P., Morbidelli, A., & Levison, H. F. 2006, *Icarus*, **184**, 39.

O'Dell, C. R., Wen, Z., & Hu, X. 1993, *Astrophysical Journal*, **410**, 696.

Ogilvie, G. I. & Lubow, S. H. 2003, *Astrophysical Journal*, **587**, 398.

Ohtsuki, K., Stewart, G. R., & Ida, S. 2002, *Icarus*, **155**, 436.

Oort, J. H. 1950, *Bulletin of the Astronomical Institutes of the Netherlands*, **11**, 91.

Ormel, C. W. 2017, in *Formation, Evolution, and Dynamics of Young Solar Systems*, Astrophysics and Space Science Library, **445**, Cham: Springer, p. 197.

Ormel, C. W. & Cuzzi, J. N. 2007, *Astronomy & Astrophysics*, **466**, 413.

Ormel, C. W. & Klahr, H. H. 2010, *Astronomy & Astrophysics*, **520**, article id. A43.

Ormel, C. W. & Okuzumi, S. 2013, *Astrophysical Journal*, **771**, 44.

Paardekooper, S.-J., Baruteau, C., & Kley, W. 2011, *Monthly Notices of the Royal Astronomical Society*, **410**, 293.

Paczynski, B. 1978, *Acta Astronomica*, **28**, 91.

Papaloizou, J. C. B. & Pringle, J. E. 1984, *Monthly Notices of the Royal Astronomical Society*, **208**, 721.

Papaloizou, J. C. B. & Terquem, C. 1999, *Astrophysical Journal*, **521**, 823.

Papaloizou, J. C. B. & Terquem, C. 2006, *Reports of Progress in Physics*, **69**, 119.

Papaloizou, J. C. B., Nelson, R. P., & Masset, F. 2001, *Astronomy & Astrophysics*, **366**, 263.

Peale, S. J. 1976, *Annual Review of Astronomy & Astrophysics*, **14**, 215.

Perez, L. M., Carpenter, J. M., Andrews, S. M., *et al.* 2016, *Science*, **353**, 1519.

Perri, F. & Cameron, A. G. W. 1974, *Icarus*, **22**, 416.

Petersen, M. R., Julien, K., & Stewart, G. R. 2007, *Astrophysical Journal*, **658**, 1236.

Petit, J.-M., Morbidelli, A. & Chambers, J. 2001, *Icarus*, **153**, 338.

Petrovich, C. & Tremaine, S. 2016, *Astrophysical Journal*, **829**, article id. 132.

Pinilla, P., Birnstiel, T., Ricci, L., *et al.* 2012, *Astronomy & Astrophysics*, **538**, article id. A114.

Pinilla, P., Pohl, A., Stammler, S. M., & Birnstiel, T. 2017, *Astrophysical Journal*, **845**, article id. 68.

Pollack, J. B., Hubickyj, O., Bodenheimer, P., *et al.* 1996, *Icarus*, **124**, 62.

Pollack, J. B., McKay, C. P., & Christofferson, B. M. 1985, *Icarus*, **64**, 471.

Pringle, J. E. 1981, *Annual Review of Astronomy & Astrophysics*, **19**, 137.

Pringle, J. E. & King, A. R. 2007, *Astrophysical Flows*, Cambridge: Cambridge University Press.

Pringle, J. E. & Rees, M. J. 1972, *Astronomy & Astrophysics*, **21**, 1.

Quillen, A. C. 2006, *Monthly Notices of the Royal Astronomical Society*, **365**, 1367.

Quirrenbach, A. 2006, in *Extrasolar Planets*. Saas-Fee Advanced Course 31, D. Queloz, S. Udry, M. Mayor, and W. Benz (eds.), Berlin: Springer-Verlag.

Ragusa, E., Dipierro, G., Lodato, G., Laibe, G., & Price, D. J. 2017, *Monthly Notices of the Royal Astronomical Society*, **474**, 4460.

Rafikov, R. R. 2003, *Astronomical Journal*, **126**, 2529.

Rafikov, R. R. 2006, *Astrophysical Journal*, **648**, 666.

Rafikov, R. R. 2009, *Astrophysical Journal*, **704**, 281.

Rasio, F. A. & Ford, E. B. 1996, *Science*, **274**, 954.

Raymond, S. N., Barnes, R., Armitage, P. J., & Gorelick, N. 2008, *Astrophysical Journal*, **687**, L107.

Raymond, S. N., Quinn, T., & Lunine, J. I. 2006, *Icarus*, **183**, 265.

Rein, H. & Liu, S.-F. 2012, *Astronomy & Astrophysics*, **537**, article id A128.

Rein, H. & Spiegel, D. S. 2015, *Monthly Notices of the Royal Astronomical Society*, **446**, 1424.

Rein, H. & Tamayo, D. 2015, *Monthly Notices of the Royal Astronomical Society*, **452**, 376.

Ricci, L., Testi, L., Natta, A., & Brooks, K. J. 2010, *Astronomy & Astrophysics*, **521**, article id. A66.

Rice, W. K. M. & Armitage, P. J. 2003, *Astrophysical Journal*, **598**, L55.

Rice, W. K. M., Armitage, P. J., Bate, M. R., & Bonnell, I. A. 2003, *Monthly Notices of the Royal Astronomical Society*, **339**, 1025.

Rice, W. K. M., Armitage, P. J., Mamatsashvili, G. R., Lodato, G., & Clarke, C. J. 2011, *Monthly Notices of the Royal Astronomical Society*, **418**, 1356.

Rice, W. K. M., Lodato, G., & Armitage, P. J. 2005, *Monthly Notices of the Royal Astronomical Society*, **364**, L56.

Rivera, E. J., Laughlin, G., Butler, R. P., *et al.* 2010, *Astrophysical Journal*, **719**, 890.

Robert, C. M. T., Crida, A., Lega, E., Méheut, H., & Morbidelli, A. 2018, *Astronomy & Astrophysics*, **617**, article id. A98.

Rogers, L. A. 2015, *Astrophysical Journal*, **801**, article id. 41.

Ros, K. & Johansen, A. 2013, *Astronomy & Astrophysics*, **552**, article id. A137.

Rossiter, R. A. 1924, *Astrophysical Journal*, **60**, 15.

Rybicki, G. B. & Lightman, A. P. 1979, *Radiative Processes in Astrophysics*, New York: Wiley-Interscience.

Safronov, V. S. 1969, *Evolution of the Protoplanetary Cloud and Formation of the Earth and the Planets*, English translation, NASA TT F-677 (1972).

Saha, P. & Tremaine, S. 1992, *Astronomical Journal*, **104**, 1633.

Saha, P. & Tremaine, S. 1994, *Astronomical Journal*, **108**, 1962.

Salmon, J. K. 1991, Parallel hierarchical N-body methods, Ph.D. thesis, California Institute of Technology, Pasadena.

Sano, T., Miyama, S. M., Umebayashi, T., & Nakano, T. 2000, *Astrophysical Journal*, **543**, 486.

Semenov, D., Henning, Th., Helling, Ch., & Sedlmayr, E. 2003, *Astronomy & Astrophysics*, **410**, 611.

Shakura, N. I. & Sunyaev, R. A. 1973, *Astronomy & Astrophysics*, **24**, 337.

Sicilia-Aguilar, A., Hartmann, L., Calvet, N., *et al.* 2006, *Astrophysical Journal*, **638**, 897.

Simon, J. B., Bai, X.-N., Armitage, P. J., Stone, J. M., & Beckwith, K. 2013, *Astrophysical Journal*, **775**, article id. 73

Simon, M. & Prato, L. 1995, *Astrophysical Journal*, **450**, 824.

Sinclair, A. T. 1972, *Monthly Notices of the Royal Astronomical Society*, **160**, 169.

Skrutskie, M. F., Dutkevitch, D., Strom, S. E., *et al.* 1990, *Astronomical Journal*, **99**, 1187.

Smoluchowski, M. V. 1916, *Physikalische Zeitschrift*, **17**, 557.

Sousa, S. G., Santos, N. C., Israelian, G., Mayor, M., & Udry, S. 2011, *Astronomy & Astrophysics*, **533**, article id. A141.

Spitzer, L. & Tomasko, M. G. 1968, *Astrophysical Journal*, **152**, 971.

Springel, V. 2005, *Monthly Notices of the Royal Astronomical Society*, **364**, 1105.

Spruit, H. C. 1996, in *Evolutionary Processes in Binary Stars*, NATO ASI Series C, **477**, Dordrecht: Kluwer, p. 249.

Stepinski, T. F. 1992, *Icarus*, **97**, 130.

Stevenson, D. J. 1982, *Planetary and Space Science*, **30**, 755.

Stevenson, D. J. & Lunine, J. I. 1988, *Icarus*, **75**, 146.

Stewart, G. R. & Ida, S. 2000, *Icarus*, **143**, 28.

Stewart, G. R. & Wetherill, G. W. 1988, *Icarus*, **74**, 542.

Struve, O. 1952, *The Observatory*, **72**, 199.

Takeuchi, T. & Lin, D. N. C. 2002, *Astrophysical Journal*, **581**, 1344.

Tanaka, H., Takeuchi, T., & Ward, W. R. 2002, *Astrophysical Journal*, **565**, 1257.

Tera, F., Papanastassiou, D. A., & Wasserburg, G. J. 1974, *Earth and Planetary Science Letters*, **22**, 1.

Thommes, E. W., Duncan, M. J., & Levison, H. F. 1999, *Nature*, **402**, 635.

Tobin, J. J., Kratter, K. M., Persson, M. V., *et al.* 2016, *Nature*, **538**, 483.

Toomre, A. 1964, *Astrophysical Journal*, **139**, 1217.

Trujillo, C. A. & Brown, M. E. 2001, *Astrophysical Journal*, **554**, L95.

Tsiganis, K., Gomes, R., Morbidelli, A., & Levison, H. F. 2005, *Nature*, **435**, 459.

Turner, N. J. & Sano, T. 2008, *Astrophysical Journal*, **679**, L131.

Turner, N. J., Sano, T., & Dziourkevitch, N. 2007, *Astrophysical Journal*, **659**, 729.

Umebayashi, T. & Nakano, T. 1981, *Publications of the Astronomical Society of Japan*, **33**, 617.

Urpin, V. & Brandenburg, A. 1998, *Monthly Notices of the Royal Astronomical Society*, **294**, 399.

van der Marel, N., van Dishoeck, E. F., Bruderer, S., *et al.* 2013, *Science*, **340**, 1199.

Velikhov, E. P. 1959, *Soviet Physics JTEP*, **36**, 995.

Veras, D. & Armitage, P. J. 2004, *Monthly Notices of the Royal Astronomical Society*, **347**, 613.

Wahl, S. M., Hubbard, W. B., Militzer, B., *et al.* 2017, *Geophysical Research Letters*, **44**, 4649.

Walsh, K. J., Morbidelli, A., Raymond, S. N., *et al.* 2011, *Nature*, **475**, 206.

Wang, J. & Fischer, D. A. 2015, *Astronomical Journal*, **149**, article id. 14.

Ward, W. R. 1991, *Abstracts of the Lunar and Planetary Science Conference*, **22**, 1463.

Ward, W. R. 1997, *Icarus*, **126**, 261.

Weidenschilling, S. J. 1977a, *Astrophysics and Space Science*, **51**, 153.

Weidenschilling, S. J. 1977b, *Monthly Notices of the Royal Astronomical Society*, **180**, 57.

Weidenschilling, S. J. & Marzari, F. 1996, *Nature*, **384**, 619.

Wetherill, G. W. 1990, *Icarus*, **88**, 336.

Wetherill, G. W. 1991, *Science*, **253**, 535.

Wetherill, G. W. 1996, *Icarus*, **119**, 219.

Whipple, F. L. 1972, in *From Plasma to Planet*, Proceedings of the Twenty-First Nobel Symposium, A. Evlius (ed.), New York: Wiley Interscience Division, p. 211.

Williams, J. P. & Cieza, L. A. 2011, *Annual Review of Astronomy & Astrophysics*, **49**, 67.

Winn, J. N. & Fabrycky, D. C. 2015, *Annual Review of Astronomy & Astrophysics*, **53**, 409.

Winn, J. N., Fabrycky, D., Albrecht, S., & Johnson, J. A. 2010, *Astrophysical Journal*, **718**, L145.

Wisdom, J. & Holman, M. 1991, *Astronomical Journal*, **102**, 1528.

Wolk, S. J. & Walter, F. M. 1996, *Astronomical Journal*, **111**, 2066.

Wolszczan, A. & Frail, D. A. 1992, *Nature*, **355**, 145.

Wright, J. T., Fakhouri, O., Marcy, G. W., *et al.* 2011a, *Publications of the Astronomical Society of the Pacific*, **123**, 412.

Wright, J. T., Veras, D., Ford, E. B., *et al.* 2011b, *Astrophysical Journal*, **730**, article id. 93.

Wu, Y. & Murray, N. 2003, *Astrophysical Journal*, **589**, 605.

Wyatt, M. C. 2008, *Annual Review of Astronomy & Astrophysics*, **46**, 339.

Wyatt, M. C., Clarke, C. J., & Booth, M. 2011, *Celestial Mechanics and Dynamical Astronomy*, **111**, 1.

Wyatt, M. C., Smith, R., Greaves, J. S., *et al.* 2007, *Astrophysical Journal*, **658**, 569.

Wu, Y. & Lithwick, Y. 2011, *Astrophysical Journal*, **735**, article id. 109.

Yang, C.-C., Johansen, A., & Carrera, D. 2017, *Astronomy & Astrophysics*, **606**, article id. A80.

Yoder, C. F. 1979, *Celestial Mechanics*, **19**, 3.

Youdin, A. N. 2010, *European Astronomical Society Publications Series*, **41**, 187.

Youdin, A. N. & Chiang, E. I. 2004, *Astrophysical Journal*, **601**, 1109.

Youdin, A. N. & Goodman, J. 2005, *Astrophysical Journal*, **620**, 459.

Youdin, A. N. & Lithwick, Y. 2007, *Icarus*, **192**, 588.

Youdin, A. N. & Shu, F. H. 2002, *Astrophysical Journal*, **580**, 494.

Zeng, Li, Sasselov, D. D., & Jacobsen, S. B. 2016, *Astrophysical Journal*, **819**, article id. 127.

Zharkov, V. N. & Trubitsyn, V. P. 1974, *Icarus*, **21**, 152.

Zhong, G. & Marsden, J. E. 1988, *Physics Letters A*, **133**, 134.

Zhu, Z., Hartmann, L., & Gammie C. 2009, *Astrophysical Journal*, **694**, 1045.

Index

accretion disk, *see* disk
adiabatic temperature gradient, 228
aerodynamic drag, 142–144
albedo, 19
Alfvén velocity, 111
aluminium (^{26}Al), in Solar Nebula, 78
angular momentum deficit, 275
angular momentum problem (star formation), 3, 50
angular momentum transport, 90, 98, 102–127
asteroids, 6
astrometry, 26
atmosphere, condition for formation, 222
azimuthal drag, 149–153
azimuthal velocity (gas disk), 60, 150

Balbus–Hawley instability, 109
barotropic disks, 108
binary systems, Kozai resonance, 272
blackbody spectrum, 20
Bode's law, 289
boundary layer, 91
brown dwarf, 1
Brownian motion, 161

carbonate–silicate cycle, 45
catastrophic disruption (threshold in collisions), 186
chaos
   Solar System, 270
   three-body problem, 285
Chiang–Goldreich model (of protoplanetary disks), 68
chondritic meteorites, 16
circulation (of resonant argument), 269
co-rotation radius, 138
coagulation equation, 207–210
   analytic solutions, 208
   discrete form, 207
   dust growth, 163
   integral form, 164
   runaway growth, 208
collisional cascade, 293–298
   size distribution, 297

collisions, inelastic, 204
condensation sequence, 73–74
conductivity, 115
continuity equation, 109
convective stability, 227
cooling time scale, of disk, 243
core accretion, 221–237
   dilute core, 239
core mass, dilute, 239
cosmic rays, flux, 78
Coulomb logarithm, 201
critical core mass, 230
   analytic expressions for, 236

dead zone, 114
debris disks, 293–300
   collisional cascade, 293–298
   time evolution, 298
   white dwarf, 299
deuterium-burning threshold, 1
diffraction limit, 20
diffusion approximation, 71, 97, 226
diffusion equation, 90
diffusivity, 114
disk
   α model, 100
   angular momentum transport, 102
   debris, 87
   dispersal, 133
   evolution equation, 88
   flaring, 59, 63–65
   frequency in young clusters, 86
   gas temperature, 67
   Green's function solution, 93
   ionization, 75
   lifetime, 87
   mass estimates, 54
   molecular line observations, 55
   nonlinear stability, 103
   photoevaporation, 134
   radiative equilibrium, 65
   ring structure, 56
   scale height, 58

disk (cont.)
  self-gravity, 103, 240
    effect on vertical structure, 59
  self-similar solution, 94
  steady-state solution, 92
  surface density profile, 4
  temperature profile, 60, 95
  time scales
    cooling, 243
    observed, 87
    viscous, 90
  transition, 56
  vertical structure, 57–59, 96
  vortices, 81
  winds, 127
  zero-torque boundary condition, 93
disk instability mode, 246
disk instability model(, 240
dispersion dominated encounters, 183, 191
dispersion relation
  magnetorotational instability, 111
  self-gravitating disk, 170
dissociative recombination, 78
Doppler method (planet detection), 21
drag coefficient, 143
drag law
  Epstein regime, 142
  Stokes regime, 143
dust
  drift-limited growth, 166
  emissivity, 66
  fragmentation-limited growth, 165
  growth, 145, 153, 160, 162
  opacity, 71
  pile-up, 155
  radial drift, 149
  settling, 144
  size distribution, 71
  sticking efficiency, 162
dust–gas interaction, 142, 158
dust–grain collisions, 162
dwarf planets, 1, 2
dynamical friction, 202

eccentric anomaly, 23
eccentricity, of extrasolar planets, 288
Einstein ring, 32
emissivity (of dust), 66
epicyclic frequency, 170
Epstein drag, 142
extrasolar planets
  16 Cyg B, 275
  detection methods, 18–34
  eccentricity distribution, 288
  GJ 876, 270
  habitability, 44
  scattering, 280

feeding zone, 193
fragmentation
  cooling-driven, 242

gas disk, 105, 240
  infall-driven, 245
  limit on particle size, 165
  particle disks, 166
  planetesimals, 186
friction time (dust particles), 144

gap depth, 260–261
gap formation, 258–260
gas drag (on planetesimals), 203
giant planets
  gravity field, 238
  interior structure, 237
Gibbs free energy, 73
Goldreich–Ward mechanism, 166
Grand Tack model (for the inner Solar System), 292
gravitational focusing, 182–183
gravitational instability
  giant planet formation, 240
  planetesimal formation, 172
  tidal downsizing, 246
gravitational lensing, 32
gravitational moments (giant
          planets), 238
gravity dominated regime (in planetesimal collisions),
          187
great inequality (between Jupiter and Saturn), 4

habitability, 44
Hamiltonian dynamics, 316
headwind regime (of pebble accretion), 213
high eccentricity migration, 275
Hill sphere, 183, 185
Hill stability, 282–286
Hill surfaces, 283
Hill's equations, 185
horseshoe orbits, 185, 253
hydrostatic equilibrium
  disk vertical structure, 57
  giant planets, 226

ice giants, 2
ice line, 14, 83
ideal magnetohydrodynamics, 110
induction equation, 110
inelastic collisions, 204
ionization
  nonthermal, 76
  thermal, 75
isochron diagram, 11
isolation mass, 197–198, 225
isotope dating, 9

Jacobi constant, 283
Jeans escape, 222
JUNO constraints (on Jovian structure, 239
Jupiter, interior structure models, 239

Kelvin–Helmholtz instability, 175
Kelvin–Helmholtz time scale (giant
          planets), 236

Kepler's equation, 23
Keplerian orbital velocity, 21
Kida vortex solution, 81
kinematic viscosity, 89
Kirkwood gaps, 6
Kozai–Lidov dynamics, 272–275
Kuiper Belt Objects (KBOs), 7
  resonant capture of, 290
  total mass, 6

Lagrange stability, 285
Late Heavy Bombardment, 292
layered accretion disk model, 115–117
lead–lead dating, 12
leapfrog scheme, 315
Ledoux criterion, 228
libration (of resonant argument), 269
Lindblad resonances, 252

magnetic braking, 127
magnetic dynamo, 113
magnetic fields
  diffusivity, 114
  disk winds, 127
  instabilities, 109
  magnetohydrodynamics (MHD), 109
  non-ideal MHD, 114
  protostellar, 137
magnetic torque, 127
magnetorotational instability, 109
magnetospheric accretion, 137–140
Maxwell stress, 114
mean anomaly, 268
mean longitude, 268
mean motion, 267
meteorites, 16
microlensing, 32
migration
  eccentricity damping/growth, 264
  gas disk, 248–265
  high eccentricity tidal, 275
  maps, 261
  particles, 149
  planetesimal disks, 276
  rate, Type 2, 262
  simulations, 258
  Solar System evidence for, 290
  stopping mechanism, 140
  torque, 249
  torque formulae, 255
  Type 1, 257
  Type 2, 257
minimum mass Solar Nebula, 4
molecular cloud cores, 49
molecular viscosity, 99
momentum equation, 60

N-body problem, 311
  leapfrog, 315
  softening, 314
  symplectic transformations, 316

Neptune, 2
  migration of, 290
Newtonian dynamics, 311
Nice model (for the outer Solar System), 291

oligarchic growth, 206, 217
opacity
  analytic approximations, 72
  giant planet envelopes, 232
  Rosseland mean, 71
  sources within disks, 71
orderly growth, 208
overlap (of resonances), 270

particle pileup, 155
particle-in-box model (of planetary growth), 190–193
passive disk, 60
Pb–Pb dating, 12
pebble accretion, 210–217
  accretion rates, 216
  headwind (or Bondi) regime, 213
  headwind speed, 212
  lower mass limit, 214
  shear (or Hill) regime, 213
  transition mass (from headwind to shear regime), 215
pendulum model (of resonance), 269
photoevaporation, 134
photoionization cross-section, 77
Planck function, 20
planet migration, *see* migration
planet traps, 261
planetary satellites, 9
planetesimals
  collisions, 186
  definition, 141
  distribution of orbital elements, 190
  formation of, 178
  formation via gravitational instability, 172–173
  gravitational collapse, 166
  growth rates, 192
  strength, 188
  velocity dispersion, 198–205
planets
  definition, 1
  habitability, 44
  interior structure, 237
  pulsar, 18
  Solar System properties, 2
Pluto, 2
  capture into resonance of, 290
Poisson equation, 167
protoplanetary disk, *see* disk

Q parameter (disk stability), 104, 167–171, 240

radial drift (of solid particles), 149
radial velocity method, 21–26
  biases, 23
  eccentric orbits, 23
  noise sources, 24

radiative diffusion, 226
radiative temperature gradient, 226
radioactive dating, 9
   lead–lead system, 12
   short-lived radionuclides, 13
ratio–ratio plot, 11
Rayleigh criterion (spatial resolution), 20
Rayleigh criterion (stability of shear flows), 103
Rayleigh distribution, 190
recombination, 78
resonances, 3, 265–270
   capture, 270–271, 290
   extrasolar planets, 270
   Kozai, 272
   Laplace, 9, 270
   libration time scale, 269
   Lindblad, 252
   mean motion, 4, 267
   overlap, 270, 285
   pendulum model, 269
   resonant argument, 268
   saturation, 264
   secondary, 269
   secular, 269, 272
   three-body, 269
   torque transfer, 251
   width, 269
restricted three-body problem, 282–283
Reynolds number, 99
   magnetic, 116
Reynolds stress, 114
Richardson number, 175
Roche lobe, 185
Rossby wave instability, 108
Rosseland mean opacity, 71
Rossiter–McLaughlin effect, 30
rotation (T Tauri stars), 138
rubble pile, 186
runaway growth (of giant planets), 230
runaway growth (of protoplanets), 192, 217

Saha equation, 75
satellites, 9
scale height, 58
scattering (of giant planets), 280
Schmidt number, 157
Schwarzschild criterion (convective stability), 228
secular chaos, 276
secular dynamics, 275–276
self-gravitating disk, *see* disk, self-gravity
settling (of dust), 144
Shakura–Sunyaev α-prescription, 98–100
shear dominated encounters, 183, 193
shear regime (of pebble accretion), 213
shot noise, 25–26
silicates, 72, 73
snow line, 14–16
softening (in N-body problem), 314
Solar Nebula, 4
sound speed, 58
spectral energy distribution (SED), 69–70
spiral arms, 243
stability (of planetary systems), 281–286

Stokes drag, 143
stopping time, *see* friction time
streaming instability, 176
strength dominated regime (in planetesimal
     collisions), 187
symplectic integration, 316

T Tauri stars, 77
   accretion rates, 52
   classification, 51
terrestrial planets, formation, 181–219
thermal time scale (giant planet envelopes), 236
three-body dynamics, 183
three-body problem, 282
Tisserand sphere, 185
Titius–Bode law, 289
Toomre Q parameter, 104, 167–171, 240
torque
   calculation in impulse approximation, 248
   formulae, 255
   horseshoe drag, 254
   Lindblad, 251
   linear co-rotation, 254
   resonant, 251
   thermal torque, 255
transition disks, 56
transits, 27–32
   secondary eclipse, 30
transmission spectroscopy, 31
tree codes, 312
Trojan asteroids, origin of, 292
true anomaly, 23
turbulence
   effect on particle settling, 147
   effect on terrestrial planet formation, 205
   hydrodynamic linear stability, 102
   hydrodynamic sources of, 102
   magnetic field instabilities, 109
   phenomenological description of, 99
   radial diffusion, 156
   self-gravitating, 102, 241

ultraviolet radiation, 134

viscosity, 90, 98, 102–114
   molecular, 99
viscous stirring (of planetesimals), 199
vortensity, 108
vortices, 81, 107
   Rossby wave instability, 108

water delivery (to Earth), 15

X-ray ionization, 77

Young Stellar Objects
   accretion rates, 52
   classification, 51
   disk masses, 54

zero-velocity surfaces, 283
zonal flows, 80

Printed in the United States
by Baker & Taylor Publisher Services